Robustness in Language and Speech Technology

Text, Speech and Language Technology

VOLUME 17

The titles published in this series are listed at the end of this volume.

Robustness in Language and Speech Technology

Edited by

Jean-Claude Junqua

Panasonic Speech Technology Laboratory,
Santa Barbara, California, U.S.A.

and

Gertjan van Noord

University of Groningen, The Netherlands

elsnet

KLUWER ACADEMIC PUBLISHERS

DORDRECHT / BOSTON / LONDON

Library of Congress Cataloging-in-Publication Data

Robustness in languages and speech technology / edited by Jean-Claude Junqua and Gertjan van Noord.
p. cm. -- (Text, speech, and language technology ; v. 17)
Includes bibliographical references.
ISBN 0-7923-6790-1 (hardbound : alk. paper)
1. Speech processing systems. I. Junqua, Jean-Claude. II. Noord, Gertjan van. III. Series.

TK7882.S65 .R65 2001
006.3'5--dc21

00-067183

ISBN 0-7923-6790-1

Published by Kluwer Academic Publishers,
P.O. Box 17, 3300 AA Dordrecht, The Netherlands.

Sold and distributed in North, Central and South America
by Kluwer Academic Publishers,
101 Philip Drive, Norwell, MA 02061, U.S.A.

In all other countries, sold and distributed
by Kluwer Academic Publishers,
P.O. Box 322, 3300 AH Dordrecht, The Netherlands.

Printed on acid-free paper

Printed in the Netherlands

Contents

Chapter 1

INTRODUCTION

Jean-Claude Junqua
Panasonic Speech Technology Laboratory, Santa Barbara
jcj@research.panasonic.com

Gertjan van Noord
Rijksuniversiteit Groningen
vannoord@let.rug.nl

Consider a computer system that you can talk to using ordinary speech (either directly or perhaps using your telephone), and that you can ask questions concerning such things as timetables for public transportation. For example, you might ask the system the departure time of a train from Brussels to Amsterdam, specifying that you wish to arrive in Amsterdam at 11. Before proceeding to answer your query such a system will probably ask you on which date you want to travel, and whether you mean '11 in the morning' or '11 in the evening'. After such a clarification, it will then give you the desired information, e.g. that you could take the '7:49' train from platform 2b at Brussels Central Station.

Such a system is capable of recognizing and understanding spoken language. It is furthermore capable of conducting a natural dialogue to obtain the information necessary for answering your questions. Moreover, the system is capable of producing spoken natural language to conduct this dialogue, and to produce the answer to your request for information.

During the last decade, systems of this kind employing language and speech technology have slowly become a reality. Robustness became a concern when technology began to be transferred from laboratory to field applications. It became clear immediately that the idealizations which could be maintained in the isolation of a laboratory setting could no longer be maintained in field

J.-C. Junqua and G. van Noord (eds.), Robustness in Languages and Speech Technology, 1–7.

applications. Systems had to be extended in various ways to make them capable of performing even if something unexpected happened. For example, part of the speech signal might have been distorted because of sudden background noise (e.g. because a dog started barking nearby); the speaker may have corrected him/herself halfway through the sentence ('I want to go to, eh no, what is the departure time of ...'), etc. Robustness in language and speech technology therefore is at the heart of many commercial applications and is essential for bridging the gap between laboratory and real-world performance. This book brings together speech and language technologies often considered separately. It also provides a good review of the field in some essential areas of robust language and speech technology.

Because applications were becoming progressively more ambitious and required increasingly complex dialogs, there was a move in the late 1980's from recognition (of the sequence of words) to understanding (of those words). In speech understanding systems, for example, synonyms can be used and the word order can be altered. Transaction systems that support this type of dialog are called *conversational systems* (Zue, Seneff, Glass, Hetherington, Hurley, Meng, Pao, Polifroni, Schloming & Schmid 1997). In the last decade a number of conversational systems have been developed, including those developed at MIT such as Voyager (Glass, Flammia, Goodine, Phillips, Polifroni, Sakai, Seneff & Zue 1995), Pegasus (Zue, Seneff, Polifroni, Phillips, Pao, Goddeau, Glass & Brill 1994), Galaxy (Goddeau, Brill, Glass, Pao, Phillips, Polifroni, Seneff & Zue 1994, Zue 1995) and Jupiter (Zue et al. 1997). In Europe, Philips Aachen developed its train timetable information system (Aust, Oerder, Seide & Steinbiss 1995), and the SUNDIAL and ARISE projects led to a number of conversational system prototypes (Peckham 1991, Lamel, Gauvain, Bennacef, Devillers, Foukia, Gangolf & Rosset 1996, Albesano, Baggia, Danieli, Gemello, Gerbino & Rullent 1997, Boros, Aretoulaki, Gallwitz, Nöth & Niemann 1997).

Figure 1.1 shows a block diagram of a typical conversational system for a database retrieval application using both audio and visual feedback. The following components need to be developed:

- a speech recognizer where the two main application-dependent components are the *acoustic and language models*. Acoustic models can be bootstrapped from other domains and refined with in-domain data. The need for in-domain data is more critical for the language model;
- a natural language parser which extracts task-dependent semantic information and deals with phenomena such as recognition errors, hesitations, ill-formed sentences and incomplete sentences. Generally, the natural language parser is customized for each application using task-dependent knowledge;
- a dialog manager which maintains the dialog context and the history of the dialog and initiates actions when a request that makes sense has been identified. To do this it uses a semantic representation of the task

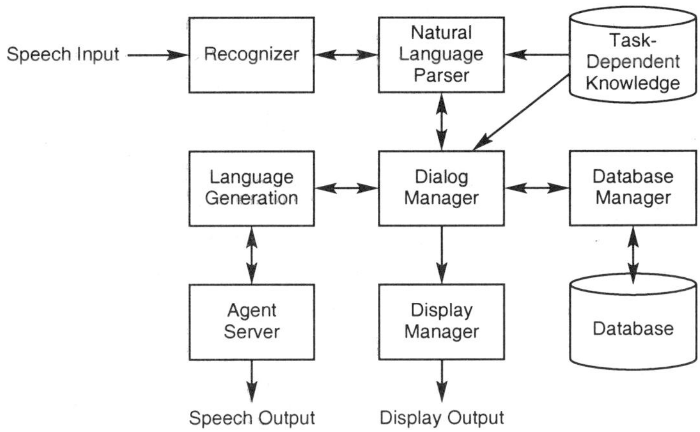

FIGURE 1.1: Generic block diagram of a conversational system with audio and visual feedback.

(in most state-of-the-art systems semantic frames are used). More details about various semantic representations can be found in Kuhn & De Mori (1997);

- a language generation module which conveys the information to the user in natural sounding sentences. In some systems this task is done by the dialog manager;
- feedback managers (the agent server and the display manager in Figure 1.1) which convey dialog cues and the displayed information to the user. The agent server gives cues to the user about the state of the system and enhances the naturalness of the conversation.

In this book we address robustness issues for the first two components, namely the speech recognizer and the natural language parser, with a focus on feature extraction and noise robust recognition, adaptive systems, language modeling, the interface between speech recognition and natural language understanding and natural language parsing. This book attempts to give a clear overview of the main technologies used in language and speech processing, along with an extensive bibliography to enable topics of interest to be pursued further. As material about state-of-the-art feature extraction and speech recognition can be very easily found in the literature, purposely we did not review these areas in this book. However, for the sake of completeness more information about these topics can be found in Junqua & Haton (1996) and Rabiner & Juang (1993).

At the feature extraction level discussed in Chapter 2, Johan De Veth, Bert Cranen and Louis Boves present several techniques that are used to alleviate the effects of unknown transmission channels and acoustic background on the acoustic features used in automatic speech recognition. They also emphasize 1) the tight coupling between feature processing and decoding and 2) missing feature theory, which provides a basis for an explanation of how humans manage to understand heavily distorted speech.

Generally speaking we can distinguish three classes of speech recognition systems: speaker independent, speaker dependent, and speaker adaptive systems. Speaker independent systems are designed to recognize speech from anyone; generally, the models are trained on data from a large number of speakers. Speaker dependent systems are designed to recognize speech from a particular individual; the models are trained on data from that individual. Error rates for speaker independent systems are roughly two to three times higher than those for speaker dependent systems, when the latter are tested on the speaker they are trained for (Lee, Lin & Juang 1991, Hazen & Glass 1997). Adaptive systems are an attempt to combine the advantages of speaker independent and speaker dependent systems. When a user first speaks to an adaptive system, the system employs speaker independent models; once speech data from this user has been obtained, the parameters of the models are updated to reflect user-specific traits.

While in the last decade speaker independent systems were the focus of most of the research, recently speaker adaptive systems have attracted much interest. Adaptive systems have been studied to compensate for a large variety of problems: speaking style, speaking rate, non-native speakers, transducers and transmission channels, noise, language, task, etc.

Speaker adaptive techniques are the focus of Chapter 3. In this chapter, Daniel Tapias first summarizes the main sources of distortion/variability and then focuses on compensation and adaptation techniques that have proven to be useful in dealing with these sources of distortion/variability. Adaptive automatic speech recognition provides a useful path to improve the robustness of our systems.

To be able to obtain reasonable recognition accuracy, constraints are placed on the utterances that can be spoken. This is the role of the language model. Robustness in language modeling refers to the need to maintain adequate speech recognition accuracy as fewer and fewer constraints are placed on the spoken utterances. This topic is covered in Chapter 4 by Jerome Bellegarda. More specifically, in this chapter a selected subset of recent approaches dealing with robust language modeling are reviewed and some perspectives for future directions are given.

Conversational systems encourage the use of spontaneous speech. However, even if spontaneous speech data are often used to train acoustic models, there is very little modeling of spontaneous speech events such as speech repairs. To deal effectively with spontaneous speech it may be necessary to model some of these events. For example, traditional language modeling techniques do not model the interactions between spontaneous speech events and speech recognition, and thus have difficulty providing accurate prediction of the words in the vicinity of these events. In Chapter 5 James Allen and Peter Heeman advocate that these spontaneous speech events should be explicitly modelled and they propose to account for them in the language model.

In Chapter 6 Mehryar Mohri and Mark-Jan Nederhof present an algorithm for approximating context-free languages with regular languages, while in Chapter 7 Mehryar Mohri describes a general grammar library that provides tools for building dynamic grammars for large vocabulary recognition. Such tools or algorithms are essential for the building of conversational/dialogue systems, because approximations of weighted context-free grammars are in general computationally less demanding than context-free grammars and more accurate than classical n-gram language models.

Traditional parsing algorithms that are designed to parse completely grammatical input sentences are unsuitable for practical spoken language applications, where disfluencies and ungrammatical sentences are common. Therefore robust parsing does not aim at providing a full syntactic analysis but rather tends to identify partial analyses. In Chapter 8 Jean-Pierre Chanod introduces the notion of robust parsing, discusses a number of linguistic phenomena which motivate a robust approach, and summarizes some pioneer and recent work in this area. Special attention is given to robust parsing techniques which work in an incremental way by resolving syntactic structures with various degrees of depth (creating a deeper analysis on top of a more superficial analysis). Such techniques provide a partial constituent structure (chunks) which might be used as input for semantic and pragmatic analysis.

There are different ways to interface a speech recognizer with a natural language understanding component. The simplest way is to connect the recognition and natural language components serially. In this case, the best recognized sentence is passed to the natural language component that interprets it. However, this type of connection does not permit good interaction between the two components. Moreover, error recovery is made difficult. Other ways to interface recognition and natural language processes include an N-best sentence interface, a word-lattice or word graph interface, or an interface where the natural language component provides a search space reduction to the speech recognizer by the ability to feed back into the speech recognition component.

In Chapter 9 Gertjan van Noord presents a general model for robust parsing of word graphs and discusses a number of search techniques aimed at optimizing the efficiency while maintaining the accuracy of concept extraction. An important property is that various partial analyses of the input word graph are considered in addition to complete analyses. This property has the desirable effect that even in case of errors the linguistic analysis component is able to extract some useful information.

The book is completed by Carolyn Penstein Rose and Alon Lavie. Rather than attempting to construct partial analyses in the context of errors (the approach in the two previous chapters) their approach aims to find the analysis of a corresponding grammatical input that is *closest* to the given input. The chapter describes a number of experiments indicating the relationship between various distance measures and efficiency. These distance measures include the possibility of word skipping, non-terminal inserting, and flexible unification.

As shown by the topics covered in this book, robustness in speech and language technology is a very wide field of research. In recent conferences an emphasis has been put on dynamic systems that incrementally learn from data and can adapt automatically to new environmental conditions as human do. While a number of research paths are currently being explored, the following areas represent what we believe are the most promising directions:

- robust feature extraction;
- rapid speaker adaptation;
- separate modeling of phonetic and speaker-dependent information;
- environment-adaptive automatic speech recognition and language/task adaptation;
- combination of syntactic knowledge, global semantic analysis and pragmatic information for better language models and robust parsing algorithms.

This book originates from the 6th European Summer School on Language and Speech Communication, which was held in the summer of 1998 in Barcelona, Spain. During two weeks slightly fewer than 100 participants followed presentations focused on the theme "Robustness in Language and Speech Technology". The quality of the courses encouraged us to bring together contributions of the lecturers and external additional contributions and to publish this material as a book. This project was initiated and organized by ELSNET, the European Network in Language and Speech, which is the main sponsor of the summer school.

Website

The book is accompanied by a website with additional material. The website is located at the following URL:

`http://www.elsnet.org/publications.html`

Acknowledgments

The editors would like to express their gratitute to the reviewers, Ted Applebaum (Panasonic), Gosse Bouma (Groningen University), Jan Daciuk (Technical University Gdansk; Groningen University), Marcello Frederico (IRST), Hans-Guenter Hirsch (Ericsson), Roland Kuhn (Panasonic), Mark-Jan Nederhof (DFKI, AT&T), Günter Neumann (DFKI), Patrick Nguyen (Panasonic; Eurecom), Giorgio Satta (Trento University), Yunxin Zhao (Missouri University). In addition, we are grateful for the technical support we received from Peter Kleiweg.

References

Albesano, D., Baggia, P., Danieli, M., Gemello, R., Gerbino, E. & Rullent, C. (1997). A robust system for human-machine dialogue in telephony-based applications, *Journal of Speech Technology* **2**(2): 99–110.

Aust, H., Oerder, M., Seide, F. & Steinbiss, V. (1995). The Philips automatic train timetable information system, *Speech Communication* **17**: 249–262.

Boros, M., Aretoulaki, M., Gallwitz, F., Nöth, E. & Niemann, H. (1997). Semantic processing of out-of-vocabulary words in a spoken dialogue system, *Proceedings of the 5th European Conference on Speech Communication and Technology (Eurospeech '97)*, pp. 1887–1890.

Glass, J., Flammia, G., Goodine, D., Phillips, M., Polifroni, J., Sakai, S., Seneff, S. & Zue, V. (1995). Multilingual spoken language understanding in the MIT Voyager system, *Speech Communication* **17**: 1–18.

Goddeau, D., Brill, E., Glass, J., Pao, C., Phillips, M., Polifroni, M., Seneff, S. & Zue, V. (1994). GALAXY: A human-language interface to on-line travel information, *International Conference on Spoken Language Processing (ICSLP)*, pp. 707–710.

Hazen, T. & Glass, J. (1997). A comparison of novel techniques for instantaneous speaker adaptation, *Proceedings of the 5th European Conference on Speech Communication and Technology (Eurospeech '97)*, pp. 2047–2050.

Junqua, J.-C. & Haton, J.-P. (1996). *Robustness in Automatic Speech Recognition*, Kluwer Academic Publishers.

Kuhn, R. & De Mori, R. (1997). Sentence interpretation, *in* R. De Mori (ed.), *Spoken Dialogues with Computers*, Academic Press, pp. 485–520.

Lamel, L., Gauvain, J.-L., Bennacef, S. K., Devillers, L., Foukia, S., Gangolf, J.-J. & Rosset, S. (1996). Field trials of a telephone service for rail travel information, *IEEE Workshop International Voice Technology for Telecommunications Applications (IVTTA)*, pp. 111–116.

Lee, C.-H., Lin, C.-H. & Juang, B. H. (1991). A study of speaker adaptation of the parameters of continuous density hidden Markov models, *IEEE Transactions on Signal Processing* **39**: 806–814.

Peckham, J. (1991). Speech understanding and dialogue over the telephone: An overview of progress in the SUNDIAL project, *Proceedings of the 2nd European Conference on Speech Communication and Technology (Eurospeech '91)*, pp. 1469–1472.

Rabiner, L. & Juang, B. H. (1993). *Fundamentals of Speech Recognition*, Prentice Hall.

Zue, V. (1995). Navigating the information superhighway using spoken language interfaces, *IEEE Expert* **10**(5): 39–43.

Zue, V., Seneff, S., Glass, J., Hetherington, L., Hurley, E., Meng, H., Pao, C., Polifroni, J., Schloming, R. & Schmid, P. (1997). From interface to content: Translingual access and delivery of on-line information, *Proceedings of the 5th European Conference on Speech Communication and Technology (Eurospeech '97)*, pp. 2227–2230.

Zue, V., Seneff, S., Polifroni, J., Phillips, M., Pao, C., Goddeau, D., Glass, J. & Brill, E. (1994). PEGASUS: A spoken language interface for on-line air travel planning, *Speech Communication* **15**: 331–340.

Chapter 2

ACOUSTIC FEATURES AND DISTANCE MEASURE TO REDUCE VULNERABILITY OF ASR PERFORMANCE DUE TO THE PRESENCE OF A COMMUNICATION CHANNEL AND/OR BACKGROUND NOISE

Johan de Veth
University of Nijmegen
J.deVeth@let.kun.nl

Bert Cranen
University of Nijmegen
B.Cranen@let.kun.nl

Louis Boves
University of Nijmegen
L.Boves@let.kun.nl

2.1 Introduction

2.1.1 Automatic Speech Recognition is pattern recognition

Saying that late 20th century automatic speech recognition (ASR) is *pattern recognition*, is something of a truism, but perhaps one of which the fundamental implications are not always fully appreciated. Essentially, a pattern recognition task boils down to measuring the distance between a physical representation of a new, as yet unknown token, and all elements of a set of preexisting patterns, of course in the same physical representation. On the one hand, the 'patterns' that can be recognized are, implicitly or explicitly, separate and invariable entities. For example, the command *open* in a Windows control application always has the same invariable and unique meaning. On the other hand, the unknown input tokens are continuous signals that typically

J.-C. Junqua and G. van Noord (eds.), Robustness in Languages and Speech Technology, 9–45.

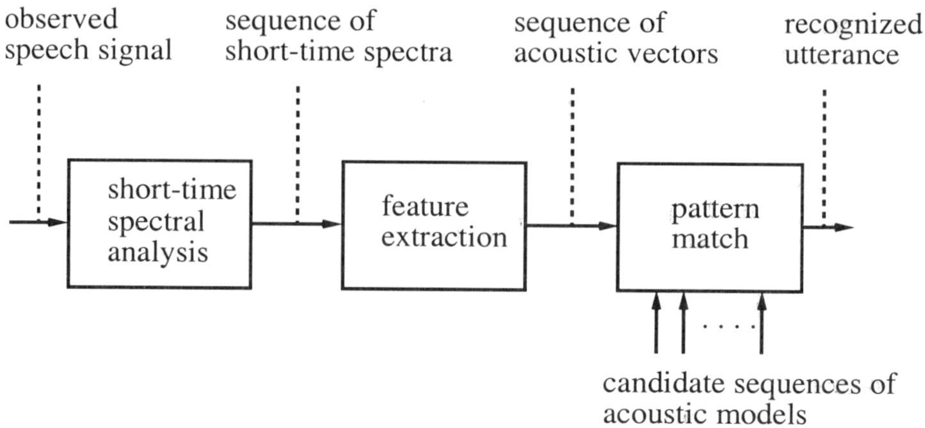

FIGURE 2.1: Automatic speech recognition as a form of pattern matching. The feature extraction module is used to compute a compact representation of the short-time spectra that describes characteristics which are best suited for recognition.

show a high degree of variability. ASR research has centered around the problem of how to map continuous, variable acoustic representations onto discrete, invariable patterns. In ASR the physical representation of the speech tokens is some kind of dynamic power spectrum, for reasons which date back to the days of Ohm and von Helmholtz, who have shown that the power spectrum explains most of the perceptual phenomena in human speech processing. Since the inception of digital signal processing dynamic spectra are approximated by a sequence of short-time spectra (Rabiner & Schafer 1978). Consequently, the pattern match in ASR is invariably implemented as the accumulation of some distance measure between the acoustic features derived from a sequence of short-time spectra of the input token and the corresponding representation of the active patterns (see Fig. 2.1). Therefore, anything which adds to the variability of the short time spectrum of a speech signal will, as it were by definition, complicate pattern matching, and consequently complicate ASR.

The basic aim of robust speech recognition is to make the pattern match insensitive to variability in the short-time spectra. It goes without saying that there is no single optimal approach to find the *holy grail*. There are too many factors which affect the spectra; in addition, there are just too many different ways in which the spectral features can be represented, in which the patterns can be cast, and in which the search for the best matching pattern can be implemented, even if we restrict the discussion to hidden Markov models (HMMs) and Viterbi search. In this contribution we investigate two factors which increase the variability of the short-time spectra in more detail, viz. the transmission channel and background noise. To clarify the discussion we first sketch a model of the speech signal at the input of an ASR device. With the

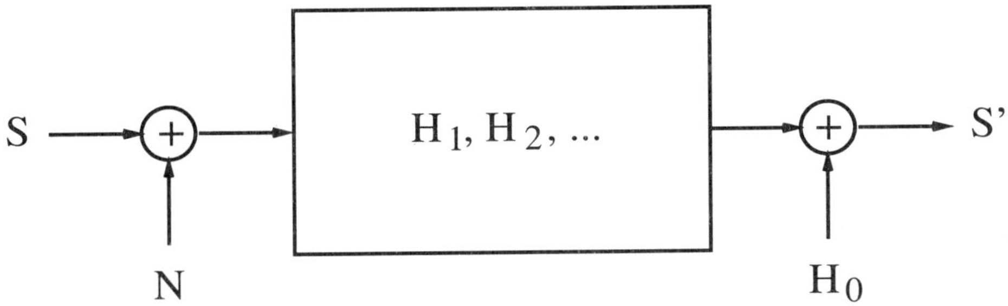

FIGURE 2.2: The observed speech spectrum S' is a mix of contributions from the original speech spectrum S, the background noise spectrum N, the linear and non-linear transfer characteristics of the channel, which can be described by a series of Volterra kernels $H_1, H_2, \ldots$ (Schetzen 1980), and the spontaneous activity of the channel H_0.

help of this model we can explain the relation between different approaches of robust speech recognition, and show how these approaches can be combined to reduce the effects of different sources that distort the short-time spectra.

2.1.2 A simple model

It is easy to draw a comprehensive, but very abstract, conceptual model of the signals at the input of an ASR device. Such a model is depicted in Fig. 2.2. We always have a speaker, who is in some physical, acoustic environment, talking into a microphone which is connected to a transmission chain that eventually delivers a signal to the input of the ASR device. Filling in the details of this abstract model is less easy and straightforward, however. How must one model the acoustic environment of the speaker? The model of an anechoic room is certainly different from the model of a car running on the highway, and also different from the model of a busy train station, and from a quiet hotel room. On top of the impact of the acoustic environment, the microphone that converts the sound into an electrical signal has several effects of its own. It is not equally sensitive to all frequencies, nor to sounds coming from all different directions. In addition, the microphone may also introduce non-linear distortions, like the plops caused by the airflow in fricative and/or plosive sounds. In short, the microphone may introduce linear and non-linear effects. The transmission chain connecting the microphone to the ASR device may introduce its own additive noise and linear filtering (e.g., caused by the cabling and amplifiers in analogue telephone transmission), as well as non-linear distortions. Digital transmission should help to alleviate channel distortions, but it is certainly no panacea. In cellular digital telephone networks radio transmission errors make a very substantial contribution to recognition errors (both human and machine, for that matter).

Of course, we build conceptual models of the signal at the input of the

ASR device with the aim to recover the original undistorted speech signal, or rather, to recover the power spectrum of the undistorted signals. The mathematical expression describing the relation between the clean speech spectrum and the spectrum at the ASR input is dependent on (1) the details of the transmission chain that are accounted for, and (2) the description of the linear and non-linear effects that were mentioned above.

For the minimalist model shown in Fig. 2.2 let S' denote the spectrum of the speech signal at the ASR input. Clearly, S' is a function of time t and frequency ω: $S' = S'(t,\omega)$. If we assume that (1) the background noise is additive and (2) the channel can be described as a linear, finite memory system, $S'(t,\omega)$ can be written as

$$S'(t,\omega) = H_0(t,\omega) + H_1(t,\omega)[S(t,\omega) + N(t,\omega)], \tag{2.1}$$

where $S(t,\omega)$ is the original speech spectrum and $N(t,\omega)$ the spectrum of the background noise. $H_0(t,\omega)$ corresponds to the spectrum of the signal spontaneously produced by the channel and $H_1(t,\omega)$ is the spectrum of the linear transfer function of the channel (Schetzen 1980). S', S, N, H_0 and H_1 are complex-valued functions of time and frequency.

The original 'clean' speech spectrum $S(t,\omega)$ is very much an abstract concept. Even under quiet, 'noise free' conditions the clean signal cannot be observed, because it is affected by the room acoustics and the recording equipment. Fortunately, the lion's share of the inevitable effects are very small, much smaller than the differences between two speakers or between two realizations of the same linguistic utterance by the same speaker. As long as the effects are similar for all utterances, they can be considered as part of the 'clean' spectrum $S(t,\omega)$. Robust speech recognition comes into play where the impact on the 'clean' speech is variable, and so strong that the contributions to $S'(t,\omega)$ can no longer be neglected.

For robust speech recognition, Eq. 2.1 can be used to tell us how the speech spectrum observed at the ASR input $S'(t,\omega)$ can be understood in terms of the original speech spectrum $S(t,\omega)$ on the one hand, and the distortion terms on the other, i.e., the spontaneous activity of the transmission channel $H_0(t,\omega)$, the linear transfer function of the channel $H_1(t,\omega)$, and the power spectrum of the additive noise $N(t,\omega)$. Any approach to robustness will need to consider (1) the relative importance of each of the three distortion terms $H_0(t,\omega)$, $H_1(t,\omega)$, and $N(t,\omega)$, and (2) the accuracy with which each term can be estimated. We are now in the position to explain why under certain conditions specific solutions are superior. The key issue here is the amount of available prior knowledge. To illustrate our point, we consider the following two scenarios.

First, take an ASR device attached to a switch in a telephone network. In this case, things look pretty hopeless. The input is a single signal, from which the set of actual parameter values of all components of the model in

Fig. 2.2 must be estimated. From a mathematical point of view this is an ill-posed problem. It is an attempt to find a unique solution based on one equation with many unknowns. Elementary algebra tells us that this is impossible. Instead of a unique solution, a whole family of solutions is possible and without additional and independent observations the 'true' parameter values cannot be reconstructed. Consequently, in this situation we are obliged to simplify the model as much as possible (to reduce the number of unknowns). As we will see below, the well-known cepstrum mean subtraction technique (Atal 1974, Furui 1981) is a typical example of this simplified approach. For a second scenario, consider an ASR device built into a Bugatti car that is almost exclusively driven on the highways of Arizona and New Mexico. In this case, we might be able to reliably estimate the parameters of most components of the model in Fig. 2.2. To be more specific, with the engine switched off and no speech input we can establish the parameter values for the model component representing the spontaneous activity of the channel. While driving the car and no speech input, we can record the typical background noise. Finally, we can measure the linear transfer characteristic of the channel by testing the microphone in an an-echoic room. Now, the most appropriate robustness approach in the first scenario will seem much more primitive than the approach in the Bugatti case. Although it may be considered more appropriate from a 'physical' point of view, the more sophisticated Bugatti model would fail miserably in the first scenario, because the parameters of the noise will vary substantially between calls, making it impossible to come up with a useful prediction for an individual call. It is common knowledge in the field of System Identification that it is better to have an overly simplistic model of which the parameters can be estimated reliably, than to try a physically more adequate model, the parameters of which cannot be estimated reliably.

In the following Section, we will discuss techniques for dealing with linear filtering effects caused by the microphone and transmission channel. In Section 2.3, we will discuss different methods to deal with additive noise. In both problems, we will stress the inter-dependence between the underlying models, the choice for parameter representations, and the eventual spectral distance computation which is at the heart of any ASR algorithm.

2.2 The presence of a transmission channel

2.2.1 Assumptions for channel robustness

As said before, it is only possible to observe the speech signal through some transmission channel. Under most practical circumstances it is reasonable to assume that the linear transfer function of the channel is time-invariant or at least varying slowly in comparison with the articulation process. Then $H_1(t, \omega)$ in Eq. 2.1 reduces to

$$H_1(t, \omega) = H_1(\omega). \tag{2.2}$$

In this Section, we want to focus the discussion on techniques for dealing with the effects of $H_1(\omega)$. To simplify the discussion, we make the additional assumption that the energy of the combination of the spontaneous activity of the channel and the background noise can be neglected, i.e.,

$$|H_0(t,\omega) + H_1(\omega)N(t,\omega)|^2 \ll |H_1(\omega)S(t,\omega)|^2. \tag{2.3}$$

The speech spectrum at the ASR input can then be approximated as

$$S^{'}(t,\omega) = H_1(\omega)S(t,\omega). \tag{2.4}$$

As can be seen, the speech spectrum at the ASR input now contains only two contributions: the time-invariant linear transfer function of the channel and the time-variant speech spectrum. As we will see below, the separation into two contributions that have different temporal characteristics is the key to many channel robustness strategies. Without the assumption expressed by Eq. 2.2 such a convenient separation is not possible.

In the log-energy domain Eq. 2.4 becomes

$$\log(|S^{'}(t,\omega)|^2) = \log(|H_1(\omega)|^2) + \log(|S(t,\omega)|^2). \tag{2.5}$$

Taking the Fourier transform, we have in the cepstral domain

$$c^{'}(t,\tau) = c_h(\tau) + c(t,\tau), \tag{2.6}$$

with $c^{'}(t,\tau)$ the cepstrum of the channel output, $c_h(\tau)$ the cepstrum of the channel and $c(t,\tau)$ the cepstrum of the original speech signal. $c^{'}(t,\tau)$, $c_h(\tau)$, and $c(t,\tau)$ are real-valued functions, because the terms in Eq. 2.5 ($\log(|S^{'}(t,\omega)|^2)$, etc.) are real and even. According to Eq. 2.6 the contribution of the unknown channel is a constant for a given quefrency τ. From Eqs. 2.4, 2.5 and 2.6 it can be understood how the channel can affect ASR: If training and testing are performed using two different channels, the cepstra used during training differ from the corresponding cepstra at recognition time. Channel normalization methods aim to reduce the differences between training and testing speech spectra caused by the channel.

2.2.2 Channel normalization techniques

Channel normalization (CN) techniques have been studied for quite different conditions. In one such condition, which is not addressed in this contribution, a recognizer is trained with speech recorded with a close talking microphone and recognition is attempted on speech recorded with a different microphone. The contribution of the channel during training does not need to be known in great detail, because it is constant. The channel conditions during test are different from those during training, but constant too. Thus, a single, fixed transformation should suffice (see for example Anastasakos, Kubala, Makhoul

& Schwartz 1994, Liu, Moreno, Stern & Acero 1994, Orloff, Gillick, Roth, Scattone & Baker 1994, Weintraub, Neumeyer & Digalakis 1994).

In ASR applications over the telephone the situation is different: the channels over which the training speech is recorded are unknown and likely to differ between all recording sessions. The same goes for the testing speech. Under these conditions a CN technique is needed both during training and testing. This is the situation that is reviewed here.

Use of many different channels

Under the assumption that the channel characteristics and the speech signal are statistically independent, the first and second order statistics of the cepstral parameters for a particular pattern (in our case a subword unit s) can be written according to Eq. 2.6 as

$$E\{c_s^{'}(t,\tau)\} = E\{c_h(\tau)\} + E\{c_s(t,\tau)\} \tag{2.7}$$

and

$$Cov\{c_s^{'}(t,\tau)\} = Cov\{c_h(\tau)\} + Cov\{c_s(t,\tau)\}, \tag{2.8}$$

where $E\{\}$ denotes the expected value and $Cov\{\}$ denotes the covariance. Thus, the emission probability density functions of the states associated with s contain a contribution due to the statistical differences between the channels. The estimates of the means are biased with the unknown function $E\{c_h(\tau)\}$, which approaches the average channel cepstrum if enough different channels are used. At the same time, the covariance estimates are increased by the factor $Cov\{c_h(\tau)\}$. If the training speech covers a sufficient number of different but representative channels, the estimated parameters of the emission probability density functions may be expected to adhere to Eqs. 2.7 and 2.8, (e.g. Hermansky, Morgan, Bayya & Kohn 1991, Hirsch, Meyer & Ruehl 1991, Aikawa, Singer, Kawahara & Tohkura 1993, Haeb-Umbach, Beyerlein & Geller 1995, Junqua, Fohr, Mari, Applebaum & Hanson 1995, Nadeu, Paches-Leal & Juang 1995, Singer, Paliwal, Beppu & Sagisaka 1995, de Veth & Boves 1996). Thus, using many different channels in training helps to reduce the impact of specific channels on the eventual models. However, it is of limited help when an unknown utterance must be recognized, because there is no guarantee that the bias due to the particular channel at hand is close to the average channel $E\{c_h(\tau)\}$.

Explicit channel estimation

Some authors have proposed to estimate the contributions of the channel explicitly (statistical channel modeling, Gish, Karnovsky, Krasner, Roucos, Schwartz & Wolf 1985, Gish, Krasner, Russell & Wolf 1986). This estimate is then used to correct the HMM parameters or the speech feature values. The well-known cepstrum mean subtraction technique (Atal 1974, Furui 1981) can be considered as a form of explicit channel estimation. The cepstrum mean $\hat{c}_{mean}(\tau)$ is computed by taking the average over all speech signal frames

$$\hat{c}_{mean}(\tau) = \frac{1}{T}\sum_{t=1}^{T}[c^{'}(t,\tau) + c_h(\tau)] = \hat{c}_{utt}(\tau) + \hat{c}_h(\tau). \quad (2.9)$$

As indicated in Eq. 2.9, the cepstrum mean consists of the average speech cepstrum of the utterance $\hat{c}_{utt}(\tau)$ (an approximation of the average speech cepstrum of the person who produced the utterance) and the contribution due to the channel $\hat{c}_h(\tau)$. As can be inferred from Eq. 2.9, cepstrum mean subtraction has two effects. Firstly, the variability that is due to differences between speakers is reduced in the statistics for a particular speech sound. Secondly, cepstrum mean subtraction reduces the variability due to differences between communication channels used when recording that sound.

Note that non-speech signal portions are not used in Eq. 2.9. The reason for this can be understood as follows. According to Eq. 2.1, the observed spectrum in non-speech signal portions is

$$S^{'}(t,\omega) = H_0(t,\omega) + H_1(t,\omega)N(t,\omega). \quad (2.10)$$

As can be seen, the non-speech spectrum consists of two contributions: the spontaneous activity of the channel $H_0(t,\omega)$ and the linearly filtered background noise $H_1(\omega)N(t,\omega)$. Without additional assumptions about the spontaneous activity of the channel $H_0(t,\omega)$ and the background noise $N(t,\omega)$, non-speech portions of the signal cannot be used to obtain reliable information about the linear filter characteristic of the channel $H_1(\omega)$ alone. As a consequence, using non-speech portions of the signal introduces bias in the estimate of the mean cepstrum in a way that cannot be predicted.

Filtering of log-energy or cepstral feature values

It is well-known that any differentiable function $f(t)$ can be recovered (up to a known constant) as follows when it is observed with a constant unknown bias k:

$$f(t) - f(t_{low}) = \int_{t_{low}}^{t} \frac{d}{dt'}[f(t') + k]dt', \quad (2.11)$$

where t_{low} satisfies $-\infty < t_{low} < t$. Due to the differentiate and re-integrate operation the unknown bias term k is replaced by the constant term $f(t_{low})$.

According to Eq. 2.6, the contribution of the channel results in a time-invariant additive bias for each cepstral coefficient, which is independent of the original speech cepstrum. When applying Eq. 2.11 to the cepstrum observed at the channel output, the cepstrum will no longer depend on the channel after such a differentiate and re-integrate operation. Recalling our assumption in Eq. 2.3 about the energy of the silent signal portions being very small compared to the energy of speech portions, we see that the contribution $f(t_{low})$

will vanish if we take t_{low} in a silent portion (e.g., in the leading silence at the beginning of the utterance). As a result, the cepstrum of the original speech signal is obtained.

The differentiate-and-integrate operation can be implemented as a linear filter. Filtering can be performed either in the log-energy domain (Hermansky et al. 1991, Hermansky & Morgan 1994) or in the cepstral domain (Haeb-Umbach et al. 1995, de Veth & Bourlard 1995, de Veth & Boves 1996). It was shown that channel robustness can also be improved if the re-integration of Eq. 2.11 is omitted (Furui 1981, Soong & Rosenberg 1986). In that case, channel robustness is improved because the constant bias term k actually corresponds to DC in the modulation spectrum and the differentiation effectively attenuates this DC component.

If a properly designed leaky integrator is used, the differentiate-and-integrate operation will also be effective if the channel transfer function is slowly time-varying. This approach of the unknown channel problem has resulted in many different proposals for filtering the observed sequence of cepstral parameters, for instance RASTA filtering (Hermansky et al. 1991, Hermansky & Morgan 1994), the Gaussian dynamic cepstrum representation (Aikawa et al. 1993, Singer et al. 1995, Boda, de Veth & Boves 1996), the high-pass filter method proposed in Hirsch et al. (1991), Slepian filters (Nadeu, Paches-Leal & Juang 1995), phase-corrected RASTA (de Veth & Boves 1996, de Veth & Boves 1997b) and combinations of these methods (Junqua et al. 1995).

The cepstrum mean subtraction technique (Atal 1974, Furui 1981) can also be formulated as a linear filter operation. If the channel estimate is calculated over the full length of the speech utterance (as can be done in off-line experiments), then cepstrum mean subtraction can be interpreted as a FIR filter operation, with the filter adjusted to the length of each utterance. If a running mean is used to obtain the channel estimate (which is the common *modus operandi* for cepstrum mean subtraction in real-time applications) the definition of the FIR filter is the same for each utterance. The Gaussian Dynamic Cepstrum Representation (Aikawa et al. 1993) and the discrete cosine transform (Milner & Vaseghi 1995) are other examples of FIR filtering, although none of these techniques was originally presented in that formalism.

2.2.3 Comparison of feature track filtering techniques

Using the linear transfer function description of Eq. 2.4, it is easy to show that the differences between the many techniques for undoing the effect of the channel relate to either the way in which the parameters of the linear filter $H_1(\omega)$ are estimated, and/or to the way in which the operation, used to undo the effect of $H_1(\omega)$ is implemented. In this Section, the focus is on the effects of the implementation of the filter. We will show that the details of the implementation are important, because these details interfere with the type of models that are used in speech recognition. Although it would have been interesting to show that the arguments developed in this Section also hold for more sophisticated models of the channel, and the attendant more complex techniques to undo these effects in a recogniser (Rahim & Juang 1996, Junqua & Haton 1996), we will limit ourselves to the techniques decribed below due to space limitations.

Effect of the filter phase response

We concentrate our discussion on a comparison of three CN techniques which can be described in terms of cepstral filtering: RASTA filtering, cepstrum mean subtraction and phase-corrected RASTA. RASTA filtering (Hermansky et al. 1991, Hermansky & Morgan 1994) is interesting for several reasons. First, it is well-known that RASTA filtering is effective with units that incorporate context dependency such as words or triphones. In addition, RASTA can be viewed as a crude model of auditory time masking, and it has been argued that this correspondence to perception accounts for much of its effectiveness (Hermansky & Pavel 1995, Hermansky 1996). However, Eq. 2.11 was the original point of departure for RASTA (Hermansky et al. 1991); its relation to auditory masking was only established later on (Hermansky & Pavel 1995, Hermansky 1996). The second CN technique discussed here is cepstrum mean subtraction (Atal 1974, Furui 1981), because this technique is very simple, yet highly effective (Steinbiss, Ney, Aubert, Besling, Dugast, Essen, Geller, Haeb-Umbach, Kneser, Meier, Oerder & Tran 1995, Haeb-Umbach et al. 1995). In two independent studies the effectiveness of RASTA filtering and cepstrum mean subtraction was compared in a recognition set-up based on context independent HMMs (CI-HMMs) (Haeb-Umbach et al. 1995, de Veth & Boves 1998a). In both studies the task was recognition of digit strings, be it that different languages were being used: Haeb-Umbach et al. (1995) used German and American English, while de Veth & Boves (1998a) used Dutch. It was found in both studies that RASTA filtering is effective as a CN technique, but that cepstrum mean subtraction is more effective. The apparent limited effectiveness of RASTA filtering can be attributed to the well-known left-context dependency introduced by the RASTA filtering (Koehler, Morgan, Hermansky, Hirsch & Tong 1994, Hermansky & Morgan 1994). To be able to better understand this left-context dependency, and to be able to verify that this is indeed the underlying cause for the limited effectiveness of RASTA filtering when compared to cepstrum mean normalization, phase-corrected RASTA was introduced in de Veth & Boves (1998a).

We start our discussion by considering the signal shown in the upper panel of Fig. 2.3. This artificial signal is intended to represent a sequence of cepstral values for quefrency τ. It models a sequence of seven time-invariant 'speech' states, preceded and followed by a rest state ('silence'). The signal contains a constant DC-component that represents the effect of the channel. The RASTA filtered version of this signal is shown in the middle panel of Fig. 2.3.

Two observations can be made. First, the DC-component has been removed (at least for times larger than, say, 70 frames). Second, the shape of the

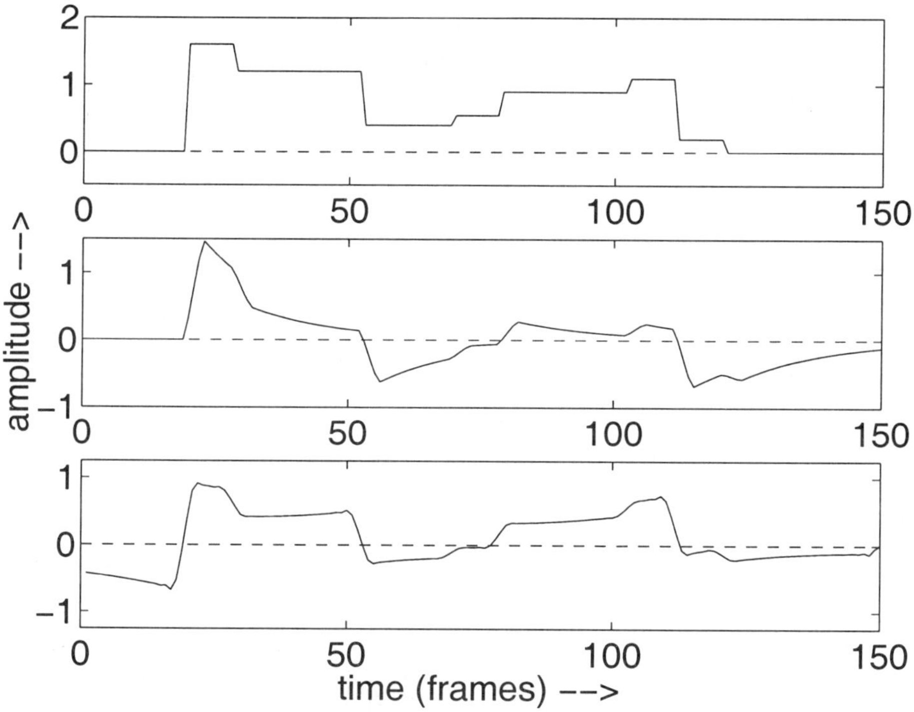

FIGURE 2.3: Synthetic signal representing one of the cepstral coefficients in the feature vector. Upper panel: Original signal containing a time-invariant DC-offset. Middle panel: RASTA filtered signal. Lower panel: Phase-corrected RASTA filtered signal.

signal has been altered. Originally, the states of the signal had a constant amplitude. After filtering, the amplitude for each state drifts towards zero, while the values immediately after an abrupt change are more or less preserved. This explains why RASTA enhances the dynamic parts in the spectrum of a speech signal (Hermansky & Morgan 1994). However, a description of the signal in terms of states with well-defined means and small variances becomes less accurate. Worse even, the mean amplitude of each state has become a function of the state itself AND of the preceding states. This is the left-context dependency in RASTA (Koehler et al. 1994, Hermansky & Morgan 1994).

To identify the origin of this left-context dependency, we take a close look at the frequency response $H_R(\nu)$ of the classical RASTA filter which can be written as

$$H_R(\nu) = |H_R(\nu)|\, e^{j\phi_R(\nu)}, \tag{2.12}$$

with ν the modulation frequency (in radians), $|H_R(\nu)|$ the magnitude response and $\phi_R(\nu)$ the phase response. The magnitude and phase response of the RASTA filter with integration factor $a = -0.94$ are shown in Figs. 2.4AB for modulation frequencies between $0 - 20$ Hz. This range includes the $2 - 16$ Hz region which has been shown to be most important for human speech recognition (Drullman, Festen & Plomp 1994). From Fig. 2.4B it can be seen that the phase response is non-linear for modulation frequencies below approximately 3 Hz. This non-linearity causes the time-domain shape distortions observed in the middle panel of Fig. 2.3.

To compensate for the phase distortion of the RASTA filter, while preserving its magnitude response, we followed a procedure suggested in Hunt (1978). After the RASTA filter an all-pass filter is applied whose phase response $\phi_{pc}(\nu)$ is exactly the inverse of the phase response of the RASTA filter

$$\phi_{pc}(\nu) = -\phi_R(\nu). \tag{2.13}$$

Thus, the frequency response $H_{pc}(\nu)$ of the phase correction filter is

$$H_{pc}(\nu) = e^{-j\phi_R(\nu)}. \tag{2.14}$$

With this phase correction, the frequency response $H_{pcR}(\nu)$ of the complete phase-corrected RASTA filter is

$$H_{pcR}(\nu) = H_R \times H_{pc} = |H_R(\nu)|\,. \tag{2.15}$$

The phase correction filter $H_{pc}(\nu)$ can be implemented as a pole-zero filter, obtained by solving for the coefficients {b,a} that satisfy

$$e^{-j\phi_R(\nu)} = \frac{b_0 + b_1 e^{-j\nu} + \ldots + b_q e^{-jq\nu}}{1 + a_1 e^{-j\nu} + \ldots + a_p e^{-jp\nu}}, \tag{2.16}$$

where q (p) is the order of the numerator (denominator) polynomial. In de Veth & Boves (1998a) a Matlab procedure with $q = 1$ and $p = 7$ was used to calculate the {b,a} coefficients (Little & Shure 1993). As it turns out, three of

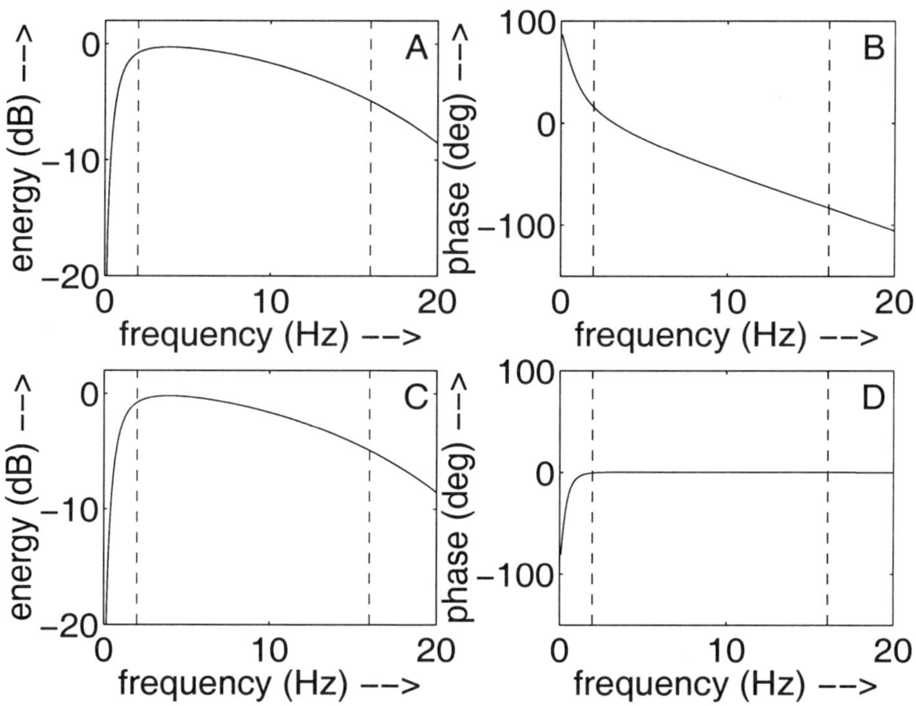

FIGURE 2.4: (A) Log-energy response of classical RASTA. (B) Phase response of classical RASTA. (C) Log-energy response of phase-corrected RASTA. (D) Phase response of phase-corrected RASTA.

the seven poles of the phase-correction filter are lying outside the unit circle, while the zero is lying inside. Due to the poles lying outside the unit circle, the phase-correction filter is unstable. Therefore, it cannot be applied directly to the RASTA filtered signal. However, the inverse of this filter is stable and, as a result, in off-line experiments the following engineering trick can be used (Hunt 1978): (1) reverse the RASTA-filtered signal in time, (2) take the inverse of the pole-zero phase-correction filter, (3) apply the inverted phase-correction filter to the time-reversed RASTA-filtered signal and (4) reverse the resulting signal in time. With a slight performance penalty the non-causal filtering can be cast in a form that allows a real-time implementation with short processing delay (de Veth & Boves 1997a). In this paper we will only discuss results for off-line experiments.

In Figs. 2.4CD the magnitude and phase response of the phase-corrected RASTA filter are shown. It can be seen that the magnitude response is almost identical to the original one and that the new phase response is flat and very close to zero in the region of important modulation frequencies. The result for phase-corrected RASTA in the time domain is shown in the lowest panel of Fig. 2.3. The shape of the phase-corrected RASTA filtered signal closely resembles that of the original signal. The phase correction effectively removes the amplitude drift towards zero in time-invariant parts of the signal and decreases the left-context dependency. Thus, phase-corrected RASTA does not model temporal masking, but it is in better agreement with the usual model of a speech utterance as a sequence of time-invariant states.

Continuous speech recognition with phase-corrected RASTA

We compared the recognition performance of classical RASTA, phase-corrected RASTA and cepstrum mean subtraction for a continuous speech recognition task, where utterances recorded over the (land-line) public switched telephone network were used. About nine hours of speech were used for training, while three hours of speech were used for testing. The recognition lexicon contained 983 words. 1.2% of the words in the test set were out-of-vocabulary. The test set perplexity of the recognition task was 36.7. For our evaluations we restricted ourselves to the single best recognized sentence. Full details of these experiments can be found in de Veth & Boves (1998b).

We trained and tested HMMs for four different channel normalization conditions, i.e., no channel normalization (NCN), classical RASTA (clR), cepstrum mean subtraction (CMS) and phase-corrected RASTA (pcR) in combination with two different recognizer set-ups, i.e. context independent phone-based HMMs (CI-HMMs) and context dependent phone-based HMMs (CD-HMMs). In these experiments, the off-line versions of CMS and pcR were used. In other words, we used the whole utterance when we computed the cepstrum mean and when we applied the time-reversal operation needed for pcR. Taking the number of substitution, deletion and insertion errors into account, we computed the word error rate for all combinations of channel normalization

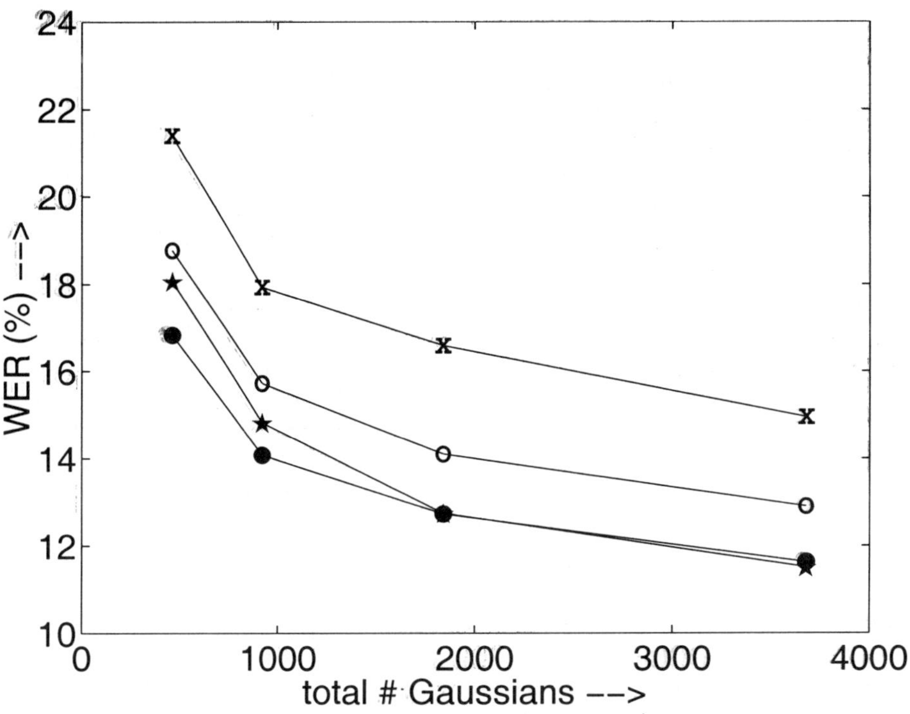

FIGURE 2.5: Recognition accuracy as a function of the total number of Gaussians in the trained HMM set for clR (×), pcR (⋆) and CMS(•), compared to the feature set without CN (○) when using CI-HMMs.

method and recognizer set-up, where we varied the number of Gaussians used to describe the emission probability density function of each state. For the CI-HMMs, mixtures with 4, 8, 16 and 32 Gaussians per state were used. This corresponds to using a total number of Gaussians of 460, 920, 1840 and 3680 respectively. The results for the different CN techniques with CI-HMMs are shown in Fig. 2.5.

From Fig. 2.5 it can be seen that clR deteriorates recognition performance compared to NCN, when CI-HMMs are used. Removing the channel bias by using clR introduces so much left-context dependency that the potential CN gain is completely annihilated. The results for pcR indicate that the poor performance of classical RASTA is a direct consequence of the phase distortion. By removing the phase distortion the recognition performance is significantly and substantially improved compared to clR. At the highest total number of Gaussians in this CI-HMM system the WER is reduced by 23% relative to clR. In addition, for the more complex acoustic models pcR recognition performance is significantly better than NCN and in fact becomes as good as CMS.

It is interesting to compare these results to the results reported in de Veth

& Boves (1998a). Whereas the continuous speech results in Fig. 2.5 show that clR actually decreases recognition performance relative to NCN, the digit experiments in Haeb-Umbach et al. (1995) and de Veth & Boves (1998a) showed that clR is viable as a CN technique. These findings may seem contradictory at first glance, but can be understood if one realizes that the main difference between these two set-ups is the number of different phone contexts. In fact, the number of different phone contexts for the continuous speech recognizer is more than 70 times as large as in the digit recognizer (de Veth & Boves 1998b). As a consequence, the loss of recognition performance due to enhancement of the left-context dependencies is likely to be much more important in the continuous speech recognizer. Apparently, this effect is so strong that it completely annihilates the potential performance gain obtained from the attenuation of modulation components near DC. In the case of the digit recognizer, the net effect of RASTA filtering is still positive, because the performance gain obtained by suppressing the DC component is less affected by the left-context effect. The number of different contexts for the digits vocabulary is apparently so small that the models are effectively context dependent.

When using clR with CD-HMMs one would expect that the loss of recognition performance due to the left-context effect is diminished, because different contexts are modeled with different states. When every individual left context could be modeled independently, the left-context effect should disappear completely, and the CN effect should remain in its full strength. However, under all practical conditions in continuous speech recognition, the amount of training data is not sufficient to model each left context independently. This lack of training data forces one to pool the data from different contexts for subword units with low occurrence counts. In our experiments the data sharing for infrequent units was implemented as a data-driven state-tying mechanism. Due to the data sharing, one can no longer expect that the loss of recognition performance caused by the left-context effect of clR is completely annihilated.

The results for CN techniques with CD-HMMs are shown in Fig. 2.6. In this set-up we used HMMs with 1, 2, 4 and 8 Gaussians per state, corresponding to a total number of Gaussians of 388, 776, 1552 and 3104 respectively. First, it can be seen that the difference between clR and NCN has become smaller than the one we observed for CI-HMMs. For the best CI-HMMs clR decreases recognition performance by 16% relative to NCN. In the case of the best CD-HMMs the performance only drops 9%. Thus, we have some gain when switching from CI- to CD-HMMs in the case of clR, but this improvement is limited due to the state-tying mechanism that is used to avoid undertraining. However, even with CD-HMMs the detrimental effect of the left-context dependency is still stronger than the beneficial effect of CN in this recognition task.

It can also be seen that introducing the phase-correction brings the recognition performance curve very close to the one for CMS (except at the models corresponding to 1 Gaussian per state). For the CD-HMMs corresponding to 8 Gaussians per state, WER is improved by 15% when clR is replaced by

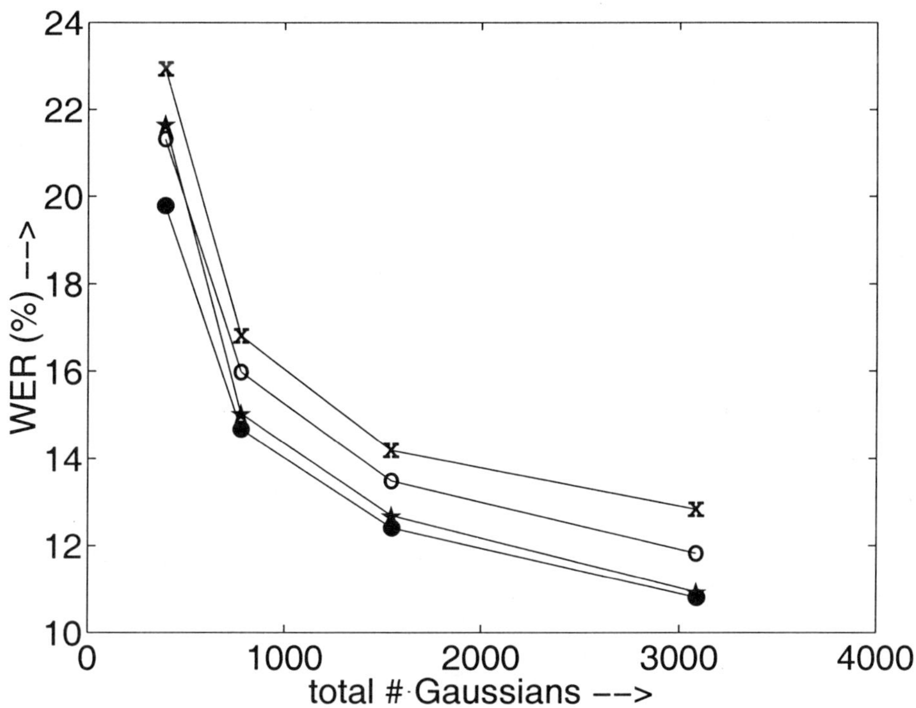

FIGURE 2.6: Recognition accuracy as a function of the total number of Gaussians in the trained HMM set for clR (×), pcR (⋆) and CMS(•), compared to the feature set without CN (○) when using CD-HMMs.

pcR. This is in good agreement with the results of pcR obtained for CI-HMMs.

2.2.4 Conclusions

Most, if not all techniques intended to eliminate the variability introduced by the communication channel imply some form of filtering of the sequence of feature vectors. The results of the comparison of three different techniques for CN and the explanation of these results show that care must be taken that the phase response of the filter is linear. In other words: The overall shape of the feature track must be preserved as much as possible. This requirement is due to the structure of the basic patterns which represent the spoken words. In most cases speech is modeled as a sequence of essentially time-invariant states, which are only dependent on a very local context. Any phase distortion caused by a filter that is applied to remove the influence of the channel by necessity interferes with the time-invariance and independence assumptions. In the particular case of classical RASTA, the signal segments represented by conventional sub-word units are much shorter than the RASTA filter memory. This results in a conflict between the intrinsic nature of the feature values after filtering and the assumptions underlying the structure of the speech model. As a consequence, the intended beneficial effect of this CN technique is completely destroyed by the negative effect of the phase-distortion.

These findings show that any technique to improve robustness can only be expected to yield improved recognition results as long as it is compatible with the basic assumptions made in the models of the speech signal and in the procedure to search for the best matching patterns. Specifically, and in a way unfortunately, this implies that findings from human speech perception cannot simply be re-used in ASR algorithms which model speech as a sequence of discrete, time-invariant, context insensitive units.

2.3 Robustness against background noise

2.3.1 Assumptions for noise robustness

To understand the effect of acoustic background noise on the feature values at the input of an ASR system, we must return to Eq. 2.1. We keep the assumption that the channel characteristics are time-invariant (cf. Eq. 2.2), but we drop the additional assumption that the magnitudes of the terms related to noise are negligible relative to the magnitudes of the terms related to the speech signal. This is equivalent to the assumption that

$$|H_0(t,\omega) + H_1(\omega)N(t,\omega)| \approx |H_1(\omega)S(t,\omega)|. \tag{2.17}$$

We then have

$$S^{'}(t,\omega) = H_0(t,\omega) + H_1(\omega)[S(t,\omega) + N(t,\omega)]. \tag{2.18}$$

Without much loss of generality, Eq. 2.18 can be simplified by lumping all additive components into a single, possibly time-varying, noise component $U(t,\omega)$:

$$S^{'}(t,\omega) = H_1(\omega)S(t,\omega) + U(t,\omega). \tag{2.19}$$

The general model in Eq. 2.19 is a good starting point for discussing different approaches for improving noise robustness. As we already discussed in Subsection 2.1.2, the choice for a particular approach depends on the assumptions that can be made about our knowledge of $U(t,\omega)$, or perhaps more accurately, on the possibility to obtain useful parametric estimates of $U(t,\omega)$ in a specific situation. The model of $U(t,\omega)$ must be more simple as the noise becomes more variable between situations and unpredictable for a specific situation. Attempts to undo the effect of additive noise can be classified according to their working domain. Popular approaches include methods (1) to clean the acoustic features, (2) to adapt the models trained on clean speech to noisy conditions and (3) to adapt the distance computation in the Dynamic Programming search. These approaches essentially try to reduce the variation in the feature values due to the noise or they try to limit the impact of this variation on the computation of the similarity between new observations and pre-existing models.

It is reasonable to assume that noise robustness of an ASR system will increase if methods developed in the different domains are properly combined. Although it might seem attractive to compare the performance of individual approaches for improved noise robustness, such a straightforward comparison is hardly fair. Some approaches may be inherently more effective with certain types of distortions. In addition, experience has shown that the effectiveness of robustness techniques may be dependent on the details of the implementation. For these reasons, we refrain from making direct comparisons between different noise robustness approaches. We will limit the discussion to summary descriptions of observation cleaning, predictive model-based compensation and model adaptation, and focus in more detail on a new approach, which is formulated in the local distance computation domain. For an extensive review of observation cleaning (and other noise robustness techniques), we refer to Gong (1995). Recently, many predictive model-based compensation schemes were reviewed in Gales (1998). Finally, we refer to Lee (1998) for an excellent review of model compensation and model adaptation techniques.

2.3.2 Three domains for noise robustness

Feature domain

We first consider a well-known noise robustness method which is defined in the feature domain. In those scenarios where it is reasonable to assume that the noise is quasi time-invariant, an obvious strategy would be to make an estimate of the noise spectrum $U(\omega)$ and to subtract it from the noisy input spectra $S^{'}(t,\omega)$. This strategy is known as *spectral subtraction* (Boll 1979, Lockwood & Boudy 1992) and has a long tradition in research in speech enhancement, i.e., processing of noisy speech to make it more pleasant and intelligible for humans. Spectral subtraction can be regarded as a classic example of the idea to try and find a feature representation for which the statistical characteristics are minimally affected by the background noise. As long as ways can be found to reliably estimate the background noise characteristics, spectral subtraction is a useful pre-processing step that will increase recognition robustness and can be combined with any of the strategies yet to be described.

Model domain

If it is reasonable to assume that the ASR system is always used in the same noise environment, probably the simplest way of handling the problem is by training models using speech recorded in that particular environment (e.g. Dautrich, Rabiner & Martin 1983), or speech corrupted by artificially adding the noise (e.g. Gales 1995). These approaches have shown good results, but their use is limited to those situations where the speech to be recognized is always picked up in the same noise environment. Moreover, with this approach new models need to be trained for each new type of noise. Finally, artificially adding noise to clean recordings is only effective as long as the noise level under actual conditions is not so high that it gives rise to the Lombard reflex. This kind of spontaneous adaptation of the speech production enhances human intelligibility, but may very well harm ASR performance (Junqua 1996).

If the noise is not easily predicted, but one can still obtain a reliable estimate of $U(t,\omega)$, one might want to use that estimate to adapt the observation distributions in the models trained on clean speech (Lee 1998, Lee & Huo 1999). Another example of a set of approaches developed for the model domain, is *predictive model combination, PMC,* also known as *parallel model combination* (Gales 1998). In this case, the idea is to train separate models of noise and speech; if necessary, different types of noise can be modelled in parallel. During recognition the most likely combination of speech sounds and noise is computed. Searching the optimal path while using both noise and speech models leads to a three dimensional dynamic programming problem (Varga & Moore 1990) with time, speech states and noise states as the three dimensions. If the noise can be described by an ergodic HMM, the three dimensional search problem can be converted into a conventional two dimensional search (Gales 1998).

Despite the good results reported for different implementations of the PMC scheme (see for example Gales 1998), such an approach is not always feasible. In particular, the usefulness of a PMC approach may be limited for two reasons, which are both growing more important with the increased use of mobile phones. Firstly, even if it is perfectly known beforehand what different noises can occur, the choice for the appropriate noise type will have to be made at recognition time. The decision will become more difficult as the number of different noise types known to the ASR system increases. Secondly, if the noise is time-variant, then the need will arise to continuously update the noise model on-line. Due to lack of observations, the noise model estimate may become poor to the extent that it limits the effectiveness of the compensation technique (Gales 1998). These difficulties have inspired people to look for approaches that make less specific assumptions on how $U(t,\omega)$ affects the features or the models. These are the approaches developed in the distance computation domain.

Distance Computation Domain

The basic mechanism in an approach that does not rely on explicit estimates of the noise in terms of features or models consists of changing the similarity measurement between the trained models and the test utterance. The moment one realizes that feature values have an inherent uncertainty due to the presence of acoustic background noise, it is only natural to try to develop decision strategies that are primarily based on feature values that are least affected by noise characteristics. In Lee & Huo (1999) a number of such robust decision methods are discussed. These methods all attempt to account explicitly for the uncertainty in the feature values.

A somewhat different starting point is taken in the approaches based on *Missing Feature Theory* (MFT) (Cooke, Morris & Green 1996, Morris, Cooke & Green 1998). According to Eq. 2.19 the signal spectrum at the ASR input can be considered as a mixture of a reliable component (i.e. the channel filtered original speech spectrum) and an unreliable component (i.e. the unknown noise contribution). Depending on the exact nature of the distortion some of the observed values in the acoustic feature vector may still be reliable, while other values may have become unreliable. In several recent proposals the key idea is to somehow disregard the unreliable information and base recognition on reliable information only. This idea can be pursued in different manners. First, let us suppose that one is working with acoustic feature vectors that are defined in the spectral domain, e.g., filter bank outputs. Then, if it can be assumed that $U(t,\omega)$ takes non-negligible values only for a limited number of time frames t or a limited range of frequencies ω, the *marginalisation approach* of MFT can be used (Cooke et al. 1996, Morris et al. 1998). With these assumptions it might be possible to explicitly detect all time-frequency regions where the observed feature values are dominated by $U(t,\omega)$, and either discard these features (Cooke et al. 1996, Dupont, Bourlard & Ris 1997, Tibrewala & Hermansky 1997, Lippmann & Carlson 1997, Morris et al. 1998) or correct them in some way or another (Cooke et al. 1996, Morris et al. 1998, Dupont 1998, Raj, Singh & Stern 1998). Of course, the problem then immediately arises how corrupted values can be reliably detected. In the spectro-temporal domain this is not an easy task, although good progress was recently reported (Vizinho, Green, Cooke & Josifovski 1999). If one is working with acoustic feature vectors defined in another domain (e.g., cepstra) then the detection task becomes even more intricate, because the components of the acoustic feature vectors that are significantly affected are not solely a function of the spectro-temporal characteristics of $U(t,\omega)$, but also of the transformations applied to the sequence of short-time spectra. We will elaborate this issue in more detail below.

Recently, a new way was suggested to handle contaminated feature values, which is not restricted to spectral features, and avoids the need to define a detector that is running independently from the decoder for identification of unreliable acoustic feature vector components. This idea, which was proposed in de Veth, Cranen & Boves (1998) and de Veth, Cranen & Boves (2000), is yet another implementation of a method where the similarity measurement has been altered to cope with the noise. It focusses on the computation of the emission probabilities in the presence of disturbed acoustic feature vectors. This approach is based on the assumption that the statistical models built for clean speech are not proper models for observations obtained in the presence of noise. By modifying the function used to evaluate the match between an input sequence of acoustic feature vectors and each one of the candidate sequences of acoustic models (cf. Fig. 2.1), a situation is created in which unlikely feature values affect the search to a lesser degree. For convenience we use the term 'local distance function' when we refer to the mathematical expression used to evaluate the cost of assuming that a given sound segment pertains to a given HMM state.

If there is noise present at recognition time that was not present when the models were trained (i.e., in a mismatched training test condition), it is *a priori* evident that not all observations were actually seen in the training phase. Therefore, some part of the total probability mass is set apart to account for the unseen observations. For recognition, a new robust local distance function can then be determined by interpolating between the contributions of the cost for 'seen' and 'unseen' observations:

$$-\log[p(O)] = -\log[(1-\epsilon).p(O|seen) + \epsilon.p(O|unseen)], \tag{2.20}$$

where $p(O)$ denotes the probability of the observation O, $p(O|seen)$ is the probability of the observation according to the data seen during during training, $p(O|unseen)$ the probability of the observation according the unknown process, and ϵ the a priori probability that an observation is generated that was not seen during training. The idea of the robust local distance function in Eq. 2.20 is in fact an attempt to incorporate the well-known Tukey-Huber distortion model (Huber 1981) in the recognition stage of an otherwise conventional HMM-based ASR (de Veth et al. 2000). What is essential here is the assumption that an observed event is the realization of a mixture of two processes. The first is the known process of which the parameters could be reliably estimated in the training phase, i.e., the process which produced the set of all seen observations. The second process is the one that produces all observations that were not seen in the training data. The only thing that is known about this second process is that some observations will be generated at recognition time that were not seen during training.

The idea that individual observations may originate from a mixture of a known and an unknown process plays an important role in the theories of Statistical Robustness (Huber 1981) and Robust Statistical Pattern Recognition (Kharin 1996). In the cases that we want to address (speaker independent recognition over the telephone), it is impossible to estimate the distortion distributions from the training speech. In addition, it is difficult to obtain a reliable estimate of the distribution of distortions from the unknown speech that is to be recognized. Under these two conditions we find ourselves in the situation (again) that it may be better to use an overly simplistic model, than to try to use a more sophisticated model. As we will see below, it is indeed possible to improve recognition performance based on an extremely simple assumption about the distribution of the observation values that were not seen during training.

In the remainder of this Section, we will first explain the robust local distance function in more detail. Next, we will introduce a topic that has not attracted much attention during the last decades, but that still might prove to be of considerable importance, viz. the way several transformations of the sequence of short-time spectra in the presence of additive noise (cf. Eq. 2.19) may affect the recognition result.

2.3.3 Disregarding unreliable information

Robust local distance function

As stated before, the pattern match in state-of-the-art ASR systems is implemented as a search through frame-state space in the form of a dynamic programming algorithm (usually a Viterbi algorithm). For each acoustic feature vector, it is decided how each candidate optimal partial path so far is best extended with any of the HMM states that are candidates for extension. For each candidate optimal partial path, that state is selected which minimizes the path extension cost. For HMMs, this path extension cost is the combination of the emission cost of the candidate extension state and the transition cost for jumping to the candidate extension state (Rabiner 1988). In what follows we will concentrate on the emission cost, since experience has shown that transition costs can actually be disregarded in a practical system without significant loss of recognition performance. Assuming that we really do not have any prior knowledge about the noise, which is not unreasonable when dealing with speech recognition over the telephone, one might reason as follows. An actually observed acoustic feature vector (or vector component) can be considered to be the realization of a mixture of two random processes: the known process as observed during training and the unknown process of all observations not previously seen. There is no need to explicitly determine by which of the two processes the observation was generated. It suffices to determine the emission cost due to the mixture of these two processes.

For an HMM state $\mathcal{S}_i$ that is described by a mixture of M Gaussian probability density functions the conventional local distance function d_{loc}, which we approximated to be equal to the emission cost, is described as

$$d_{loc}(\mathcal{S}_i, \mathbf{x}(t)) = -\log\{\sum_{m=1}^{M} w_{im} \prod_{k=1}^{K} G_{imk}(x_k(t))\}, \tag{2.21}$$

where $\mathbf{x}(t)$ denotes the acoustic observation vector at time t, w_{im} denotes the m-th mixture weight for state $\mathcal{S}_i$, K denotes the dimension of the acoustic observation vector, $x_k(t)$ the k-th component of $\mathbf{x}(t)$, and G_{imk} the k-th component of the m-th Gaussian probability density function for state $\mathcal{S}_i$. The robust local distance function d_{robust} defined in de Veth et al. (1998) and de Veth et al. (2000) is

$$d_{robust}(\mathcal{S}_i, \mathbf{x}(t)) = -\log\{\sum_{m=1}^{M} w_{im} \prod_{k=1}^{K} [(1-\epsilon)G_{imk}(x_k(t))+\epsilon\hat{p}_0(x_k(t))]\}, \tag{2.22}$$

where ϵ denotes the a priori probability that a feature value originates from the distribution of disturbed, unreliable speech values ($0 \leq \epsilon < 1$) and $\hat{p}_0(x_k(t))$

denotes the unknown probability density function used to compute the probability for observing an outlier with value $x_k(t)$. It can be seen that Eq. 2.22 reduces to Eq. 2.21 if we choose $\epsilon = 0$.

Having reached this point, we still need to decide how the unknown process is best statistically described, where 'best' means optimal according to the principles of Robust Statistical Pattern Recognition. For the particular problem we study (i.e., how to make the computation of the local cost in the search robust) the best description of the unknown process is, as yet, an open question. However, this does not mean that one cannot make a sensible choice based on practical considerations. In de Veth et al. (1998) and de Veth et al. (2000) it was proposed to model the unknown distribution as a uniform distribution, because this choice reflects our assumption best that we do not have any prior knowledge about the unknown process.

Another decision that remains to be made is how to choose the a priori probability ϵ that a feature value originates from the distribution of values not seen during training. Without additional assumptions about the noise distortion there is no obvious way in which the 'optimal' value of ϵ can be found. According to the experience gained so far, it appears to be reasonable to choose the Acoustic Backing-off parameter ϵ such that the recognition performance in the matched training-test condition does not suffer too much, while in the mismatched condition the word error rate is maximally decreased (de Veth, de Wet, Cranen & Boves 1999, de Veth, Cranen, de Wet & Boves 1999).

We will now explain the effect of using the robust local distance function as defined in Eq. 2.22 and why the choice to model the unknown distribution as a uniform distribution is already convenient. In Fig. 2.7 the local distance functions corresponding to two competing, active HMM states (marked i and j) are shown for the conventional local distance computation. For illustration purposes, we have assumed that the emission probability density function $p(x_k|S_i)$ is modeled as a single Gaussian. Then the local distance becomes a quadratic function of the difference between the value of the observed feature vector component and the mean value of the given distribution. We consider three different observation values: one undisturbed value corresponding to the clean condition (marked 'clean') and two different disturbed ones (marked 'd1' and 'd2'). We assume that the frame vector actually 'belongs' to state i. It can be seen that the contribution to the emission cost is lower for state i than for state j for the clean observation value. Now consider disturbed observation values 'd1' and 'd2'. In both cases the contribution to the emission cost for state i is (much) higher than the one for state j, thereby increasing the risk of recognition errors.

In Fig. 2.8 the same situation is depicted, but now the conventional local distance functions have been replaced by their robust versions. As can be seen, state i is being preferred over state j in the clean condition as before. However, for the distorted observation value 'd1' the contributions to the emission cost for states i and j have become identical. As a result the corrupted value will

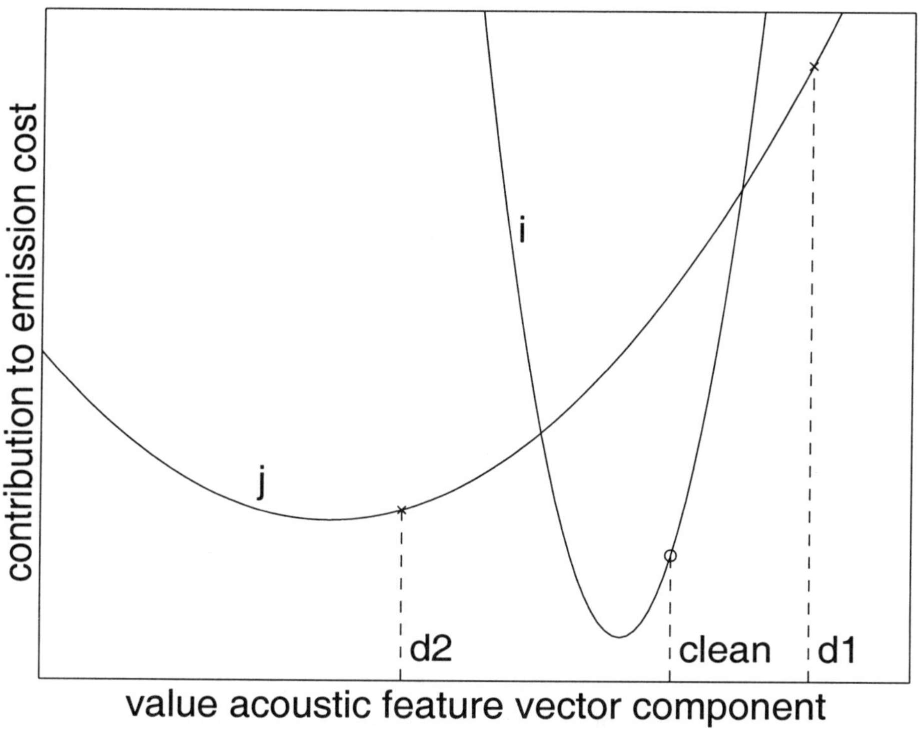

FIGURE 2.7: The contribution to the emission cost as a function of the observation value of one acoustic feature vector component for two competing states (indicated as 'i' and 'j'), when a conventional local distance function is used. Assuming that the current observation actually corresponds to state 'i', three observation values are considered: a reliable, undistorted observation value ('clean') and two different types of unreliable, distorted observation values ('d1' and 'd2'). For the conventional local distance function, the contribution to the emission cost due to a distorted value may lead to an unreliable assignment of the most probable state.

no longer favor the wrong state j. Obviously, it will not favor the right state i either. But if the corrupted value lies in the tail of all (or most) distributions for the active states, its contribution to the decision how to extend the candidate optimal partial path best is effectively canceled. If sufficient components of the acoustic feature vector of this frame contain uncorrupted values, they will discriminate between the active states and weigh in favor of the correct one. Obviously, this approach is not capable of removing the detrimental influence of distributional outliers of the type 'd2'. Here we are even more dependent on the presence of a sufficient number of undistorted values to compensate for the incorrect boost of the likelihood of state j.

The robust local distance function shown in Fig. 2.8 can also handle frames in which all values are corrupted, as long as the values are affected in the same manner as the 'd1' type of distortion. In this case, the emission cost for all competing states becomes essentially the same. When this happens, the frame makes no contribution to the decision of what is the best path and thus is effectively eliminated.

The effect of dispersion of unreliable information

As already indicated in the general scheme depicted in Fig. 2.1, in typical ASR systems the raw short-time spectra are not directly used for pattern matching. Most of the time, various normalization (e.g., gain normalization, channel normalization) and orthogonalization and dimensionality reducing transforms (e.g., Discrete Cosine Transform, Linear Discriminant Analysis) are applied. By using normalizing transforms, acoustic feature vectors are obtained that mainly represent the statistics of individual speech sounds and represent much less the variation due to differences in voice effort between different speakers or the variation due to different telephone channels. Orthogonalization transforms are used because they allow for more efficient modeling. For instance, only if the features are orthogonal, it is safe to assume that the covariance matrix is diagonal.

With clean speech data, normalization and orthogonalization transforms generally improve recognition performance significantly. However, a complication may arise when a subset of the components in the short-time spectrum are disturbed. In this case, corrupted values in a restricted number of short-time spectral components will be smeared out over the entire transformed vector. If this happens, the effectiveness of any strategy based on disregarding unreliable information might be jeopardized. This is readily illustrated for the case of MFT. The basic presupposition in MFT is that disturbances affect only part of the acoustic feature vector components and leave the rest intact. The idea of MFT is that recognition will be based only on those intact components. If some transform causes dispersion of the distortions over all acoustic feature vector components, none of the components are completely intact any more. The extent to which the effectiveness of MFT is undermined will then depend on how severely individual components are disturbed. In short, it is important to limit

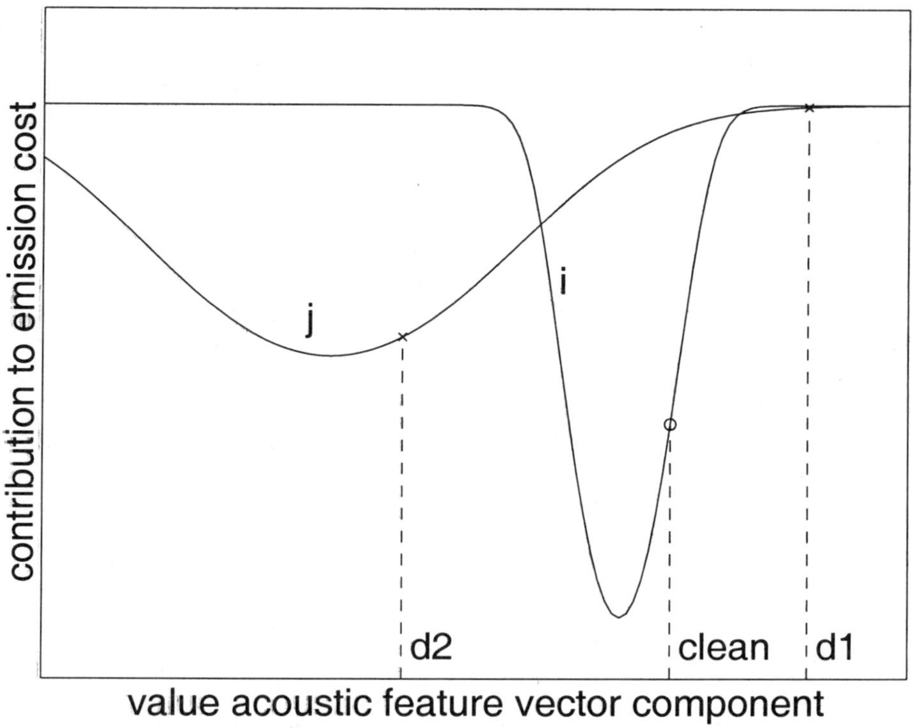

FIGURE 2.8: The contribution to the emission cost as a function of the observation value of one acoustic feature vector component for the same two competing states 'i' and 'j' as shown in Fig. 2.7, now using a robust local distance function. With the robust local distance function, the contributions to the emission cost due to distorted feature values of type 'd1' become identical for the two competing states 'i' and 'j'. As a result, the assignment of the most probable state becomes independent of this type of distorted values. For a distortion of type 'd2', however, the robust local distance function is not effective.

the spread of unreliable information in the acoustic feature vectors as much as possible, to keep the full effect of a strategy based on disregarding unreliable information.

2.3.4 Connected digit recognition with additive band-limited noise

We studied the effect of the spread of unreliable information due to acoustic feature vector transformations and the effect of using a robust local distance function in the context of connected digit recognition over the telephone. In all experiments we started with mel-frequency log-energy coefficients as the basic representation of the short-time spectrum. These are the raw features. We compared the recognition performance for two types of acoustic feature representations. The first type of features are obtained by a full-smearing transformation of the raw features, i.e., a linear combinations of *all* raw features. For ease of reference, such feature representations are called F-type. The second type of feature representations are obtained by feature transforms of the raw features that only partly smear distortions over all feature vector components (P-type features). In particular, we used within-vector mean normalized mel-frequency log-energy coefficients (in short: F1) and mel-frequency cepstral coefficients (F2) and compared these full-smearing transforms to sub-band mel-frequency cepstral coefficients (Okawa, Bocchieri & Potamianos 1998) (P1) and within-vector filtered mel-frequency log-energy coefficients (Nadeu, Hernando & Gorricho 1995) (P2). To study the effect of the type of local distance function, we conducted two sets of experiments with connected digit recognition, one set with the conventional and the other with the robust local distance function. As a distortion we used additive band-limited Gaussian noise. The cut-off frequencies of the band-pass filter were chosen such that approximately one quarter of the energy bands that we used would be contaminated by noise ($F_{low} = 395Hz$ and $F_{high} = 880Hz$). We used three different signal-to-noise ratios of 20, 10 and 5 dBA respectively, i.e., both the speech and noise energy levels were weighted according to the A-scale (Hassall & Zaveri 1979). The ten words used for the digits in Dutch were modeled using 3-state, context independent phone-based HMMs with 16 Gaussians per state. In all experiments reported here the robust local distance function was computed using $\epsilon = 0.1$. The uniform distribution that we used was defined independently for each component $k, k = 1, \ldots, K$ of the acoustic feature vector. Using all available observations in the training data, we determined a lower and upper bound ($T_{k,low}$ and $T_{k,high}$) such that 99.9% of all observations $x_k(t)$ fell within the range between $T_{k,low}$ and $T_{k,high}$. The uniform distribution for feature component k was defined to be equal to $\frac{1}{T_{k,high} - T_{k,low}}$ inside this range and zero everywhere else. More details about the robust local distance function are given in de Veth et al. (2000). Additional details about the experimental set-up can be found in de Veth, de Wet, Cranen & Boves (1999) and de Veth, Cranen, de Wet & Boves (1999). The recognition results using the conventional local distance function

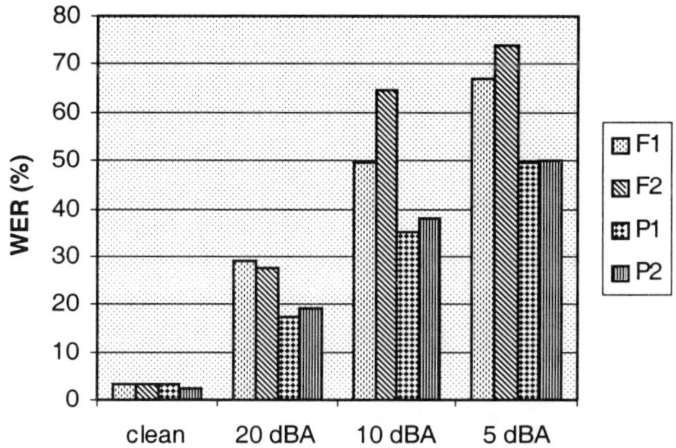

FIGURE 2.9: Recognition results as a function of signal-to-noise ratio when using the conventional local distance function. F1: within-vector mean normalized Mel-frequency log-energy coefficients. F2: Mel-frequency cepstral coefficients. P1: sub-band Mel-frequency cepstral coefficients. P2: within-vector filtereded Mel-frequency log-energy coefficients.

for the clean and noisy conditions are shown in Fig. 2.9. The results using the robust local distance function are shown in Fig. 2.10A, and the WER difference $\Delta WER = WER_{robust} - WER_{conventional}$ is shown in Fig. 2.10B.

Looking at the clean conditions first, it can be seen that all four feature representations essentially perform at the same level and that recognition performance in the clean condition is affected only slightly when switching from the conventional to the robust local distance function. Focusing on the conditions where noise was added to the speech signals, two effects are clearly visible. Firstly, recognition performance is better for the two feature representations that only partially smear distortions (i.e., P1 and P2 (two rightmost bars)) than for the representations that smear distortions over all feature components (i.e., F1 and F2 (two leftmost bars)). This observation holds for the recognizer with the conventional as well as for the recognizer with the robust local distance function. Secondly, it can be seen that the recognizer based on the robust local distance function yields better results than the recognizer based on the conventional local distance function when noise is present in all cases, but one. The single exception occurs at SNR = 20 dBA for P1: The WER increases from 17.1% to 18.6% when switching from the conventional to the robust local distance function.

Given the data shown in Fig. 2.9 and Figs 2.10AB, two remarks are in place. Firstly, application of the robust local distance function in the clean condition consistently leads to a slight loss of recognition performance. This is probably due to the fact that the distributions of the observations to be recognized in the clean condition are better represented by the distributions found during training than by the mixture of distributions used in the robust local

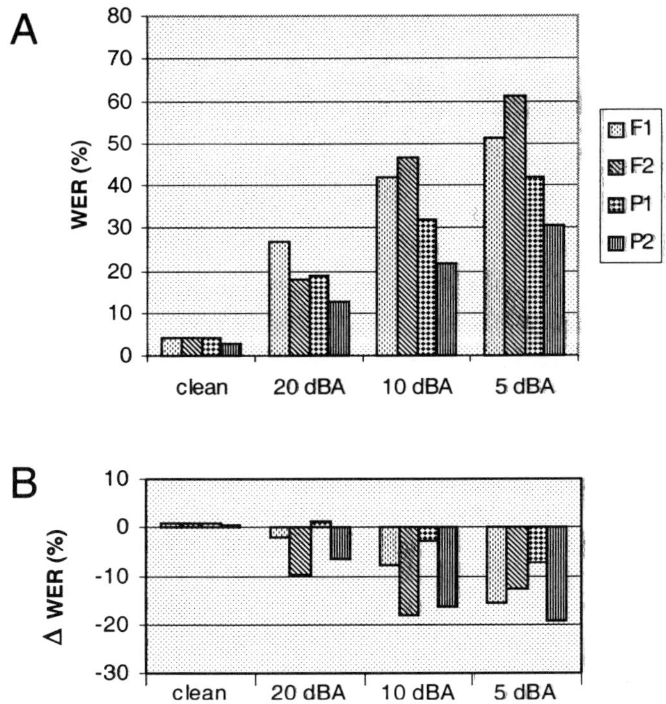

FIGURE 2.10: (A) Recognition results as a function of signal-to-noise ratio when using the robust local distance function. Same abbreviations as in Fig. 2.9. (B) Corresponding Δ WER results.

distance computation. A mismatch between the mixture of distributions used in the robust local distance computation and the actual distribution of observations could also explain the slight loss of recognition performance observed in one of the noise conditions. Secondly, the results show that performance improvements are observed for all types of features that were tested. In other words: Even for a feature representation that fully spreads spectrally local distortions over all feature vector components, the robust local distance function is capable of improving recognition performance. Apparently, the detrimental effect of the noise can be partially repaired by the robust local distance function, albeit that the improvement is not equally large for all feature types.

The results discussed in this Section cannot be readily generalized, because it must be expected that each specific noise type will affect different features differently. Consequently, it must be expected that the amount of success that our robust local distance function can offer, will depend on a complex interaction between feature types, noise types and model characteristics. Additional research is needed to fully come to grips with this matter.

2.3.5 Discussion and Conclusions

In this Section, we have presented a simple model of speech corrupted by additive noise that can be used as a framework to compare and understand several different approaches to making ASR more robust to noise. Noise robustness can be pursued in the acoustic feature domain, in the acoustic model domain, or in the distance measure domain. According to this scheme, we mentioned observation cleaning methods, of which spectral subtraction is the classic example. Next, we mentioned predictive model-based compensation (viz. by assuming that useful estimates of the parameters of the noise can be obtained, which in their turn can be used to adapt the models to better fit the conditions present in the new signal). As another approach in the model domain, we referred to work in model adaptation that attempts to make corrections based on observations as they are received at recognition time.

In our contribution, we have focused on conditions in which no dependable estimates of the noise can be made, so that we are left with the assumption that observed acoustic feature vectors can be considered as realizations of a mixture of two different processes. The first process is known and corresponds to the 'speech process' as observed during training. No knowledge about the second process is available. This unknown process corresponds to observations that were not seen during training. We have argued that this description allows to make a link with the theory of Robust Statistical Pattern Recognition (Kharin 1996) and also to Missing Feature Theory (Cooke et al. 1996, Morris et al. 1998). These links deserve (and need) further research.

From a speech science point of view, two possible ways can be identified to extend the work on the robust local distance function. Firstly, our implementation of Missing Feature Theory might open new alleys towards the deployment of phonetic and auditory knowledge in automatic speech recognition. For example, it could be possible to introduce an estimate of fundamental frequency as one of the elements in the acoustic feature vector. Fundamental frequency is only defined for voiced speech segments and undefined for other signal portions. With Acoustic Backing-off, it should not be difficult to consider the value for fundamental frequency missing in signal portions that do not contain voiced speech. Secondly, the physical and perceptual basis under Missing Feature Theory might help to determine the best way for the incorporation of recent results from Robust Statistical Pattern Recognition in ASR.

We have also drawn attention to an issue which has not been widely discussed in the literature, viz. the potential interaction between transformations of the components of the short-time spectra and robustness against additive noise. With few exceptions additive noise will not affect all components of the short-time spectrum equally. We argued that a transformation may be suboptimal when it smears distortions which are local in the input short-time spectra over (almost) all components of the acoustic feature vectors. For Missing Feature Theory this is evident, since smearing violates the basic assumption underlying Missing Feature Theory, i.e., that part of the observation values are undistorted. The results from our experiments with feature representations that do not smear local distortion over the full feature vector have shown convincingly that it pays to minimize smearing. However, our research has also shown that it is not always straightforward to predict how a given distortion in the spectro-temporal domain will be smeared out in another domain under a given transformation. For instance, the fact that our robust LDF has a positive effect even for full-smearing features like within-vector mean normalized mel-frequency log-energy coefficients and mel-frequency cepstral coefficients can at least in part be explained by the fact that not all transformed features have suffered equally from the smearing of the low frequency spectral distortion (de Veth, Cranen, de Wet & Boves 1999). More research is needed with respect to this subject.

As a final subject for further research, we recall that the robust local distance function in the form of Acoustic Backing-off is not capable to handle the 'd2' type distortions, shown in Figs 2.7 and 2.8. In general, the combination of the characteristics of the additive noise and the feature transform will result in a mixture of 'd1' and 'd2' type distortions. It is reasonable to expect that Acoustic Backing-off will be more effective if the proportion of 'd1' distortions in such a mixture is larger. However, it is still an open question how to handle a mixture with a large proportion of 'd2' type distortions to improve recognition robustness.

2.4 Concluding Remarks

In this chapter we have discussed two environmental factors which contribute to variation in speech signals and which therefore make automatic speech recognition difficult. The first factor is the effect of the transmission channel on the speech signal observed at the input of the ASR device; the second is the effect of additive noise. Both factors play their role in almost every recognition task, be it small vocabulary isolated word recognition or the recognition of unconstrained spontaneous speech.

Throughout this contribution, the discussion was based on a physical and mathematical model of the signals. It was argued that a model which encompasses all physical effects in great detail (including possible non-linear distortions) is far too complex to be handled. We have discussed how a simplified model can be adopted. Some simplifications are quite realistic, e.g., the assumption that the transmission channel is time-invariant (or varies only very slowly) during a human-machine interaction session. Yet, in some specific situations the simplifications may become physically irrealistic. For instance, it is very unlikely that radio transmission errors in digital cellular networks are adequately represented by Eq. 2.19. However, it should be stressed that the simplifications we addressed in this contribution are motivated by the important finding that a simplistic model of which the parameters can be reliably estimated is always to be preferred over a physically more realistic model, if the parameters of the latter cannot be reliably obtained.

Another issue which has been emphasized throughout the paper is the interdependence of the modules of state-of-the-art ASR devices. Thus, an 'improvement' in one module, even one which is perfectly motivated by solid theory, may prove to deteriorate recognition accuracy, because it violates essential assumptions underlying other modules. This helps to explain why it has proven to be so difficult to harness conventional and recent knowledge from phonetics and auditory perception to improve ASR: Until we have a viable alternative for the Dynamic Programming search through a frame-state space in which the frames constitute observations at equidistant time points with a single fixed frequency resolution, only the most basic phonetic and auditory perception knowledge can be brought to bear. We have illustrated this issue by means of RASTA filtering: It is precisely its relation to human temporal masking – and the attendant conflict with basic assumptions underlying HMM recognizers (like the assumption that speech can be modelled as a sequence of relatively invariant and static sub-word units) – which restricts the usefulness of classical RASTA to the realm of recognition based on whole-word or triphone units, and prevents its generalization to sub-word model systems based on units other than triphones.

Finally, we have pointed out how several different approaches to robust speech recognition can be unified or at least be related to one another. Again, the point of departure was a simplistic model, in which the signal at the input of an ASR device is considered as the sum of the 'clean' speech signal and some noise signal. Different approaches can be developed depending on the choice of the working domain (see Fig. 2.1): the acoustic feature domain, the model domain or the distance computation domain. Examples for the different domains are spectral subtraction, predictive model compensation, and Missing Feature Theory, respectively. We argued that the choice for any particular method depends on the assumptions about the parameters of the noise, and on the possibilities to reliably estimate these parameters. We have elaborated a recently emerged approach which has relations to the theory of Robust Statistical Pattern Recognition (Kharin 1996) and more in particular to Missing Feature Theory (Cooke et al. 1996, Morris et al. 1998). Departing from the bottom line assumption that we have no prior knowledge about the noise, we have introduced 'Acoustic Backing-off' as a means for handling observations that are potentially corrupt and do not correspond to the distribution of observations seen during training. To that end we have introduced a new, robust local distance function. In doing so, we have uncovered a new issue, viz. the impact of feature transformations on the local distance function and the attendant search.

We hope that the presentations and discussions in this paper help to provide a framework to compare and unify the increasing stream of research papers on robust ASR. At the same time, it should help to guide future research and to focus it on those aspects which are most promising, given the full context of the models and assumptions implied in a speech recogniser. Finally, this framework should help to prevent disappointments by showing how local improvements can be turned counterproductive because of the way in which they violate critical assumptions in other components of a full-fledged ASR system.

Acknowledgment

The contribution of Johan de Veth to this research was funded through the Priority Programme Language and Speech Technology (TST). The TST-Programme is sponsored by NWO (Dutch Organization for Scientific Research).

References

Aikawa, K., Singer, H., Kawahara, H. & Tohkura, Y. (1993). A dynamic cepstrum incorporating time-frequency masking and its application to continuous speech recognition, *Proceedings International Conference on Acoustics, Speech, and Signal Processing*, pp. 668–671.

Anastasakos, A., Kubala, F., Makhoul, J. & Schwartz, R. (1994). Adaptation to new microphones using tied-mixture normalization, *Proceedings ARPA Spoken Language Technology Workshop*, pp. 89–93.

Atal, B. (1974). Automatic recognition of speakers from their voices, *Proceedings IEEE*, Vol. 64, pp. 460–475.

Boda, P.-P., de Veth, J. & Boves, L. (1996). Channel normalisation by using RASTA filtering and the dynamic cepstrum for automatic speech recognition over the phone, *Proceedings ESCA Workshop on the Auditory Basis of Speech Perception*, Keele, UK, pp. 317–320.

Boll, S. (1979). Suppression of acoustic noise in speech using spectral subtraction, *IEEE Transactions on Acoustics, Speech, and Signal Processing* **27**: 113–120.

Cooke, M., Morris, A. & Green, P. (1996). Recognising occluded speech, *Proceedings ESCA Workshop on the Auditory Basis of Speech Perception*, Keele, UK, pp. 297–300.

Dautrich, B., Rabiner, L. & Martin, T. (1983). On the effect of varying filter bank parameters on isolated word recognition, *IEEE Transactions on Acoustics, Speech, and Signal Processing* **31**: 793–806.

de Veth, J. & Bourlard, H. (1995). Comparison of hidden Markov model techniques for automatic speaker verification in real-world conditions, *Speech Communication* **17**: 81–90.

de Veth, J. & Boves, L. (1996). Comparison of channel normalisation techniques for automatic speech recognition over the phone, *Proceedings International Conference on Spoken Language Processing*, pp. 2332–2335.

de Veth, J. & Boves, L. (1997a). Channel normalisation using phase-corrected RASTA, *Proceedings ESCA-NATO Workshop on Robust Speech Recognition for Unknown Communication Channels*, Pont-a-Mousson, France, pp. 119–122.

de Veth, J. & Boves, L. (1997b). Phase-corrected RASTA for automatic speech recognition over the phone, *Proceedings International Conference on Acoustics, Speech, and Signal Processing*, pp. 1239–1242.

de Veth, J. & Boves, L. (1998a). Channel normalization techniques for automatic speech recognition over the telephone, *Speech Communication* **25**: 149–164.

de Veth, J. & Boves, L. (1998b). Effectiveness of phase-corrected RASTA for continuous speech recognition, *Proceedings International Conference on Spoken Language Processing*, pp. 963–966.

de Veth, J., Cranen, B. & Boves, L. (1998). Acoustic backing-off in the local distance computation for robust automatic speech recognition, *Proceedings International Conference on Spoken Language Processing*, pp. 1427–1430.

de Veth, J., Cranen, B. & Boves, L. (2000). Acoustic backing-off as an implementation of missing feature theory. Accepted for publication in Speech Communication.

de Veth, J., Cranen, B., de Wet, F. & Boves, L. (1999). Acoustic pre-processing for optimal effectivity of missing feature theory, *Proceedings of the 6th European Conference on Speech Communication and Technology (Eurospeech '99)*, pp. 65–68.

de Veth, J., de Wet, F., Cranen, B. & Boves, L. (1999). Missing feature theory in ASR: Make sure you miss the right type of features, *Proceedings Workshop on Robust Methods for ASR in Adverse Conditions*, pp. 231–234.

Drullman, R., Festen, J. & Plomp, R. (1994). Effect of temporal envelope smearing on speech reception, *Journal of the Acoustical Society of America* **95**: 1053–1064.

Dupont, S. (1998). Missing data reconstruction for robust automatic speech recognition in the framework of hybrid HMM/ANN systems, *Proceedings International Conference on Spoken Language Processing*, pp. 1439–1442.

Dupont, S., Bourlard, H. & Ris, C. (1997). Robust speech recognition based on multi-stream features, *Proceedings ESCA-NATO Workshop on Robust Speech Recognition for Unknown Communication Channels*, Pont-a-Mousson, France, pp. 95–98.

Furui, S. (1981). Cepstral analysis technique for automatic speaker verification, *IEEE Transactions on Acoustics, Speech, and Signal Processing* **29**: 254–272.

Gales, M. (1995). *Model-Based Techniques for Noise Robust Speech Recognition*, PhD thesis, Cambridge University.

Gales, M. (1998). Predictive model-based compensation schemes for robust speech recognition, *Speech Communication* **25**: 49–75.

Gish, H., Karnovsky, K., Krasner, M., Roucos, S., Schwartz, R. & Wolf, J. (1985). Investigation of text-independent speaker identification over telephone channels, *Proceedings International Conference on Acoustics, Speech, and Signal Processing*, pp. 379–382.

Gish, H., Krasner, M., Russell, W. & Wolf, J. (1986). Methods and experiments for text-independent speaker recognition over telephone channels, *Proceedings International Conference on Acoustics, Speech, and Signal Processing*, pp. 865–868.

Gong, Y. (1995). Speech recognition in noisy environments: A survey, *Speech Communication* **16**: 261–291.

Haeb-Umbach, R., Beyerlein, P. & Geller, D. (1995). Speech recognition algorithms for voice control interfaces, *Philips Journal of Research* **49**: 381–397.

Hassall, J. & Zaveri, K. (1979). *Acoustic Noise Measurements*, Brüel & Kjær, Denmark.

Hermansky, H. (1996). Auditory modeling in automatic recognition of speech, *Proceedings ESCA Workshop on the Auditory Basis of Speech Perception*, Keele, UK.

Hermansky, H. & Morgan, N. (1994). RASTA processing of speech, *IEEE Transactions on Speech and Audio Processing* **2**: 578–589.

Hermansky, H., Morgan, N., Bayya, A. & Kohn, P. (1991). Compensation for the effect of the communication channel in auditory-like analysis of speech, *Proceedings of the 2nd European Conference on Speech Communication and Technology (Eurospeech '91)*, pp. 1367–1370.

Hermansky, H. & Pavel, M. (1995). Psychophysics of speech engineering systems, *Proceedings International Conference on Phonetic Sciences*, pp. 3.42–3.49.

Hirsch, H.-G., Meyer, P. & Ruehl, H. (1991). Improved speech recognition using high-pass filtering of subband envelopes, *Proceedings of the 2nd European Conference on Speech Communication and Technology (Eurospeech '91)*, pp. 413–416.

Huber, P. (1981). *Robust Statistics*, Wiley, New York.

Hunt, M. (1978). Automatic correction of low-frequency phase distortion in analogue magnetic recordings, *Acoustic Letters* **32**: 6–10.

Junqua, J.-C. (1996). The influence of acoustics on speech production: A noise-induced stress phenomenon known as the Lombard reflex, *Speech Communication* **20**: 13–22.

Junqua, J.-C., Fohr, D., Mari, J.-F., Applebaum, T. & Hanson, B. (1995). Time derivatives, cepstral normalisation and spectral parameter filtering for continuously spelled names over the telephone, *Proceedings of the 4th European Conference on Speech Communication and Technology (Eurospeech '95)*, pp. 1385–1388.

Junqua, J.-C. & Haton, J.-P. (1996). *Robustness in Automatic Speech Recognition: Fundamentals and Applications*, Kluwer Academic Publishers, Boston.

Kharin, Y. (1996). *Robustness in Statistical Pattern Recognition*, Kluwer Academic Publishers, Dordrecht.

Koehler, J., Morgan, N., Hermansky, H., Hirsch, H.-G. & Tong, G. (1994). Integrating RASTA-PLP into speech recognition, *Proceedings International Conference on Acoustics, Speech, and Signal Processing*, pp. 421–424.

Lee, C.-H. (1998). On stochastic feature and model compensation approaches to robust speech recognition, *Speech Communication* **25**: 29–47.

Lee, C.-H. & Huo, Q. (1999). Adaptive classification and decision strategies for robust speech recognition, *Proceedings Workshop on Robust Methods for ASR in Adverse Conditions*, pp. 45–52.

Lippmann, R. & Carlson, B. (1997). Using missing feature theory to actively select features for robust speech recognition with interruptions, filtering, and noise, *Proceedings of the 5th European Conference on Speech Communication and Technology (Eurospeech '97)*, pp. 37–40.

Little, J. & Shure, L. (1993). *Matlab Signal Processing Toolbox Users Guide*, The MathWorks, Inc., Natick.

Liu, F.-H., Moreno, P., Stern, R. & Acero, A. (1994). Signal processing for robust speech recognition, *Proceedings ARPA Spoken Language Technology Workshop*, pp. 110–115.

Lockwood, P. & Boudy, J. (1992). Experiments with a non-linear spectral subtractor (NSS), hidden Markov models and the projection, for robust speech recognition in cars, *Speech Communication* **11**: 215–228.

Milner, B. & Vaseghi, S. (1995). An analysis of cepstral-time matrices for noise and channel robust speech recognition, *Proceedings of the 4th European Conference on Speech Communication and Technology (Eurospeech '95)*, pp. 519–522.

Morris, A., Cooke, M. & Green, P. (1998). Some solutions to the missing feature problem in data classification, with applications to noise robust ASR, *Proceedings International Conference on Acoustics, Speech, and Signal Processing*, pp. 737–740.

Nadeu, C., Hernando, J. & Gorricho, M. (1995). On the decorrelation of filter-bank energies in speech recognition, *Proceedings of the 4th European Conference on Speech Communication and Technology (Eurospeech '95)*, pp. 1381–1384.

Nadeu, C., Paches-Leal, P. & Juang, B.-H. (1995). Filtering the time sequence of spectral parameters for speaker-independent CDHMM word recognition, *Proceedings of the 4th European Conference on Speech Communication and Technology (Eurospeech '95)*, pp. 923–926.

Okawa, S., Bocchieri, E. & Potamianos, A. (1998). Multi-band speech recognition in noisy environments, *Proceedings International Conference on Acoustics, Speech, and Signal Processing*, pp. 641–644.

Orloff, J., Gillick, L., Roth, R., Scattone, F. & Baker, J. (1994). Adaptation of acoustic models in large vocabulary speaker independent continuous speech recognition, *Proceedings ARPA Spoken Language Technology Workshop*, pp. 119–122.

Rabiner, L. (1988). Mathematical foundations of hidden Markov models, *Recent advances in speech understanding and dialog systems*, Vol. F46 of *NATO ASI Series*, Springer-Verlag, Berlin, pp. 183–205.

Rabiner, L. & Schafer, R. (1978). *Digital Processing of Speech Signals*, Prentice-Hall, Englewood Cliffs.

Rahim, M. & Juang, B.-H. (1996). Signal bias removal by maximum likelihood estimation for robust telephone speech recognition, *IEEE Transactions on Speech and Audio Processing* **4**: 19–30.

Raj, B., Singh, R. & Stern, R. (1998). Inference of missing spectrographic features for robust automatic speech recognition, *Proceedings International Conference on Spoken Language Processing*, pp. 1491–1494.

Schetzen, M. (1980). *The Volterra and Wiener Theories of Nonlinear Systems*, Wiley, New York.

Singer, H., Paliwal, K., Beppu, T. & Sagisaka, Y. (1995). Effect of RASTA-type processing for speech recognition with speaking-rate mismatches, *Proceedings of the 4th European Conference on Speech Communication and Technology (Eurospeech '95)*, pp. 487–490.

Soong, F. & Rosenberg, A. (1986). On the use of instantaneous and transitional spectral information in speaker recognition, *Proceedings International Conference on Acoustics, Speech, and Signal Processing*, pp. 877–880.

Steinbiss, V., Ney, H., Aubert, X., Besling, S., Dugast, C., Essen, U., Geller, D., Haeb-Umbach, R., Kneser, R., Meier, H.-G., Oerder, M. & Tran, B.-H. (1995). The Philips research system for continuous-speech recognition, *Philips Journal of Research* **49**: 317–352.

Tibrewala, S. & Hermansky, H. (1997). Sub-band based recognition of noisy speech, *Proceedings International Conference on Acoustics, Speech, and Signal Processing*, pp. 1255–1258.

Varga, A. & Moore, R. (1990). Hidden Markov model decomposition of speech and noise, *Proceedings International Conference on Acoustics, Speech, and Signal Processing*, pp. 845–848.

Vizinho, A., Green, P., Cooke, M. & Josifovski, L. (1999). Missing data theory, spectral subtraction and signal-to-noise estimation for robust ASR: An integrated study, *Proceedings of the 6th European Conference on Speech Communication and Technology (Eurospeech '99)*, pp. 2407–2410.

Weintraub, M., Neumeyer, L. & Digalakis, V. (1994). SRI november 1993 CSR spoke evaluation, *Proceedings ARPA Spoken Language Technology Workshop*, pp. 135–144.

Chapter 3

SPEAKER COMPENSATION IN AUTOMATIC SPEECH RECOGNITON

Daniel Tapias Merino
Telefónica Investigación y Desarrollo S.A.
dtapias@tid.es

Abstract Current state-of-the-art speech recognition technology allows the implementation of speaker independent recognisers which have very low error rates when the training and testing conditions are similar. However, the word accuracy dramatically degrades if the training and testing conditions differ: Well-known factors that affect the word error rate (WER) are background noise and channel and speaker variability.

This chapter deals with the latter (inter/intraspeaker variability) which is a key problem in most current commercial systems and real applications. The chapter is divided into four parts: The first part analyses the sources of inter/intraspeaker variability, shows the trend of current research in this area and also give some directions to select the best adaptation technique for a particular problem. The second, third and fourth parts concentrate on compensation techniques of the current state-of-the-art technology. In particular, the second part focuses on speaker adaptation techniques and provides a summary of them by describing some of the current methods for speaker adaptation: clustering techniques, transformation based methods, bayesian estimation and pronunciation modelling. The third part focuses on speaker normalisation techniques and finally, the fourth part centres on speaker adaptive training.

3.1 Introduction

Human beings are continuously adapting to the environment characteristics, i.e.: the background noise, the channel and the speaker. Our capacity of adaptation is so impressive compared to automatic systems (Cox, Linford, Hill & Johnston 1998, Collado, Concejero, Tapias, Castellanos & Hernández 1999) that only in very extreme conditions there is a significant degradation of the word accuracy. Therefore, complex tasks like language identification, speaker

J.-C. Junqua and G. van Noord (eds.), Robustness in Languages and Speech Technology, 47–100.

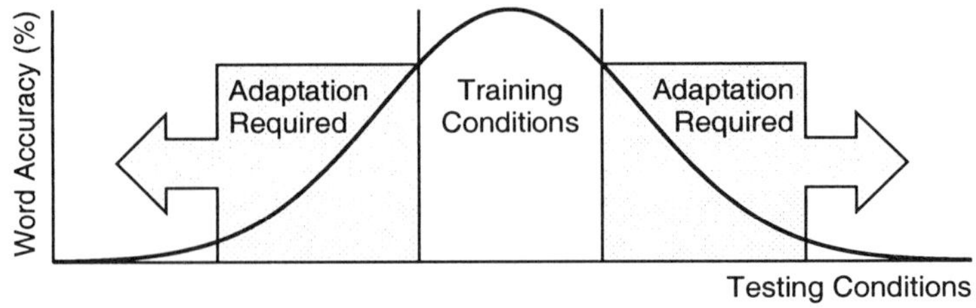

FIGURE 3.1: Variation of the word accuracy for current state-of-the-art speech recognisers.

recognition or speech recognition in adverse noise conditions, different channels (telephone, digital and analog mobile telephone channels, reverberant rooms, etc.), different accent/dialects and voices as well as combinations of all these sources of variability in many different proportions and intensities (for example: broadcast news recognition) are carried out with apparently no effort by humans.

At the other end, the behaviour of machines is strongly dependent on the testing conditions. Figure 3.1 graphically shows this fact: if the training and testing conditions are similar, the word accuracy is high and, in many cases, enough to allow the use of a speech recogniser in real applications. On the other hand, the larger the differences between the training and testing conditions, the lower the word accuracy, and consequently, the use of some adaptation technique is required.

Figure 3.2 exemplifies the effect of the mismatch between the training and testing conditions as far as speaker variability is concerned. It represents part of the results of the evaluation of the 1996 version of the Automatic Telephone Operator Service (ATOS) developed at Telefonica Investigacion y Desarrollo. The ATOS conversational system provides users with personal agenda, telephone directory, mail box and PABX services (recall, call transfer, etc.) and is able to recognise a vocabulary of 2,000 words (Álvarez, Tapias, Crespo, Cortazar & Martínez 1997). The horizontal axis identifies ten of the users that participated in the evaluation of the system and the vertical axis represents the average WER. Although the average WER is low, some speakers have recognition rates considerably worse than others: It can be observed that while the speech recogniser worked well for some of the speakers (speakers B and I), there were others for which the speech recogniser did not behave properly (speakers E and G). The variability of the WER in Figure 3.2 is due to the effect of the speaker variability and the characteristics of the spontaneous speech, and represents the typical performance of a speaker independent speech recogniser.

In this chapter we will overview current methods for speaker variability compensation. We will not focus just on the techniques that are currently being used in the state-of-the-art recognisers, but we will review many different

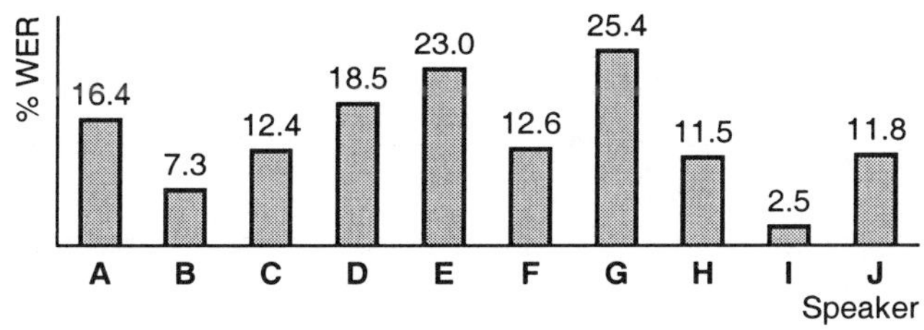

FIGURE 3.2: Word error rate of the ATOS conversational system (1996 version).

solutions to reach a double goal:

- Provide as many ideas, strategies and approaches to speaker variability compensation as possible with the aim of helping to develop new and more powerful compensation techniques.
- Show possible ways of improving the performance of the speech recognisers that are currently being used in real applications by adding speaker compensation techniques: Most speech recognisers used in real applications are real time and are not state-of-the-art. On the other hand, most popular compensation techniques are non-real time, are time consuming and require from moderate to substantial amounts of adaptation data, which makes their use in a large number of real applications difficult. For this reason, we will also focus on other approaches to speaker variability compensation that are not used in state-of-the-art systems but still can successfully be applied in these 'other' speech recognisers.

Additionally, some MATLAB programs can be downloaded at http://www.elsnet.org/. To graphically see how the Maximum Likelihood (ML) and Maximum A Posteriori (MAP) techniques work and to compare their performance when there is sufficient and insufficient adaptation data. Finally, the web site also contains an appendix with proofs of some algorithms and some distance measures.

3.2 The speaker compensation problem

At present, the capacity of machines to perceive their environment is very limited and therefore the automatic interaction with the environment is not a trivial problem unless the system works under controlled conditions and the signals are easy to interpret.

In general, current computers are able to automate reliably and rapidly many tasks that for a person are difficult, repetitive or time consuming, while they show up their limitations to perform apparently simple everyday human tasks like hand writing characters reading, speech recognition or image identification. The reason for this is that in order to accurately do, for example, speech recognition, the computer should go from relatively simple tasks like

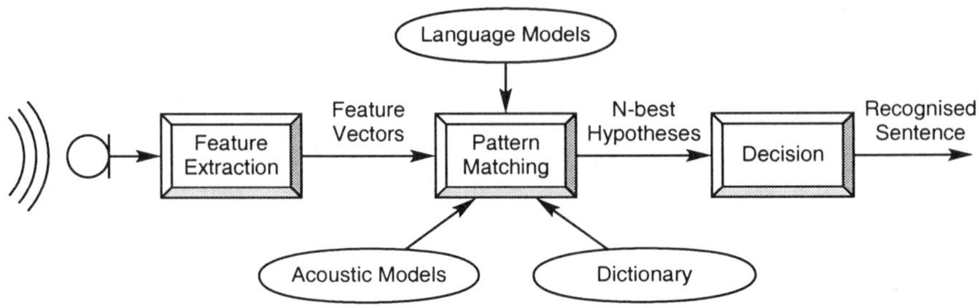

FIGURE 3.3: Traditional block diagram of a speech recogniser.

signal measurement, data ordering, data retrieval and mathematical computation to the complex task of data interpretation, which additionally requires complex processes of reasoning, knowledge databases and learning capacity.

The latter point, learning capacity, is what would allow the system to adapt to new conditions and would convert a traditional static speech recogniser into an adaptive speech recognition system. Two major design considerations appear when we want to provide the speech recogniser with adaptation capabilities:

1. How to include speaker adaptation techniques into the complex structure of an automatic speech recognition (ASR) system.
2. What sources of information have to be considered in the adaptation algorithm.

3.2.1 Architecture for compensating the inter/intraspeaker variation

Figure 3.3 shows the traditional block diagram of a speech recogniser, that is composed of three modules:

The feature extraction module, which computes the feature vectors.

The pattern matching module, which computes the probability of the observation vectors given the models and generates either a list with the N-best candidates or a lattice containing all the possible recognition hypotheses.

The decision module, which uses information of the task and/or additional sources of knowledge to select the recognised sentence.

On the other hand, Figure 3.4 represents a speech recogniser with speaker adaptation capability. It has one more block than the one in Figure 3.3, which is in charge of the speaker adaptation task. The dotted lines that connect the speaker adaptation module with the other components of the diagram indicate that the adaptation could be applied to one or several parts of the speech recogniser even though currently, most of the adaptation techniques are applied to the acoustic models or just to a part of them.

Figure 3.5 goes one step further since it integrates understanding in the continuous speech recognition process by means of a semantic parser and a dialog manager. Additionally, it includes an environment analyser that analyses

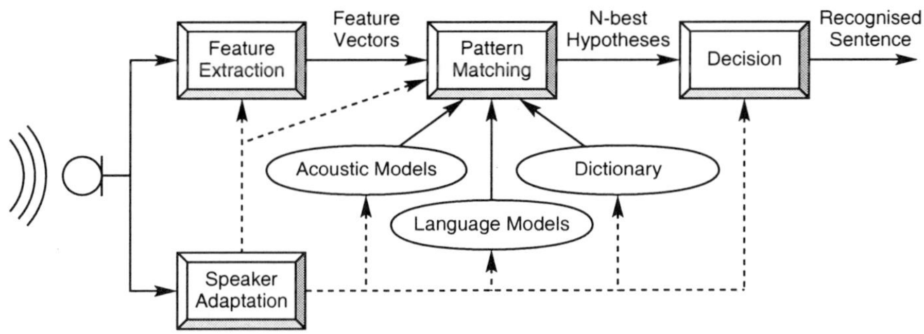

FIGURE 3.4: Block diagram of a speech recogniser with speaker adaptation.

all the factors that can affect the speech recogniser performance, i.e., characterises the environment in the broadest sense of the word: signal-to-noise ratio, kind of noise, kind of speaker, etc.

The information produced by the environment analyser is passed to the agent and to the dialogue manager. The agent stores all the relevant information that can be used next time the same person uses the system and provides the dialogue manager with the available information about the user. Finally, the dialog manager utilises the incoming information from the environment analyser and the information from the agent to adapt the whole system to the current characteristics of the environment. Since the dialog manager also knows the state of the conversation with the user, it can go one step further on adaptation providing the speech recogniser with information about the expected information from the user in the next dialogue turn.

3.2.2 Sources of information for compensating the inter/intraspeaker variation

Human beings have an impressive capability to understand any person that speaks our native language even through monophonic channels (e.g. telephone channels). This is mainly due to three reasons:

- The capability of the brain for processing the acoustic signal.
- The capability we have to predict the word or sequence of words that are going to be uttered by the speaker based on our knowledge about the context of the conversation, the language, the world and the person who is speaking.
- Our adaptation capability, which allows us to rapidly adapt ourselves to the specific features of the speaker's voice and his/her speaking style.

Using these capabilities we can extract the maximum information from the acoustic signal, which suggests that the speaker independent speech recognisers could increase their recognition accuracy if they could use more knowledge and adaptation techniques.

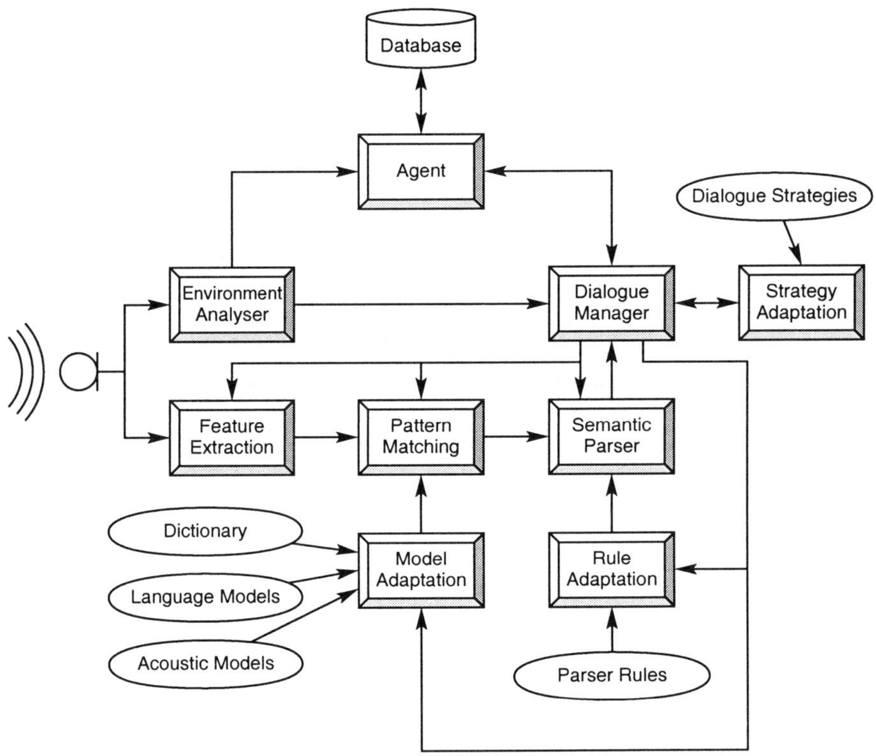

FIGURE 3.5: Future block diagram of a speech recogniser.

Additionally, human beings perceive the differences between voices belonging to different people very easily. This fact allows us to do speaker identification by just listening to the speaker's voice or even get information about the other person like sex, age, education, manners, state of mind, region where the person grew up or usually lives in, etc. Therefore, the speech signal provides human beings with more information than just its orthographic transcription. We all use this additional, non-orthographic information to improve the communication process, but most current state-of-the-art speech recognisers hardly use this 'additional' information. In fact, the 'additional' speech information is usually responsible for a dramatic increase of the WER.

Causes of inter/intraspeaker variation

It is well known that the same phoneme can be articulated in different ways (Olive, Greenwood & Coleman 1993, Quilis 1988). The phonetic realisation of a phoneme depends on its phonetic context, but also on the environment, the physiological characteristics of the speaker and his speaking style. Phonemes serve to distinguish meaning and therefore they are usually the units that appear in the pronunciation dictionary of the speech recognisers. On the other hand, allophones are variations of a phoneme pronunciation and, in many systems, their specific characteristics are modelled by triphone models (a triphone is a phoneme together with its right and left contexts).

The problem of adapting a speech recogniser to all possible causes of inter/intraspeaker variation is not only the large number of different sources of variability of the speech signal, but also the large number of combinations of all of them and the large number of different degrees of intensity with which each source of variability can affect the speech signal.

In the following, a simple classification is presented for the causes of inter/intraspeaker variation that even though neither claims to be complete nor totally accurate, will give us a clearer idea of the complexity of the speaker compensation problem (see also section 3.2.1). This classification shows the sources of variability of the speech signal that may also be used as sources of information for compensating the effects of inter/intraspeaker variation. The sources of variability have been divided into three groups: cultural differences, physiological differences and environmental differences:

Cultural differences In this section we discuss sources of variability caused by differences in the environments in which individuals grew up or in which they usually live. Some of these sources of variability are not always due to cultural differences, but they have been grouped in the same section for the sake of simplicity. For example, there are speakers who usually speak very slow or very fast, but speech rate is also a resource of the spoken language to emphasise a part of the message. In the same way, there are users that usually turn down the speech production loudness when they speak over the phone, but this phenomenon can also occur if the user does not want to be heard, etc.

Taking this into account, the cultural sources of variability are:

- Speech production loudness (volume),
- Set of sounds (phones and allophones),
- Sound duration,
- Intonation,
- Way of building sentences,
- Speech rate,
- Vocabulary.

It is worth mentioning that in some cases the pronunciation of a word by a particular speaker can lead to predictions about the pronunciation of other words. For example, in Madrid and its area of influence, the aspiration phonetic phenomenon is typical in the voiceless fricative phoneme /s/ at the end of a syllable, e.g. the words *es que* are usually pronounced as '/e/ /h/ /k/ /e/' instead of being pronounced as '/e/ /s/ /k/ /e/'. Therefore, if a speaker utilises this kind of pronunciation, he/she will be consistent with the /s/ pronunciation in similar cases like *escalera* or *esqueleto*. In English, for example, the production of vowel /a/ in *bath* is usually coherent with the production of the vowel in words like *grass* (Leggetter 1995). In other cases the pronunciation of a word will not give any indication of how the person will pronounce other words.

There are many other phonetic phenomena that affect the sound production in spontaneous speech, for example:

Assimilation, which consists of a characteristic of a sound being shared by adjacent sounds. In English, the negative prefix /in-/ always assimilates in place of articulation to the following consonant when the consonant is a bilabial (e.g.: impatient) (Olive et al. 1993). In Spanish, the nasal /n/ is assimilated to its right context sound, i.e. its usual articulation place (alveolar) is replaced with the articulation place of the next sound (e.g.: *cancion* → [n] interdental; *blanco* → [n] velar).

Deletion, which occurs when a sound is not pronounced by the speaker. For example, in English in some cases the unstressed vowels are not pronounced (e.g.: *memory* → memry) (Olive et al. 1993). In Spanish, the intervocalic /d/ in the verbal word ending /-ado/ is usually not pronounced (e.g.: *estado* → estao)

Addition, which consists of including a sound between two sounds, probably to increase the acoustic differences between them (Olive et al. 1993).

Substitution is another interesting phenomenon of spontaneous speech, it consists of replacing a sound with another (for example, sometimes the Spanish group /sT/ becomes /rT/: *doscientos* → dorcientos) (León 1996).

Reduction, which affects unstressed vowels (Olive et al. 1993). When a vowel becomes reduced it is pronounced with the timbre of another vowel with less aperture of the oral cavity. In English, when a vowel becomes reduced, it is pronounced like the schwa allophone. This phenomenon appears also in some dialects of Spanish like for example the ones of Galicia and Asturias (e.g. /o/ > /u/: *comprado* → comprau, where there is a deletion of the intervocalic /d/ and a vowel reduction of phoneme /o/). Some other authors (Martínez Celdrán 1984) refer to vowel reduction when two equal and contiguous vowels are converted into one (e.g. *alcohol* → alcol).

Aspiration, which in English affects the voiceless stops when they are at the beginning of a word and before a stressed vowel (e.g.: *park*) (Olive et al. 1993). In Spanish it affects the voiceless fricative phoneme /s/ at the end of a syllable and the voiceless fricative phoneme /x/ and consists of a relaxation of the vocal tract constriction (e.g. *escalera* → /e/ /h/ /k/ /a/ /l/ /e/ /r/ /a/). The aperture of the oral cavity is intermediate between the vowels and the voiceless consonants and the place of articulation is not fixed and depends on the context.

Assibilation, which consists of pronouncing a sound as sibilant. This phenomenon occurs in some dialects of Spanish, like the ones in La Rioja (Spain) or the inland area of Argentina, where the phoneme /R/ becomes sibilant by converting the total obstruction of the airstream into a constriction of the vocal tract.

Physiological differences In this section we include the differences inherent to the anatomical characteristics, health and physical condition of the speaker:

- Vocal tract shape and length,
- Vibration frequency of the vocal cords,
- Glottal pulse shape,
- Age group,
- Tiredness,
- Loss of Voice, and
- Nasal congestion.

Loss of voice, tiredness and nasal congestion are more transitory in nature. Loss of voice affects the vibration of the vocal cords and consequently the speech production loudness and the quality of the speech signal, tiredness usually leads to slower speech and nasal congestion leads to constriction or even obstruction of the nasal cavity, so that the speech becomes partly or fully non-nasal.

Environmental differences Human beings, like the other living creatures, interact with the environment and therefore are influenced by the environment and vice versa. The influence of the environment can be determinant as far as quality and characteristics of the speech signal is concerned. Consequently, it is important to know the environment in which the speaker is going to be immersed in order to predict the characteristics of the expected speech signal.

Next, there is a list of different effects of the environment on the speech production process:

- Background noise forces the vocal apparatus to increase the effort during the utterance production process, which modifies the characteristics of the acoustic signal. This phenomenon is called Lombard effect (Pisoni, Bernacki, Nusbaum & Yuchtman 1985, Stanton, Jamieson & Allen 1988, Summers, Pisoni, Bernacki, Pedlow & Stockes 1984, Hansen 1988, Junqua 1993).
- Mechanical factors like acceleration or vibration also affect the resulting acoustic signal (e.g. fighter pilots).
- Emotional states like fear, anger, surprise, state of mind, stress, etc. are the response of people to different environments and can dramatically affect the speech signal characteristics.
- Cognitive load is another important environmental factor (e.g.: speech produced in a telephone conversation while working with a computer).
- Room acoustics effects like reverberation or distortion can also influence the speech production process as well as the resulting speech signal.

Additionally, the background noise and the channel mask and distort the speech signal once it has been produced, but the compensation of their effects is out of the scope of this chapter.

3.2.3 Speaker dependent versus speaker independent speech recognisers

The combination of subsets of the factors in the three described groups (cultural, physiological and environmental differences) can lead to different speaking styles like careful, clear, formal, informal, spontaneous, read, scripted, vulgar, stressed, etc. (Llisterri 1992, Eskenazi 1993) and are responsible for the fact that the same sentence pronounced by different speakers can give rise to quite distinct signals. For this reason, the speaker independent systems are trained with large speech databases composed of many utterances from a large number of speakers. This procedure ideally allows the training process to model all the phonetic variations as well as the inter/intraspeaker variability. Nevertheless, this procedure leads to a set of models with a large amount of variance and therefore reduces the modelling accuracy for each individual speaker and sound.

On the contrary, the speaker dependent recognisers are able to better discriminate between different sounds of the same speaker given that they have just to model the speech particularities of one speaker. This fact can be easily proved by a comparison of a speaker dependent and a speaker independent system tested on the same speaker with the same amount of training data. For example, it is quite evident that the performance of a dictation system improves when it is adapted to a particular speaker. Finally, it is worth mentioning that the distinction between a speaker independent (SI) and a speaker dependent (SD) speech recogniser refers to the training database only since both approaches use the same parameter estimation techniques.

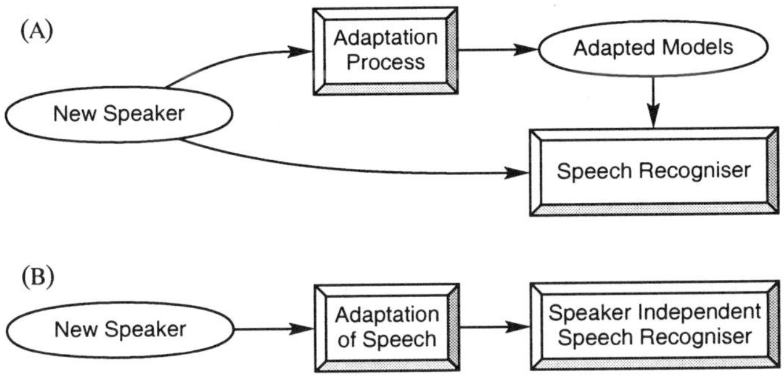

FIGURE 3.6: Block diagrams of two typical speaker adaptation approaches.

At this point it seems that speaker dependent systems should always be used, but these systems have the drawback that it is not always possible to have a large enough speaker dependent training database. This is the case, for example, of the speech recognition based telephone services. They can be accessed by a large number of different speakers and in most of these applications there are no more than two or three interactions between the user and the machine, which reduces the availability of training data to two or three isolated words or short sentences.

This problem can be partially overcome by using speaker compensation techniques, which attempt to reduce the mismatch between the training and the testing conditions as far as inter/intraspeaker variability is concerned. These techniques require smaller databases to do the adaptation than the full retraining process but they provide worse results. However, the fast evolution of these techniques both in improved performance and velocity of adaptation will allow us in the near future to have 'real' speaker independent speech recognisers by using instantaneous adaptation techniques: The baseline speech recogniser of any application, like human beings, will instantaneously adapt to every new speaker, becoming a real speaker independent system.

3.2.4 Strategies for compensating the inter/intraspeaker variation

In Section 3.1 we showed how differences between the training and testing conditions affected the performance of a speech recogniser. In particular, Figure 3.1 mirrored the general behaviour of a speech recogniser against variations of the testing conditions and Figure 3.2 showed the word error rate of a continuous speech recogniser tested with ten different speakers.

Most current compensation techniques try to reduce the mismatch between the training and testing conditions by introducing modifications into the feature extraction module, into the models or in both of them and can be divided into three groups:

Speaker adaptation: The goal of the speaker adaptation techniques is to reduce the mismatch between the feature space of the training data and

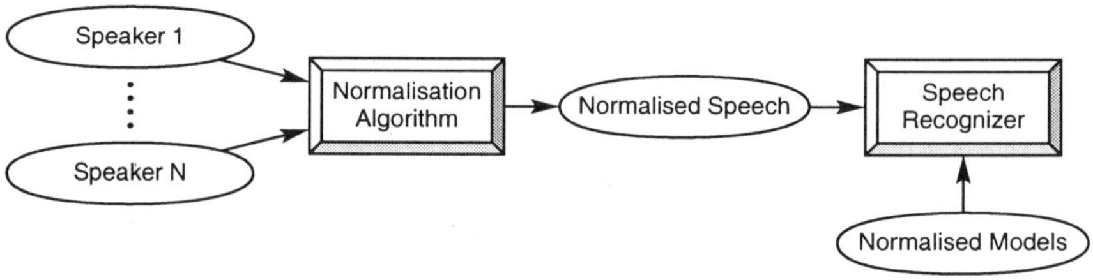

FIGURE 3.7: Block diagram of a typical speaker normalization based speech recognizer.

the feature space of the new speaker. This group of techniques is usually applied to the speech recogniser so that for each new speaker a new speech recogniser is constructed by adapting the acoustic models to the specific characteristics of the speaker (Figure 3.6a). There are some other techniques based on transformations that are applied to the incoming speech signal (Figure 3.6b). Their goal is to reduce the distance between the characteristics of the incoming speech and the characteristics of the training data. Later in this chapter we present several acoustic model adaptation techniques that can be divided into three different groups:

1. Clustering based techniques.
2. Transformation based techniques.
3. Bayesian estimation based techniques.

The speaker adaptation techniques usually provide larger reductions of the word error rate than the speaker normalisation ones, which are presented next.

Speaker normalisation: The goal of the speaker normalisation techniques is to find a transformation to map speech from any person to a common normalised speaker space in which the interspeaker differences are minimal. During the training process, these techniques first normalise the training speech database by using some transformation previously computed. Then, the estimation of the acoustic models for the normalised speech is carried out. At recognition time, the incoming speech is mapped into the normalised speaker space using the above-mentioned transformation and later the resulting speech is recognised using the normalised acoustic models. Figure 3.7 shows a speech recogniser using this approach.

Speaker normalisation requires finding a unique transformation that works properly for all speakers, which may be difficult. Among the attempts that have been made there are techniques based on estimating the vocal tract length and then computing a transformation according to the length obtained. Later in this chapter two different approaches based on this idea are presented: the first one is based on a linear transformation of the frequency axis so that speech from different speakers seems to be produced by a unique vocal tract (Wegmann, McAllaster, Orloff & Peskin 1996), the second one is based on the estimation of the positions of the first and second formants to obtain the conditional probability density functions of the position of the formants and later computation of a normalisation factor that maximises the joint probability density function (Lincoln, Cox & Ringland 1997). These two methods are just two examples of speaker normalisation techniques. There are many others that have been reported such as Ono, Wakita & Zhao (1993), Ariki, Tagashira & Nishijima (1996), Eide & Gish (1996), Lee & Rose (1996), Ishii & Tonomura (1997), and Westphal, Schultz & Waibel (1998).

At this time the speaker normalisation techniques have obtained a moderate relative decrease of the WER that depends on the task and the normalisation procedure and usually ranges from 3 to 10%.

Speaker adaptive training (SAT): In this technique, the HMM parameters and the parameters of an affine transformation of the HMM means are jointly estimated in a unified Maximum Likelihood framework. This process leads to the generation of a set of speaker independent models that are less dependent on the speaker specific characteristics than the traditional speaker independent models. At recognition time, for each new speaker, a transformation to the speaker independent models is applied to convert them into speaker dependent models.

Even though most speaker variation compensation techniques are applied either to the acoustic models or to the feature extraction module, there are other ways of performing compensation. For example, the pronunciation dictionary, the language models, the dialogue strategy, the HMM topology (e.g. increasing or decreasing the number of states and/or transitions depending on the speech rate), the vocabulary, etc. can be modified to increase the performance of the speech recogniser for a particular speaker or group of speakers. In some cases the modification of these parameters can be classified as task adaptation, but if we think of a particular task or semantic domain, there are still many ways of pronouncing words, asking a question, building a sentence or expressing an idea since the specific way a speaker expresses himself depends on factors like education, state of mind, etc. In the same way, there are users with different degrees of expertise in the use of automatic speech recognition systems. Therefore in these cases, the usual task adaptation techniques could be used to do speaker adaptation.

This chapter mainly focuses on adaptation of acoustic models, though in order to illustrate other kinds of speaker adaptation techniques, a couple of pronunciation modelling techniques will be presented at the end of the chapter.

Types of speaker adaptation

The adaptation process can be carried out in several different ways depending on factors like amount of adaptation data, computational or memory constraints and availability of the orthographic transcriptions of the adaptation utterances:

Supervised and unsupervised adaptation: The adaptation is named supervised if the right orthographic transcriptions of the adaptation data are available, otherwise the adaptation is termed unsupervised.

Block or static adaptation and dynamic or incremental adaptation: When all the adaptation data is available before the beginning of the adaptation process and it is used at once, the adaptation technique is called block or static. When the adapted system is produced by using the available adaptation data and later the adaptation is iteratively refined using more adaptation data, the adaptation is named dynamic or incremental. The advantage of incremental adaptation over batch algorithms is not usually in the final result, but in computational efficiency, reduced storage requirements, and the fact that an outcome may be provided without having to wait for all the adaptation data to be available and processed.

Complexity of the inter/intraspeaker variation compensation problem

In the development of inter/intraspeaker variation compensation techniques it would be very useful to have more knowledge about the speech signal and the recogniser. In particular, a deeper knowledge about:

- how each speech parameter affects the speech recogniser performance when its value is out of its range of variation in the training database,
- how the variation of a speech parameter affects the other speech parameters,
- what parameters of the HMMs are affected by the variation of each speech parameter, and
- how the HMMs are affected,

would provide us with clues to develop faster and more accurate compensation techniques.

The acquisition of this knowledge is very difficult because the generation of speech databases where just one speech parameter changes is not a simple task: If we create artificial databases, the artificial model of speech production and the synthesis method may introduce important differences between the artificial and the natural speech, mainly for extreme values of the speech parameter under study. On the other hand, the production of semi-artificial speech using analysis by synthesis methods, would be appropriate just to introduce changes in the fundamental frequency, the sound duration and the volume. Additionally, the problem of the interaction between speech parameters is even more difficult to address given the large amount of possible parameter combinations and possible values each parameter can be set to. There are some studies that have been done in this area to study the sensitivity of speech recognisers to speaker variability in general (Thomas,

Peckham & Frangoulis 1989, Thomas, Peckham, Frangoulis & Cove 1989), or concentrated in just a parameter of the speech like speech rate (Siegler & Stern 1995, Martínez, Tapias, Álvarez & León 1997, Richardson, Hwang, Acero & Huang 1999), the pitch (García & Tapias 2000, Thomas, Peckham & Frangoulis 1989, Thomas, Peckham, Frangoulis & Cove 1989) or the speech production loudness phenomena (Tapias, García & Cazassus 1999).

The current lack of information about the four above mentioned questions leads to the development of techniques that try to compensate all the effects of the mismatch between the training and the testing speech assuming a model that tries to describe such a mismatch. Most popular adaptation techniques employ different sophisticated and rather complex mathematical algorithms to re-estimate some of the parameters of the models from the adaptation data. These methods are usually non-real time and require from moderate to large amounts of adaptation data to work properly. Additionally, they work under the assumption that the interspeaker differences affect just the mean vectors or the mean vectors and the covariance matrices of the gaussian distributions.

Information about these four questions could be used to design compensation techniques that use *a priori* knowledge about the problem to be overcome. For instance, compensation of the speech rate is an example of this approach: It is well known that the speech rate within a dialogue varies both globally and locally among speakers due to various factors like emotion (Vroomen, Collier & Mozziconacci 1993), state of mind, emphasis, lexical stress, dialogue status, etc. This phenomenon dramatically affects the performance of large vocabulary continuous speech recognisers as to double to triple the WER for fast and slow speakers (Martínez et al. 1997). Additionally, we have found out that in real telephone applications, if the sentence is misrecognised, users usually repeat the sentence very slowly to make it more understandable. Consequently, since all the components of the system are adjusted to do speech recognition at the average speech rate, the word accuracy degrades dramatically. Finally, there are other users that usually speak fast and, for the same reason, the WER also increases for them.

Several studies have been carried out (Siegler & Stern 1995, Mirghafori, Fosler & Morgan 1995, Mirghafori, Fosler & Morgan 1996) to better understand this phenomenon, to obtain a reliable measure for the speech rate and to detect and compensate the speech rate effect in continuous speech recognition. Based on these studies and after the collection of a database composed of a large number of utterances for slow, normal and fast speech, we carried out a study to develop a fast and simple adaptation method based on previous knowledge about this phenomenon (Martínez et al. 1997). The method was based on two ideas:

Transition probabilities adaptation, that modifies the state-transition probabilities to adapt them to fast and slow speech: It assumes that the state transition probabilities are the most affected parameters when the speech rate changes from normal to fast or slow and was first pointed out in Siegler & Stern (1995) and Mirghafori et al. (1995).

Word insertion penalty and language weight adaptation, that deals with the lack of balance between insertions and deletions when the speech rate is fast or slow.

This ideas were tested on the TRESVEL database (Martínez et al. 1997), that is composed of 3200 utterances for each speech rate (slow, average and fast speech rate). The speech recogniser was based on semicontinuous HMMs and it is a modified version of the Sphinx-II system developed at Carnegie Mellon University (Ravishankar 1996). The experimental results showed a 21.5% relative improvement for slow speech while the improvement for fast speech was just 7.8%. The improvement for fast speech is lower than for slow speech due to several reasons:

1. Some triphone models cannot be properly time-aligned with the speech signal because the duration of some phones is lower than the minimum one allowed by the HMM topology.
2. It is difficult to predict phonetic phenomena like phone elision, that are very usual in fast speech.
3. The transient nature of the fast speech spectra, that is not properly modelled by the acoustic models.

There are other methods for compensating the effects of speaking rate in speech recognition like the ones presented in Martínez, Tapias & Álvarez (1998) and Richardson et al. (1999).

Selection of the most appropriate compensation technique

It is clear that in order to obtain the highest recognition accuracy a static supervised adaptation mode together with a large adaptation database is almost always desirable. Nevertheless, there are many real applications where these conditions do not occur and therefore other modes of adaptation have to be used. For example, there are cases where a very rapid adaptation is required and consequently adaptation has to be done as soon as some data is available. Additionally, in these cases, an accurate orthographic transcription will not be available most times and therefore a dynamic unsupervised mode will be the only choice for adaptation.

In the following, several considerations for carefully analysing each situation and selecting the most appropriate compensation technique are presented:

Problem Identification: This is an important step since in some cases very simple and effective ideas can be used to do adaptation. Therefore, the identification of the differences between the training and the testing speakers as well as the part of the speech recogniser that could be affected by such differences (feature extraction, acoustic models, etc.) may be helpful.

Task Constraints: There are many constraints that can be imposed by the characteristics of the task:

Availability of adaptation data: This point will help us to decide both the approaches to adapt the recogniser and the mode of speaker adaptation. Sparse data, moderate amount of data or more than enough data situations lead to different solutions. In the same way, the fact of having all the adaptation data at once or getting adaptation data every time the user uses the system also leads to different approaches. Finally, the availability or non-availability of the orthographic transcriptions of the adaptation data is also determinant. There are two alternatives for treating the sparse adaptation data problem. The first one is to reduce the number of parameters to estimate, thus reducing the need for adaptation data. However this entails a loss in modelling accuracy. The second alternative is to use speaker transformation techniques to create the required adaptation data from the training database. This can be achieved, for example, by applying speaker normalisation techniques (Ono et al. 1993, Ariki et al. 1996, Eide & Gish 1996, Lee & Rose 1996, Ishii & Tonomura 1997, Lincoln et al. 1997, Westphal et al. 1998) or using the metamorphic algorithm (Bellegarda, De Souza, Nadas, Nahamoo, Picheny & Bahl 1994).

Speech recogniser: Isolated or continuous speech recognition applications have different characteristics. For example, isolated speech recognition will typically provide us with less adaptation data in each utterance, which can be a drawback as far as the availability of adaptation data is concerned.

Real time: Some tasks require real time adaptation, which dramatically limits the number of techniques that can be applied and reduces the improvement in terms of word accuracy with respect to other non-real time techniques.

Number and kind of potential users: Speech recognition based applications that are going to be accessed by any person will not benefit from this point. Nevertheless, if we are implementing a speech recogniser that is going to be used by a population that shares some common characteristics like sex, age, dialect, etc., we can take advantage of this previous knowledge to improve the performance of the system by preadapting it to the testing conditions.

Hardware Constraints: Computational and/or memory limitations and other parameters like data transfer rate or storage capacity strongly affect the final implementation.

3.3 Speaker adaptation techniques

3.3.1 Clustering based adaptation

This set of algorithms is based first on the computation of different sets of models and then at recognition time the best set of models or the best combination of models is selected. The problem of *some* clustering based methods is that they fragment the training data, that can lead to sparse data problems in some clusters unless there is sufficient training data or there is a way to get the required training data.

The clustering based adaptation techniques can be classified into four groups:

- Speaker clustering.
- Speaker mixture model.
- Condition dependent utterance clustering.
- Eigenvoice approach.

Speaker clustering

This consists of clustering speakers whose voices are similar with respect to some distance measure and then generating a specific model for each cluster. At recognition time the system looks for the speaker cluster that is the closest to the new speaker's voice and the models corresponding to that cluster are used to recognise the utterances of the new speaker. A typical example of this approach is gender dependent modelling, which consists of dividing the training database into female and male clusters and then generating a set of models for each sex. Next, two different implementations of this technique are presented:

Padmanabhan, Bahl, Namahoo & Picheny (1996) and Padmanabhan, Bahl, Namahoo & Picheny (1998) presented a block unsupervised adaptation technique that is non-real time and is based on two assumptions:

- The training database contains a subset of speakers who are close, acoustically, to the test speaker.
- If the model parameters are re-estimated from the subset of training speakers who are close to the test speaker, the resulting parameters should be reasonably close to the ones of the speaker dependent recogniser trained on sufficiently large amounts of data from the test speaker.

In order to determine the closest training speakers to the test speaker the following procedure is carried out:

1. Compute speaker dependent models for each speaker of the training database.
2. Decode the adaptation data using a speaker independent system.
3. Viterbi align the adaptation data against the transcriptions obtained in the previous step. Tag each acoustic observation with a state identification.
4. Compute the acoustic likelihood of the adaptation data conditioned on the alignment using each speaker dependent model obtained in step 1.

5. Rank the training speakers in the order of these likelihoods.
6. Pick the top N speakers up from the ordered list as the closest to the test speaker.

Once the list of closest speakers to the test speaker is obtained, the Maximum Likelihood Linear Regression (MLLR) algorithm (see section 3.3.2) is used to map the acoustic space of each selected training speaker into the acoustic space of the test speaker. Finally, the gaussians of the speaker independent model are re-estimated by using the transformed data from the selected training speakers.

In this approach, the authors apply the MLLR transforms to the observation vectors rather than to the models, which makes it necessary to have the training data available in the adaptation process. Additionally, since multiple transformations can be applied to each training speaker, it is necessary to know what state an observation vector corresponds to in order to apply the appropriate transformation.

Later, Gao, Padmanabhan & Picheny (1997) modified this technique to avoid the need of having the training data available in the adaptation process, and Huang & Padmanabhan (1999) improved it by computing the transformations in a joint manner and combining them so as to maximise the likelihood of the adaptation data.

This method was tested on the WSJ-SI 37 database, whose vocabulary size is 20,000 words (Padmanabhan et al. 1998). The training database was WSJ-SI 284, which is composed of 35,000 utterances uttered by 284 speakers. Each training speaker pronounced from 100 to 200 sentences. The number of training speakers used in the experiments to generate the speaker dependent models was 50 and just three adaptation utterances (about 20 seconds of speech) were utilised. In these conditions, a 19.5% relative improvement with respect to the speaker independent models was obtained, which is a good result compared with the 30% relative improvement obtained with speaker dependent models.

The key idea presented by Imamura (1991) is that "spectral features of different categories uttered by different speakers sometimes overlap". This fact would be responsible for the lower capability of the speaker independent models to discriminate between sounds with respect to the capability of the speaker dependent models. Therefore, a way to divide the feature space into non-overlapped subspaces would be the estimation of an acoustic model per cluster of the training database. This is carried out by estimating a discrete HMM for each speaker in the training corpus using fuzzy vector quantisation. Later, the training speakers are clustered into K speaker classes by using a cross-entropy distance measure and finally a discrete HMM is created for each cluster.

In the recognition process a stochastic classifier is used to determine the closest classes to the new speaker and perform the recognition using a weighted combination of the models corresponding to these classes (Figure 3.8):

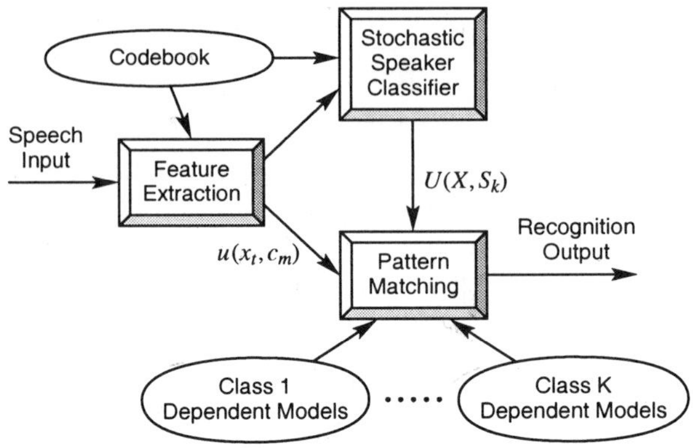

FIGURE 3.8: Block diagram of the stochastic speaker classifier based speech recognizer.

Let T be the number of frames of the utterance, K be the number of speaker classes, M be the total number of vectors in the codebook, X be the set of feature vectors of the utterance, x_t be the feature vector at time t and:

$u(x_t|c_j)$ be the fuzzy membership function.

$P(c_j|c_i, S_k)$ be the conditional co-occurrence probability of the codebook vectors (c_i, c_j) for speaker class S_k.

S_k be the kth speaker class.

The combination of the models is carried out with the following equation:

$$P(x_t, s_t = i) = \sum_{m=1}^{M} \sum_{k=1}^{K} u(x_t, c_m) \cdot U(X, S_k) \cdot b_i^k(c_m)$$

where $b_i^k(c_m)$ is the discrete observation emission probability for acoustic symbol c_m and speaker class S_k at state 'i' and $U(X, S_k)$ is the speaker classifier output, that is given by:

$$U(X, S_k) = \frac{P(X|S_k) \cdot P(S_k)}{\sum_{n=1}^{K} P(X|S_n) \cdot P(S_n)}$$

$$P(X|S_k) = \prod_{t=1}^{T} \sum_{i=1}^{M} \sum_{j=1}^{M} u(x_{t-1}|c_i) \cdot u(x_t|c_j) \cdot P(c_j|c_i, S_k)$$

This block supervised adaptation method was tested on a Japanese isolated digits task and obtained a 51.3% relative improvement (Imamura 1991). The adaptation data was just one isolated digit. The training database was composed of 6,960 utterances produced by 116 speakers that were clustered into four speaker classes. Each training speaker uttered 50 isolated words and 10 isolated digits.

Speaker mixture model

The speaker mixture models consist of the combination of speaker dependent models or models corresponding to several clusters of speakers to create a new set of adapted models. For instance, the method of Imamura (1991) presented in last section is a hybrid of speaker clustering and speaker mixture model. Next, we are presenting two examples of techniques of this kind:

Gao et al. (1997) proposed a method that is based on the one by Padmanabhan et al. (1996) and Padmanabhan et al. (1998) shown in the previous section. The main difference is that Gao works directly with models, so that he does not need to store the speech files of the training database. The sets of models are obtained from clusters of acoustically similar speakers in the training database.

Kosaka, Takami & Sagayama (1993) developed a method that consists of estimating a set of speaker dependent models and then combining them to produce a new speaker dependent model adapted to the new speaker. This combination is carried out by means of a weighted sum of the distributions of the different speaker dependent models in each state. The whole procedure can be summarised as follows:

1. Construct speaker dependent CDHMMs for each training speaker.
2. Create a speaker mixture model as a linear combination of speaker dependent models:

$$b_j(x_t) = \sum_{s=1}^{S} w_j^s \cdot b_j^s(x_t)$$

   where s is the identification of the speaker in the training database, S is the number of training speakers, w_j^s is the weight for speaker 's' and state 'j' and $b_j^s(x_t)$ is the probability density function associated to speaker 's' and state 'j'.
3. Retrain the speaker weights using the adaptation data of the new speaker and the Baum-Welch algorithm. In this way, the initial equal weights are estimated, so that the models that are closest to the new speaker will have the highest values. The retraining constraint is that just one global weight per speaker is estimated because in this way:
   - Both the number of parameters to estimate and therefore the amount of required adaptation data are reduced.
   - The problem of estimating weights for those states for which there is no adaptation data is avoided.
4. Speaker pruning: Models belonging to speakers with low weights are deleted from the model mixture. This allows to reduce the computational cost with no increase of the word error rate. The speaker weights are later re-normalised.

This block supervised method was tested on a Japanese continuous speech task with a perplexity of 5.9 and a vocabulary size of 1,000 words (Kosaka et al. 1993). The adaptation data used consisted of the utterance of a word (about 0.6 seconds of speech). The training database was composed of 16,664 utterances of isolated words where each of the 25 training speakers uttered 736 words. The estimated acoustic models were HMNets and the experimental results of this method provided a 17% relative improvement.

Condition dependent utterance clustering

While the previous methods based on speaker clustering and speaker mixture models have the goal of producing a new set of speaker dependent models, the utterance clustering techniques pursue the creation of condition dependent models (stress, speech rate, speech loudness, etc.).

Womack & Hansen (1996) use a classifier of speech under stress based on neural networks. Once the kind of stress has been detected, the speech recogniser utilises speaker independent stress dependent models. The approach is based on three points:

- Studies carried out with simulated and real speech under stress suggest that the stress classification is possible.
- "Feature analysis suggests that perturbations in speech production under stress are reflected to varying degrees across multiple feature domains depending upon the stress condition and phoneme group."
- "For a given stress condition, there are degrees of stress. Hence, it is necessary to estimate a stress probability response score to assess the stress level. However, a stress score for one stress condition alone is not sufficient, since it does not model mixtures of stress states." Therefore, it is necessary to model stress state combinations (stress mixtures), which is carried out by computing a stress score for each stress condition.

The adaptation procedure is composed of five steps:

1. Build a pronunciation dictionary based on eight phoneme groups (fricatives, vowels, affricates, etc.).
2. Estimate CDHMMs for the eight context dependent phoneme groups using data from a neutral database (TIMIT) and a stress speech database (SUSAS). Each model is composed of five states and two gaussians per state.
3. For each phoneme group and stress condition select the subset of features that classify the stress condition the best. The reported set of features is: cross sectional vocal tract areas, pitch, duration, autocorrelation, mel-cepstral features and their statistics.
4. Train a stress detector for each stress condition and a neural network to estimate the stress score:

   $$p(\varepsilon_k|w_i)$$

   where k is the stress condition and w_i is the ith word in the vocabulary.
5. Construct a set of CDHMMs for each stress condition.

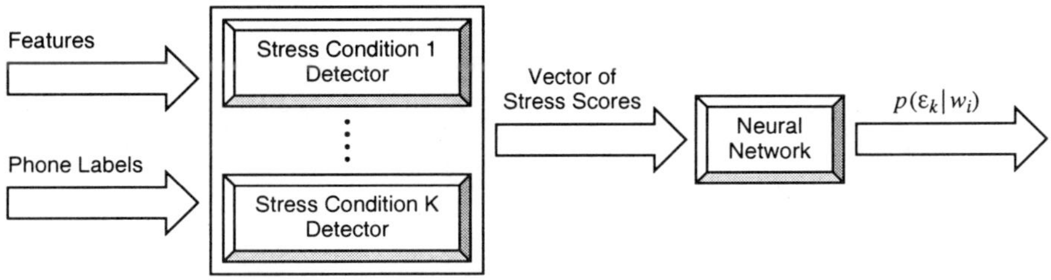

FIGURE 3.9: Block diagram of the stress detector.

The recognition procedure consists of detecting the stress condition and then using the most appropriate set of models. This is carried out in four steps:

1. Use Viterbi decoding, the CDHMMs for the eight context dependent phoneme groups and the pronunciation dictionary based on the phoneme groups to partition the incoming speech into a sequence of phoneme groups.
2. Extract the targeted features of the incoming speech and pass them to the set of stress detectors (Figure 3.9) together with the phoneme labels.
3. Pass the stress scores vector that contains the scores for each stress condition, to the neural network to obtain the maximum word stress score:

$$p(\varepsilon_{k_{max}}|w_i) = \max_k p(\varepsilon_k|w_i)$$

4. Recognise the incoming speech using the most appropriate stress dependent recogniser, which will be the one trained for the k_{max} stress condition.

This static unsupervised adaptation method was tested on the SUSAS database (Womack & Hansen 1996), which contains ten different stress conditions and its vocabulary size is 35 words. The training databases were TIMIT and SUSAS and the adaptation data was just one isolated word. In these conditions, a 34% relative improvement was obtained.

Another example of utterance clustering is the one developed to deal with the speech rate phenomenon (Martínez et al. 1998). This method, like the previous one, uses a specific database to adapt the speech recogniser, and utilises a speech rate classifier to later use the most appropriate set of models during the recognition phase.

The study was performed on the TRESVEL database, that was designed to study, evaluate and compensate the effects of speech rate in Large Vocabulary Continuous Speech Recognition systems. The study led to four conclusions:

1. Previous supervised measures of the rate of speech (ROS) had a drawback: They can provide different values of the ROS for two sentences uttered at the same speaking rate. The reason for this is that the expected duration of a phone is different from one phone to another, so that the ROS slightly depends on the set of phones that compose the

sentence. For this reason, a new supervised measure for the ROS was developed.
2. In general, slow speech is usually properly pronounced while the average and fast speech usually have effects like phone deletion or weakening, aspiration, etc.
3. The duration of phones is related to lexical stress, their contexts, the position of the phone inside the word and the sentence, the phonetic characteristics of the phone and the speech rate.
4. The effect of the speech rate in large vocabulary continuous speech recognition can be summarised as follows: while most errors for fast speech are due to substitutions (62%) and deletions (31%), for slow speech most errors are due to insertions (55.8%) and substitutions (42%).

The adaptation procedure can be summarised in two steps:

1. Clustering of the adaptation data, which is carried out in a supervised mode:
   - The Viterbi algorithm is used to force align the adaptation data and determine the phone segmentation.
   - The average speech rate of each utterance is computed as follows:

$$ROS = \frac{N}{\sum_{i=1}^{N} \frac{d_i}{E[d_i]} \cdot \bar{d}}$$

     where N is the total number of phones, d_i is the duration of the ith phone of the utterance, $E[d_i]$ is its expected duration and $\bar{d}$ is the mean duration of all phones.
   - The adaptation database is divided into three clusters, one per speech rate (slow, average and fast).
2. Training of the adapted speech recogniser:
   - Speech rate classifier training: The classification procedure is based on the dynamic coefficients of the feature vector, since the delta and delta-delta Mel Frequency Cepstrum Coefficients are the most affected by the speech rate changes and therefore, are the most appropriate to perform the classification. The classification method is based on a gaussian classifier that is composed of 64 gaussians per speech rate.
   - Speech Rate Dependent Models (SRDM) training: Two sets of models are built: one for slow speech rate and the other for fast speech rate. Both sets of models are constructed starting with the set of speaker independent models and applying the Baum-Welch algorithm and deleted interpolation.

During the recognition phase, the speech rate classifier hypothesises the speech rate using the first frames of the incoming speech. This information is passed to the model selection module that chooses the most appropriate set of models and passes them to the speech recogniser.

This block supervised method was tested on the TRESVEL database (Martínez et al. 1998). It is composed of 9,600 utterances produced by 40 speakers (20 female and 20 male). Each speaker uttered 80 sentences at slow, average and fast speech rate. The TRESVEL database was divided into an adaptation set composed of 1,000 utterances for slow speech rate and 1,000 for fast speech rate and a testing set to perform these experiments. The vocabulary size of this task is 4,700 words and the relative improvement is 64% for slow speech rate and 19% for fast speech rate.

Eigenvoices approach

This is a new approach to fast speaker adaptation (Kuhn, Nguyen, Junqua, Goldwasser, Niedzielski, Fincke, Field & Contolini 1998, Kuhn, Nguyen, Junqua, Boman, Niedzielski, Fincke, Field & Contolini 1999, Kuhn, Junqua, Nguyen & Niedzielski 1999) that is based on recent research on face representation and recognition by Kirby & Sirovich (1990). It basically consists of creating, from speaker dependent models, a reduced set of models (called eigenvoices) that are representative of the average speaker space and its main sources of variation. In this way, adaptation to a new speaker is carried out by means of a linear combination of these eigenvoices.

The eigenvoice approach starts from a set of R speaker dependent models which have been properly trained. Then, for each speaker dependent model, a supervector is created that is composed of the parameters to be adapted. All the parameters of the acoustic models can be included in this supervector, even though the reported experiments have been done using just the mean vectors. Once the R D-dimensional supervectors have been created, they are processed by a dimensionality reduction technique like Principal Component Analysis (PCA), that leads to obtaining R D-dimensional eigenvectors. The eigenvector 0 is the mean supervector. The eigenvectors 1 to R-1 are the principal components, which are ranked in the order of the decreasing magnitude of their contribution to variation in the reference supervectors. Finally, the higher order eigenvectors are removed, keeping only the $K+1$ most important ones, that are called eigenvoices ($K < R \ll D$).

Thus, the adaptation to a new speaker is computationally cheap since the supervector for the new speaker is assumed to be a linear combination of the eigenvoices:

$$P = e(0) + \sum_{k=1}^{K} \omega(k) \cdot e(k)$$

This implies that the adaptation process consists of the estimation of just some weights (5 to 10 weights), which dramatically reduces the required amount of adaptation data at the expense of assuming that all the new speakers can be represented by a linear combination of these eigenvoices. It is therefore necessary to use a properly selected set of reference speakers to make sure that the assumption can be maintained.

The estimation of the weights, $\omega(i)$, from the adaptation data is carried out using a maximum likelihood estimator called MLED (Maximum Likelihood Eigen-Decomposition). This estimation algorithm is fully described in Kuhn et al. (1998), Kuhn, Nguyen, Junqua, Boman, Niedzielski, Fincke, Field & Contolini (1999), and Kuhn, Junqua, Nguyen & Niedzielski (1999) and leads to a set of K equations and K unknowns: Let

x_t be the feature vector at time 't'.

$\sum_{s,m}$ be the covariance matrix for gaussian 'm' in state 's'.

$e^s_m(j)$ be the subvector of eigenvoice 'j' ($j = 1, \ldots, K$) corresponding to the mean vector of mixture gaussian 'm' in state 's'.

$\gamma^s_m(t)$ be the probability of gaussian 'm' and state 's' given the observation vector x_t and the model λ.

Assuming the eigenvalues are independent, the K equations are:

$$\sum_s \sum_m \sum_t \gamma^s_m(t) \cdot [e^s_m(j)]^t \cdot \sum_{s,m}^{-1} \cdot x_t =$$
$$\sum_s \sum_m \sum_t \gamma^s_m(t) \cdot \left\{ \sum_{k=1}^{K} \omega(k) \cdot [e^s_m(k)]^t \cdot \sum_{s,m}^{-1} \cdot e^s_m(j) \right\}$$

The speaker dependent model parameters that are not adapted, because they are not included in the supervector, can be obtained from a speaker independent model.

This block supervised adaptation method was tested on the Isolet database, which contains 5 sets of 30 speakers that pronounced two times the whole alphabet (Kuhn, Junqua, Nguyen & Niedzielski 1999). The database was organised into five groups. Each group was composed of four sets of speakers (120 speakers) that were used as reference data and one set of speakers (30 speakers) that were used for adaptation and testing. The results that are presented are the average of the results in the five groups. The relative word error rate reduction obtained with 6 eigenvoices is 16.6% using one letter utterance for adaptation and 26.7% using four letter utterances for adaptation. If the number of eigenvoices is increased up to 11, the performance improves a little bit for more than ten adaptation letter utterances. In all cases, the performance of the eigenvoice approach is better than the performance of MAP and MLLR for these amounts of adaptation data.

3.3.2 Transformation based methods

The transformation based methods work under the assumption that there is a transform that is able to map the feature space of the training database into the feature space of the new speaker. The main problem of this approach is that it usually computes the adapted models using a transformation that is estimated just from the observation vectors of the new speaker, i.e.: it does not use *a priori* knowledge. Therefore, when there is sparse adaptation data, the estimated transform can be very unreliable, leading to poor performance of the speech recogniser. The use of techniques that allow the prediction of regions of the feature space for which there is not sufficient adaptation data or just there is no adaptation data, usually reduces the impact of this problem.

These methods use linear transformations that are applied either to the input vector or to the models, and the main difference between them is the approach used to estimate the best transformation. In this section we focus on several approaches to model transformations. Nevertheless, it is worth mentioning the work that was carried out by Choukri, Chollet & Grenier (1986) in the spectral domain, where two spectral transformations were estimated: one was applied to the new speaker and the other to the reference speaker to map both spaces into a common space.

The four model transformation techniques that are presented in this section are:

- Predictive adaptation.
- Maximum likelihood linear regression (MLLR).
- Transfer vector field smoothing (VFS).
- Spatial relation based transforms.

Predictive speaker adaptation

This method directly addresses the problem of sparse adaptation data given that it uses *a priori* information of the relation between sounds in the training database. The implementation that is presented in this chapter was developed by Cox (1993) and (1995) and is based on the following ideas:

- An utterance of sound 'S' from the new speaker contains information about the other sounds that the speaker can produce. Under this assumption, if the speaker produces an example of sound 'S', it is possible to update the parameters of several other models and the parameters of the model for 'S'.
- The prediction of poorly represented or unseen sounds of the new speaker can be done by means of regression models based on the relation between the sounds in the training database.
- Multiple linear regression models are generally more powerful than simple linear regression models. However, multiple linear regression has a drawback in this case because we do not know, *a priori*, what sounds from a new speaker will be available to the system at recognition time. If we restrict the method to use simple linear regression, we would need to have just V(V-1)/2 regression models (V is the number of sounds in the vocabulary) to be able to predict any sound from any other sound.

The adaptation procedure assumes that we know *a priori* what the unseen sounds of the adaptation data are, even though the extension of this method is straightforward for the case where this information is not available. It basically consists of computing linear regression models between all the couples of sounds together with the correlation between them using the training database.

Then, the adaptation data is used to re-estimate the models of the 'seen' sounds and the linear regression models are used to update the models of the 'unseen' sounds.

The adaptation procedure is therefore as follows:

1. Training of speaker independent models for each sound in the vocabulary using the training database.
2. Forced alignment and segmentation of the training data using the Viterbi algorithm, so that each observation vector of each utterance is mapped to a single state in its associated HMM.
3. Compute the sample mean vector, $\bar{x}_{ij}^k$, for each speaker 'k' in each state 'j' of model 'i'.

$$\bar{x}_{ij}^k = \frac{1}{N_{ij}^k} \cdot \sum_{l=1}^{N_{ij}^k} x_{ij}^{kl}$$

   where N_{ij}^k is the number of observation vectors, of speaker 'k', assigned to state 'j' of model 'i'.
4. For every speaker of the training database, see Figure 3.10, each component of the sample mean vector of each state of each 'seen' model is paired with the same component of each state of the 'unseen' models (seen and unseen refer to the adaptation data).

$$\left\{ \left(\bar{x}_{ij}^k(c), \bar{x}_{nm}^k(c)\right) | i \neq n \right\}$$

   where 'c' is the component of the vector.
5. Form a scattergram using the S_{tr} pairs of datapoints for each couple of states 'j' and 'm', where S_{tr} is the number of training speakers:

$$\left\{ \left(\bar{x}_{ij}^0, \bar{x}_{nm}^0\right), \left(\bar{x}_{ij}^1, \bar{x}_{nm}^1\right), \left(\bar{x}_{ij}^2, \bar{x}_{nm}^2\right), \ldots, \left(\bar{x}_{ij}^{S_{tr}}, \bar{x}_{nm}^{S_{tr}}\right) \right\}$$

6. Compute the best-fit line, $\bar{x}_{nm} = a_{ijnm} \cdot \bar{x}_{ij} + b_{ijnm}$, through the scattergram and store the regression coefficients a_{ijnm} and b_{ijnm}.
7. Compute and store the correlation coefficient r_{ijnm}.

At recognition time, it is assumed that each new speaker gives labelled utterances which contain a subset of 'N' different sounds. If we call these 'N' sounds $c(1), c(2), \ldots, c(N)$, the adaptation to the new speaker is done as follows:

1. Forced alignment and segmentation of the adaptation utterances.
2. Computation of the sample mean of the observation vectors assigned to each state.
3. Prediction of the mean values of the unseen sounds by applying the transformation:

$$\bar{x}_{nm} = a_{ijnm} \cdot \bar{x}_{ij} + b_{ijnm}$$

   where a_{ijnm} and b_{ijnm} are the regression coefficients corresponding to the highest correlation coefficient r_{ijnm} between the state 'j' of the seen sound 'i' and the state 'm' of the unseen sound 'n'.

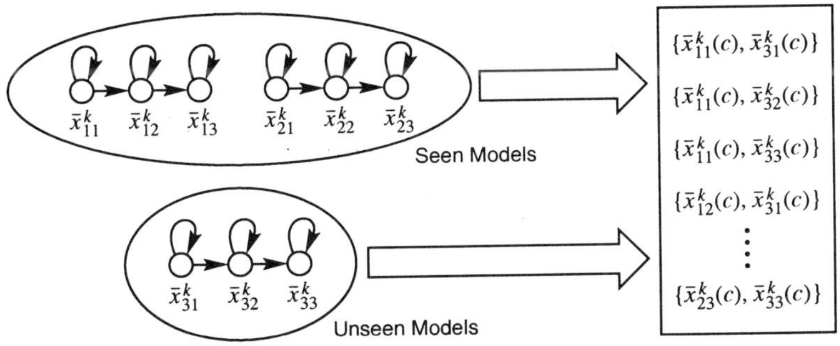

FIGURE 3.10: Seen and unseen models 'pairing' for speaker k.

If the correlation coefficient is high, we could be confident of the reliability of the predicted mean vector. However, if the correlation coefficient, r_{ijnm}, is low, we would not be so confident of this prediction and would rather use the existing speaker independent mean of the state of the HMM. Therefore, in these conditions, a maximum likelihood or a Bayesian approach are the most appropriate strategies (Duda & Hart 1973).

Maximum likelihood approach Let's denote:

N: Number of 'seen' sounds in the adaptation database.
H: Number of states of each HMM.
$\bar{x}^{kij}_{nm}$: Prediction of the mean of state 'm' of model 'n' for the new speaker 'k', that has been obtained from the mean of state 'j' of seen sound 'i'.
$\bar{x}^k_{ij}$: Adapted mean for state 'j' of model 'i' and new speaker 'k'.
$m, j = 1, \ldots, H$
$i = 1, \ldots, N$
$\bar{X}^k_{nm} = \{\bar{x}^{k11}_{nm}, \bar{x}^{k12}_{nm}, \ldots, \bar{x}^{kNH}_{nm}\}$: Set of predictions of the mean for state 'm' of model 'n' and new speaker 'k'.

If each prediction, $\bar{x}^{kij}_{nm}$, is assumed to be drawn from a normal distribution with mean and variance:

$$v_{kij}, \sigma^2_{kij} = \sigma^2_{\bar{x}_{nm}} \cdot \left(1 - r^2_{ijnm}\right)$$

then the ML estimate of $\bar{x}^k_{nm}$ is:

$$\left(\bar{x}^k_{nm}\right)ML = \frac{\sum_{i=1}^{N} \sum_{j=1}^{H} \frac{\bar{x}^{kij}_{nm}}{1-r^2_{ijnm}}}{\sum_{i=1}^{N} \sum_{j=1}^{H} \frac{1}{1-r^2_{ijnm}}}$$

Bayesian Approach This approach is applied under the assumption that the prior distribution of $\bar{x}^k_{nm}$ is gaussian with mean and variance μ_{SI}, σ^2_{SI}. In these conditions:

$$p\left(\bar{x}_{nm}^{k}|\bar{X}_{nm}^{k}\right)=\frac{p\left(\bar{X}_{nm}^{k}|\bar{x}_{nm}^{k}\right)\cdot P\left(\bar{x}_{nm}^{k}\right)}{P\left(\bar{X}_{nm}^{k}\right)}=\frac{P\left(\bar{x}_{nm}^{k}\right)}{P\left(\bar{X}_{nm}^{k}\right)}\cdot\prod_{i=1}^{N}\prod_{j=1}^{H}p\left(\bar{x}_{nm}^{kij}|\bar{x}_{nm}^{k}\right)$$

Calculating the logarithm, differentiating w.r.t. $\bar{x}_{nm}^{k}$ and setting to zero:

$$\left(\bar{x}_{nm}^{k}\right)_{MAP}=\frac{\frac{\mu_{SI}^{2}}{\sigma_{SI}^{2}}+\sum_{i=1}^{N}\sum_{j=1}^{H}\frac{x_{nm}^{kij}}{\sigma_{\bar{x}_{nm}}^{2}\left(1-r_{ijnm}^{2}\right)}}{\frac{1}{\sigma_{SI}^{2}}+\sum_{i=1}^{N}\sum_{j=1}^{H}\frac{1}{\sigma_{\bar{x}_{nm}}^{2}\left(1-r_{ijnm}^{2}\right)}}$$

This block supervised method was tested on a task of isolated utterances of letters (Cox 1995). The training database was composed of 4,056 utterances that were pronounced by 52 different speakers. The speech recognition system was adapted with 39 utterances of letters, i.e.: half alphabet (13 letters) was repeated three times by the new speaker. The vocabulary size for the isolated letters recognition task was 26 words and the experimental results showed a 76% relative improvement using the linear regression models and MAP.

Ahadi & Woodland (1995) extended this method to the case of continuous speech using context dependent continuous density HMMs. The first part of this new method uses MAP (see section 3.3.3) adaptation to estimate the mean vectors of the distributions of the 'seen' sounds. In the second part, a regression model is used to estimate the mean vectors of the distributions of the 'unseen' sounds. Therefore, the mean vectors of the distributions of the 'unseen' sounds are obtained from the re-estimated mean vectors of the 'seen' sounds by means of the following equation:

$$y=a_{0}+\sum_{l=1}^{P}a_{l}\cdot x_{l}$$

where y is the re-estimated mean of the target distribution, a_l is the regression coefficient for the mean of distribution l, x_l is the mean of the distribution l and P is just the regression order that is equal to the number of properly estimated distributions.

This approach provides a significant improvement with respect to the MAP estimate when there is few adaptation data and converges to the MAP estimate when there is enough adaptation data.

Maximum likelihood linear regression (MLLR)

This method was proposed in Digalakis, Rtischev & Neumeyer (1995), Leggetter & Woodland (1994a), Leggetter & Woodland (1994b), Leggetter (1995), and Leggetter & Woodland (1995). It is based on a group of linear transformations which transform the model parameters and maximise the model likelihood on the adaptation data. Each transformation is designed to adapt a certain portion of the model, since the acoustic mismatch may exhibit significant differences among the model units.

It is a flexible method in the sense that it can estimate just one global transformation if we have sparse adaptation data or many different transformations if there is enough adaptation data. It starts with speaker independent models and then computes statistics about the differences between the features modelled by the system and the ones of the new speaker. Finally, the statistics are used to update the means of the models.

Next, several key ideas of this method are presented:

- The main differences between speakers are assumed to be characterised by the distribution means.
- The adaptation of the mean vector is achieved by applying a transformation matrix W_s to the extended mean vector ε_s to obtain the adapted mean vector $\hat{\mu}_s$:

  $$\hat{\mu}_s = W_s \cdot \varepsilon_s$$

  The transformation matrix W_s carries out scaling and rotation of the extended mean vectors of the gaussian distributions so that the probability of the adaptation data given the adapted models is maximised.
- The same transformation matrix is shared by several distributions, so if there is no adaptation data for some distribution, a transformation may still be applied.

The adaptation procedure can be summarised in six steps:

1. Training of speaker independent continuous density HMMs.
2. Forced alignment of the adaptation data using the forward-backward algorithm.
3. Regression classes definition: The number of regression classes (and therefore, the number of transformations) depends on the amount of adaptation data. Two approaches for regression classes definition have been proposed:
   - Broad phonetic classes: all the model gaussians that belong to the same broad phonetic class (vowels, nasals, ...) are grouped into the same regression class.
   - Automatic clustering approach: the model gaussians are compared using a likelihood measure so that similar components are grouped in the same regression class. This approach has the advantage that it allows customising the number of classes to the available adaptation data.
4. Frame assignment: frames of the segmented adaptation data are assigned to their corresponding regression classes.
5. Transformation matrices computation (appendix at http://www.elsnet.org/, section 2).
6. Mixture components update using the above mentioned equation:

$$\hat{\mu}_s = W_s \cdot \varepsilon_s$$

The MLLR approach was later extended to adapt the covariance matrices by Gales & Woodland (1996a, 1996b). A drawback of MLLR is that it uses linear transformations that could be restrictive and inappropriate to model the differences between the characteristics of the training data and the characteristics of the new speaker. Diakoloukas & Digalakis (1997) presented a method that tries to solve this problem. It is based on MLLR and uses a weighted combination of linear transformations for each gaussian mixture of the model. The correct estimation of the parameters of the transformations requires a moderate amount of adaptation data (500 utterances), that could be a drawback for some applications.

Another drawback of MLLR is that maximum likelihood estimation does not introduce any constraints on the possible values of the parameters and relies only on the adaptation data and the original acoustic models. It would be therefore desirable to constrain the possible values of the parameters to avoid getting a transformed model that might misrepresent the underlying structure of the acoustic space. Chou (1999) and Chesta, Siohan & Lee (1999) propose the use of maximum a posteriori estimation (section 3.3.3) instead of maximum likelihood. This approach called MAPLR, requires a smaller amount of adaptation utterances since in MAP some values of the parameters, described by the prior density, are more likely than others, which constrains the estimation process.

One of the reported adaptation experiments carried out with MLLR was tested on the Resource Management task (RM1) (Leggetter & Woodland 1995), whose vocabulary size is 1,000 words. A word-pair grammar whose perplexity was 60 and 10 regression classes were used. The system was adapted using 40 utterances and the speaker independent models were estimated using 3,990 utterances. In these conditions, the experimental results showed a relative improvement of 37% compared with the 58% relative improvement of the speaker dependent models.

Transfer vector field smoothing (VFS)

VFS (Ohkura, Sugiyama & Sagayama 1992) addresses the two problems derived from retraining with insufficient adaptation data: Adaptation of models for which there is no adaptation data and estimation of parameters for which there is insufficient adaptation data.

VFS assumes that the gaussian means are the most affected parameters in the HMMs and therefore it is applied just to the means. It consists of a mapping between the initial speaker dependent HMMs and the adapted HMMs by means of a transfer vector field. The transfer vector field is composed of vectors calculated as the difference between the mean vectors of the initial CDHMMs and the adapted mean vectors. The transfer vectors corresponding to mean vectors for which there is no adaptation data are obtained by interpolation of the 'k' transfer vectors whose corresponding means are the closest to the untrained mean.

The adaptation procedure can be divided into four steps that are described below:

1. Embedded training: using the models of the reference speaker as initial models and the adaptation data of the new speaker.
2. Transfer vectors computation: Each transfer vector, V_k, is calculated as the difference between a mean vector of the initial CDHMM, μ_k^I, and the corresponding mean vector obtained in the embedded training, μ_k^R:

$$V_k = \mu_k^R - \mu_k^I$$

3. Interpolation: Each untrained mean vector is mapped to the new speaker's voice space by using an interpolated transfer vector that is computed as follows:

$$V_n = \sum_{k \in KNN(\mu_n^I)} \Gamma_{nk}(f) \cdot V_k$$

where f is a parameter called fuzziness ($f = 1.4$), and:

$k \in KNN(\mu_n^I)$: represents the set of the k retrained mean vectors which are the nearest neighbours of μ_n^I in the initial voice space (the voice space of the reference speaker).

$\Gamma_{nk}(f)$: is a fuzzy membership function and it is the weight of the transfer vector V_k.

The fuzzy membership function is given by:

$$\Gamma_{nk}(f) = \frac{1}{\sum_{j \in k_R} \left(\frac{d_{nk}}{d_{nj}} \right)^{1/(f-1)}}$$

where k_R represents the group of retrained mean vectors and d_{nk} is the distance between μ_n^I and μ_k^I.

4. Smoothing: Each transfer vector is modified in accordance with the nearest transfer vectors. This is carried out as follows:
   - The fuzzy membership function, $\Gamma_{km}(f)$, is calculated for each mean vector.
   - The auxiliary function $\lambda_{km}(f)$ is constructed:

$$\lambda_{km}(f) = \begin{cases} 1 & \Leftrightarrow \quad k = m \\ \Gamma_{km}(f) & \Leftrightarrow \quad k \neq m \end{cases}$$

   - The smoothed transfer vector V_k^s is obtained using the equation:

$$V_k^s = \frac{\sum_{m \in KNN(\mu_k^I)} \alpha_m \cdot \lambda_{km}(f) \cdot V_m}{\sum_{m \in KNN(\mu_k^I)} \alpha_m \cdot \lambda_{km}(f)}$$

   Where α_m represents the reliability of transfer vector V_m ($\alpha_m = 1$ if $m \in k_R$, otherwise $\alpha_m = 0$).

- The mean vector is mapped to the voice space of the new speaker:

$$\mu_k^s = \mu_k^I + V_k^s$$

Transfer vector field smoothing is a block supervised adaptation technique that was tested on a Japanese isolated phonemes task (Ohkura et al. 1992) to adapt speaker dependent models to a new speaker. The initial speaker dependent models were trained using 5,500 utterances of isolated words and the adaptation was carried out with 66 seconds of speech (11 sentences). The size of the vocabulary was 23 phonemes and the relative improvement was 74% compared with 85% improvement obtained using speaker dependent training.

Spatial Relation Based Transformation

Komori, Kosaka, Yamada & Yamamoto (1997) presented an adaptation method based on HMM transformations that are applied to the means of the gaussian distributions. These transformations are based on two different relations:

- Spatial relation between context dependent and context independent HMMs, and
- Spatial relation between speaker dependent and speaker independent HMMs.

This method starts from three sets of models to calculate the speaker dependent context dependent models:

1. Speaker independent context independent models (SICI-HMM), that are estimated using the training database.
2. Speaker independent context dependent models (SICD-HMM), that are also obtained using the training database.
3. Speaker dependent context independent models (SDCI-HMM), that are calculated using the adaptation database.

Based on the three sets of models and the two spatial relations, two kinds of transformations can be applied: Speaker direction adaptation and context direction adaptation. In context direction adaptation, the transformation that is illustrated in Figure 3.11 is performed in the following way:

$$\hat{\mu}_m = \mu_s + (\mu_m - \mu_a) \cdot \frac{\sigma_s}{\sigma_a}$$

When the transformation is done in the speaker direction adaptation, the following equation is applied:

$$\hat{\mu}_m = \mu_m + (\mu_s - \mu_a) \cdot \frac{\sigma_m}{\sigma_a}$$

These methods can also be applied to the case of mixture gaussian distributions. Komori et al. (1997) presents three different ways of adapting the models based on their spatial relations. The main idea is to calculate the mean of

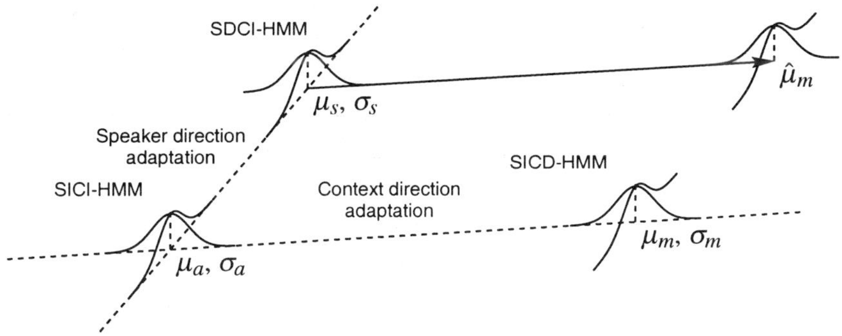

FIGURE 3.11: Context direction transformation.

each state and then compute the transformations based on the spatial relation of the state means.

Finally, all the adapted means are smoothed:

$$\tilde{\mu}^{SDCD} = \alpha \cdot \hat{\mu}^{SDCD} + (1 - \alpha) \cdot \mu^{SICD}$$

where $\tilde{\mu}^{SDCD}$ is the final speaker dependent context dependent mean, $\hat{\mu}^{SDCD}$ is the speaker dependent context dependent transformed mean, μ^{SICD} is the speaker independent context dependent mean, and α is the adaptation ratio $(0 \leq \alpha \leq 1.0)$.

This is a block supervised adaptation method that was tested on a Japanese isolated words task with a vocabulary size of 520 words (Komori et al. 1997). In the experiments, 25 isolated words were used as adaptation data. The training database was composed of 1,500 Japanese isolated words that were recorded by 60 speakers. Each training speaker uttered 25 words. The word accuracy improved from 95.6% to 97%, that represents a 31.8% reduction of the word error rate.

3.3.3 Bayesian estimation

Maximum a posteriori estimation (MAP)

Bayesian estimation was first used by Brown, Lee & Spohrer (1983) for adapting a digits recogniser based on continuous density HMMs. This method was later extended to HMM parameters adaptation for the case of gaussian distributions by Lee, Lin & Juang (1990) and Gauvain & Lee (1994).

The difference between maximum likelihood estimation (MLE) and maximum a posteriori (MAP) is that MLE assumes that the vector of parameters, λ, is fixed but unknown, while for MAP, λ is a random variable whose prior distribution is known. Therefore, the MAP estimation framework is particularly useful for dealing with the sparse adaptation data problem, where MLE would lead to inaccurate estimates. The reason for this is that MAP integrates prior information in the adaptation process and, consequently, less adaptation data is required. In MLE, since there is no information about λ, the amount of adaptation data has to be large enough to provide robust estimates of the parameters.

MAP estimation has two sources of information: the prior distribution of the parameters, $p(\lambda)$, and the adaptation data, X. The MAP estimate, λ_{MAP}, is defined as:

$$\lambda_{MAP} = \underset{\lambda}{\arg max}\, p(\lambda|X) = \underset{\lambda}{\arg max}\, \frac{p(X|\lambda) \cdot p(\lambda)}{p(X)} \propto \underset{\lambda}{\arg max}\, p(X|\lambda) \cdot p(\lambda)$$

where $p(X)$ is the probability distribution of the adaptation data, X.

When the prior distribution of the parameters, $p(\lambda)$, is uninformative, i.e. $p(\lambda)$ is uniform, MAP estimation reduces to MLE.

The MAP technique presents four main problems: the selection of the prior distributions and the estimation of their parameters, the estimation of the adapted HMM parameters, the errors of the orthographic transcriptions when MAP is used in unsupervised adaptation mode and the inability of MAP to estimate the parameters for which there is no adaptation data.

The selection of the appropriate prior densities is a key problem in MAP estimation: the computation of the posteriori density $p(\lambda|X)$ can be simplified if we assume that observations are drawn from a family of distributions for which there is a sufficient statistic of fixed dimension. In this case, the prior distributions can be chosen from a conjugate family and then the MAP estimation is reduced to obtaining the mode of the posteriori density. Nevertheless, in the case of HMMs adaptation, no sufficient statistic of fixed dimension exists due to the underlying hidden process, i.e. the observations are drawn from a probabilistic function of a Markov chain. A solution pointed out by Gauvain & Lee (1994) is to use the EM algorithm, which is based on the fact that maximising the likelihood of complete data is simpler than maximising the likelihood of the incomplete data when the complete data has sufficient statistics of fixed dimension. Therefore, with this approach, the prior densities are selected from the conjugate family of the complete data density.

Following this approach, the formulae for the estimation of the adapted HMM parameters are obtained from the EM algorithm, as shown in the appendix at http://www.elsnet.org/.

Additionally, a solution for compensating the problem of the orthographic transcription errors in unsupervised adaptation is described in next section.

Concerning the inability of MAP to estimate parameters for which there is no adaptation data, it can be overcome using Extended MAP (EMAP) estimation. This method uses correlations between the model parameters, so that adaptation data for a given gaussian mean may contribute to the adaptation of other means. The first work on this idea was carried out by Lasry & Stern (1984) and Stern & Lasry (1987), which incorporated the correlation between phonemes in the bayesian estimation framework. Later we give more detailed information about this method.

Bayesian learning can be modified to work in incremental adaptation mode. For example, Matsuoka & Lee (1993) used the segmental MAP to perform on-line adaptation. Another interesting example is the work presented by Huo & Lee (1997), who developed a quasi-Bayes framework for estimating the parameters of continuous density HMMs with gaussian mixture state observation densities. This algorithm was designed to incrementally update the parameters and used a forgetting mechanism to adjust the contribution of previous utterances. This on-line adaptive method can, thus, be used to compensate the mismatches caused by changing speakers, channels and microphones.

In summary, the regular MAP adaptation procedure can be divided into four steps:

1. Assumptions:
   - The adaptation samples $X = \{x_1, \ldots, x_T\}$ were generated by a HMM with parameters $\lambda = (\Pi_i, a_{ij}, c_{jk}, \mu_{jk}, \Sigma_{jk})$.
   - All the parameters are unknown but their prior distributions are known (Gauvain & Lee 1994, DeGroot 1970, Álvarez 1995):

$$\begin{aligned} P(\Pi_i) &\equiv \text{Dirichlet with parameter } \eta_i \\ P(a_{ij}) &\equiv \text{Dirichlet with parameter } \eta_{ij} \\ P(c_{jk}) &\equiv \text{Dirichlet with parameter } \upsilon_{jk} \\ P(\mu_{jk}) &\equiv \text{Normal with parameters } m_{jk}, \tau_{jk}, \Sigma_{jk} \\ P(\Sigma_{jk}) &\equiv \text{Wishart with parameters } \alpha_{jk}, U_{jk} \end{aligned}$$

2. Given the initial model λ^0, compute the initial values of η_i, η_{ij}, υ_{jk}, m_{jk}, α_{jk}, U_{jk}.
   Let

   L be the total number of gaussians per mixture or state.
   i, j be states of the HMM.
   τ_{jk} be the number of observations generated by gaussian k of state j.
   μ^0_{jk} be the mean of the initial model for gaussian k of state j.
   Σ^0_{jk} be the covariance matrix of the initial model for gaussian k at state j.
   c^0_{jk} be the weight of the initial model for gaussian k at state j.
   D be the dimension of the observations or feature vectors.

   Then (Álvarez 1995):

$$\begin{aligned} \eta_i &= \pi^0_i + 1 \\ \eta_{ij} &= 1 + a^0_{ij} \cdot \left[\sum_{k=1}^{L} \tau_{ik} \right] \end{aligned}$$

$$\begin{aligned} v_{jk} &= c^0_{jk} \cdot \sum_{k=1}^{L} \tau_{jk} + 1 \\ m_{jk} &= \mu^0_{jk} \\ U_{jk} &= (\tau_{jk} - 1) \cdot \Sigma^0_{jk} \\ \alpha_{jk} &= \tau_{jk} + D \end{aligned}$$

3. Calculate the set of parameters λ that maximise the posterior probability, i.e. $\lambda_{MAP} = \underset{\lambda}{\arg\max}\, p(\lambda|X)$ Gauvain & Lee (1994) and Álvarez (1995) (see section 1 of the appendix at http://www.elsnet.org/):

$$\bar{\pi}_i = \frac{(\eta_i - 1) + \gamma_1(i)}{1 + \sum_{i=1}^{N} (\eta_i - 1)}$$

$$\bar{a}_{ij} = \frac{(\eta_{ij} - 1) + \sum_{t=1}^{T} \gamma_t(i,j)}{\sum_{j=1}^{N} (\eta_{ij} - 1) + \sum_{t=1}^{T} \gamma(j)}$$

$$\bar{c}_{jk} = \frac{(v_{jk} - 1) + \sum_{t=1}^{T} \xi_t(j,k)}{\sum_{k=1}^{L} (v_{jk} - 1) + \sum_{t=1}^{T} \gamma_t(j)}$$

$$\bar{\mu}_{jk} = \frac{\tau_{jk} \cdot m_{jk} + \sum_{t=1}^{T} \xi_t(j,k) \cdot x_t}{\tau_{jk} + \sum_{t=1}^{T} \xi_t(j,k)}$$

$$\bar{\Sigma}_{jk} = \frac{U_{jk} + \tau_{jk} \cdot (\bar{\mu}_{jk} - m_{jk}) \cdot (\bar{\mu}_{jk} - m_{jk})^t + \sum_{t=1}^{T} \xi_t(j,k) \cdot (x_t - \bar{\mu}_{jk}) \cdot (x_t - \bar{\mu}_{jk})^t}{(\alpha_{jk} - D) + \sum_{t=1}^{T} \xi_t(j,k)}$$

Where:

$\gamma_t(i) = p(s_t = i|X, \lambda)$
$\gamma_1(i) = p(s_1 = i|X, \lambda)$
$\gamma_t(i,j) = p(s_t = i, s_{t+1} = j|X, \lambda)$
$\xi_t(j,k) = p(s_t = j, k_t = k|X, \lambda)$
T is the number of feature vectors.

4. Go back to step 2 until a convergence criterion is reached.

The reported experiments (Gauvain & Lee 1994) show the WER is reduced from 31.5% when the system is adapted using MLE with 2 minutes of adaptation data to 8.7% when the system is adapted using MAP and gender independent models with the same amount of adaptation data. The speaker independent models were trained using the standard RM SI-109 training set consisting of 3,990 utterances from 109 speakers, each providing from 30 to 40 utterances. The test database was RM FEB91-SD, that consists of 25 utterances from each of the 12 testing speakers. As expected, when the amount of adaptation data is large (30 min.), the results for both MLE and MAP are the same.

N-Best Based Unsupervised MAP Adaptation

In the previous section it was mentioned that the lack of orthographic transcriptions for the adaptation data is a drawback in MAP estimation. This is due to the fact that in these cases the transcriptions are generated using a speaker independent speech recogniser, that always has an associated word error rate. Since the word error rates can be high for some speakers, the adaptation algorithm could estimate wrong models for those speakers who require adaptation the most.

Matsui, Matsuoka & Furui (1997) addressed this problem by using the N-best candidates obtained during the recognition of the adaptation utterances. The method, which combines clustering and MAP estimation, uses a smoothing estimation technique that takes into account the N highest likelihood recognised word sequences. The value of each gaussian mean is calculated as a weighted sum of the values of the adapted means of the N-best recognised sentences and the weights are proportional to a confidence measure. The means for each recognised sentence are adapted with a bias vector that allows a translation of the corresponding mean.

The whole adaptation procedure can be summarised in six steps:

1. Cluster the model means using a distance measure.
2. Obtain multiple word sequence hypotheses $\{w_1, w_2, \ldots, w_N\}$ for each utterance of the adaptation data by using N-best decoding and speaker independent phoneme models.
3. For each utterance, use utterance verification to decide whether it is necessary to use N-best decoding or not.
4. For each word sequence w_n, obtain the mixture bias term, $B_n = \{b_n^1, b_n^2, \ldots, b_n^P\}$, for each cluster using MAP estimation:

$$\hat{\mu}_{jk}^n = \mu_{jk} + b_n^{\Phi(j,k)}$$

where:

μ_{jk} is the speaker independent mean for gaussian k of state j.
$\Phi(j,k)$ is a membership function that maps gaussians to clusters.
$b_n^{\Phi(j,k)}$ is the bias term.
$\hat{\mu}_{jk}^n$ is the estimated mean of gaussian k of state j for the $\mathrm{n^{th}}$ decoded sequence of words.

5. Compute the confidence measure, c_n, for each decoded sequence of words, w_n:

$$c_n = \frac{1}{e^{\alpha\lfloor L_{max} - L_n \rfloor}}$$

where L_n is the log-likelihood for the recognised sequence of words w_n after adaptation, L_{max} is the highest log-likelihood after adaptation and α is an experimental parameter.

6. The adapted means of the models are obtained using the following equation:

$$\hat{\mu}_{jk} = \frac{\sum_{n=1}^{N} c_n \cdot \hat{\mu}_{jk}^{n}}{\sum_{n=1}^{N} c_n}$$

This block unsupervised adaptation method was tested on a Japanese four-digit string task. The training database contained 24,194 utterances from 177 male speakers, each pronouncing approximately 136 utterances. The adaptation data consisted of six four-digit strings per speaker and the relative improvement was 18.7%.

Extended MAP adaptation (EMAP)

The extended MAP algorithm was first proposed by Lasry & Stern (1984) to improve the estimation of a set of gaussian classifiers in a static pattern recognition problem. This algorithm incorporates correlation information across the various speech recogniser parameters into the MAP parameter estimation process, which overcomes the problem of adapting the parameters for which there is no adaptation data. This technique was successfully used as a speaker adaptation method by Stern & Lasry (1987) in the FEATURE recognition system, Rozzi & Stern (1991) in the SPHINX semicontinuous speech recogniser and Zavaliagkos, Schwartz, McDonough & Makhoul (1995) in the BYBLOS continuous density speech recogniser.

The correlation information is incorporated in the prior distributions of the gaussian means: in MAP estimation the means of the gaussians are assumed to be independent random variables while in the EMAP approach we consider their joint distribution.

Let's consider a set of C D-dimensional gaussians. Let's call $N = C \cdot D$ and define the mean vector μ as $\mu = [\mu_1^t, \ldots, \mu_j^t, \ldots, \mu_C^t]^t$. If we assume the N-dimensional mean vector μ to be normally distributed with mean vector μ_0 and NxN-dimensional covariance matrix Σ_0 and denote the set of adaptation observation vectors as $X = \{x_{1,1}, \ldots, x_{1,n_1}, \ldots, x_{j,1}, \ldots, x_{j,n_j}, \ldots, x_{C,1}, \ldots, x_{C,n_C}\}$, the EMAP estimate of the mean μ is given by (Lasry & Stern 1984):

$$\hat{\mu}_{EMAP} = \Sigma \cdot (\Sigma + B \cdot \Sigma_0)^{-1} \mu_0 + \Sigma_0 \cdot (\Sigma + B \cdot \Sigma_0)^{-1} B \cdot \mu_{ML}$$

where:

B is a NxN block diagonal matrix which diagonal blocks are given by $n_j \cdot I$.

Σ is a NxN block diagonal matrix which diagonal blocks are given by Σ_j.

μ_{ML} is the maximum likelihood estimate of μ (the sample mean).

μ_0^t is the vector of speaker independent mean vectors.

$\Sigma_0 = E\lfloor(\mu - \mu_0) \cdot (\mu - \mu_0)^t\rfloor$

n_j is the number of observation vectors for gaussian j $(1 \leq j \leq C)$.

Zavaliagkos et al. (1995) tested a modified version of this adaptation method to adapt the BYBLOS speech recogniser to English non-native speakers. The experiments were performed on various 5K-word tasks of the Wall Street Journal corpus. The results showed 50% word error rate reduction with respect to the speaker independent system.

3.3.4 Pronunciation Modelling

Speaking Mode Dependent Pronunciation Modelling

It is well known that pronunciation differences in spontaneous conversational speech represent an important source of variability that is not well accounted for by current speech recognition systems. Initially, we would think that this problem could be solved by increasing the number of phonetic transcriptions in the dictionary. Nevertheless, the simple increase of the pronunciation set often is not useful since the gain of increasing the number of phonetic transcriptions may be offset by the loss due to increased confusability. For this reason and given that there are predictable pronunciation changes which depend on the speaking mode and/or the speaker characteristics, we could just change dynamically the set of active phonetic transcriptions to avoid the above mentioned loss in recognition accuracy.

The method presented in this section (Finke & Waibel 1997) deals with this problem by assigning a probability to each alternative pronunciation that is a function of the speaking style. The probability function is learned through decision trees from rule based generated alternative pronunciations.

The adaptation procedure is carried out as follows:

1. Define a set of context dependent pronunciation transformation rule. The rule set is defined to account for the typical pronunciation transformations of the task where the speech recogniser is going to be tested. A typical example of this kind could be the intervocalic /d/ in the past participle verb declination in Spanish: /-ado/ $\rightarrow$ /-ao/ (comprado $\rightarrow$ comprao).
2. The set of rules is used to automatically produce alternative pronunciations in the recognition dictionary.
3. Given that current state-of-the-art speech recognisers are not able to rewrite the pronunciation of phones that depend on word context, the crossword pronunciation phenomenon is captured using multiwords in the dictionary (for example, in English *kind of* is usually reduced to *kinda*).
4. The acoustic models are adapted to each training speaker by using maximum likelihood linear regression (MLLR, see section 3.3.2).
5. The training database is force-aligned (Viterbi) using the expanded dictionary, the adapted models for each speaker and the flexible transcription alignment (FTA) procedure that consists of allowing for:
   - All the alternative pronunciations.

- Multiwords.
- Optional beginning and ending words.
- Optional silence or breathing models between words.
- Optional noise words at the beginning and the end of the utterance.

6. The results of the flexible transcription alignment are used to estimate the prior probabilities of the transformation rules, $p(r)$. These probabilities are therefore obtained from the frequency of each rule in the aligned corpus.
7. A decision tree is built for each rule (Breiman, Friedman, Olshen & Stone 1993) to predict the probability, $p(r|w,m)$, of applying a transformation rule, r, given the word, w, and the speaking mode, m, measured in terms of speaking rate and phone and word duration.
8. The rule probabilities are used to compute the probabilities of each alternative pronunciation, $q_i(w)$, in the dictionary:

$$P(q_i(w)) = \frac{\prod_{\forall r^+} p(r^+) \cdot \prod_{\forall r^-} [1 - p(r^-)]}{Z}$$

   where r^+ are the rules that were used to obtain the alternative pronunciation $q_i(w)$, r^- are the rules that match with the baseform, w, but were not used to obtain $q_i(w)$ and Z is a normalisation constant.

Finally, the probability of each alternative pronunciation is used as a weight during the recognition process.

This block supervised adaptation technique was tested on the Switchboard and CallHome partitions of the 1996 NIST Hub-5e evaluation set (Finke & Waibel 1997). The experiments were carried out with the Janus recognition toolkit (JRTk). The expanded dictionary included 1.78 alternative pronunciations per word on average. The best results were obtained when the alternative pronunciations were weighted using the tree polyphonic decision tree based on the speaking mode and the FTA labels. In these conditions, the word error rate was reduced from 32.2% to 26.7% in the Switchboard task and from 43.7% to 36.1% in the CallHome task.

Accent specific pronunciation modelling

This other approach to pronunciation modelling deals with the problem of regional accent specific pronunciations and has been developed to be used in large vocabulary continuous speech recognition. It focuses on phone insertion, deletion and substitution that occur because of the regional accent specific pronunciations and also because of the spontaneous speech characteristics.

As we have already mentioned, most successful current adaptation techniques adapt just the acoustic models, so that they work under the assumption that there is no pronunciation variability among speakers and, consequently, the phonetic transcriptions of the dictionary words do not have to be adapted to new speakers. Therefore they do not account for pronunciation specific phenomena that in some cases might be an important source of recognition errors.

This adaptation technique (Humphries & Woodland 1997) was applied to use a British English speech recogniser to recognise American English speakers and can be summarised in six steps:

1. The non-native (American English) pronunciations are transcribed using a triphone-based speaker independent phone recogniser. In this way, a set of errorful phone level transcriptions, T_1, are obtained.
2. The non-native pronunciations are force-aligned using the real orthographic transcriptions and a word-based speaker independent speech recogniser. The forced alignment process is carried out by allowing all the alternative pronunciations in the dictionary and then let the speech recogniser select the most appropriate. At the end of this process, there will be a second set of transcriptions, T_2, available.
3. The two sets of transcriptions T_1 and T_2 are aligned using dynamic programming to obtain a list of context dependent phone substitutions, insertions and deletions. The list describes the differences between the non-native pronunciations and the ones used by by the native speech recognition system and each entry has the following form:

$$m(l, r) \rightarrow s$$

   where (l, r) are the left and right contexts of phone 'm' respectively and s is what replaces phone 'm'. If the replacement s, is another phone, it represents a substitution, if it is a null, it represents a phone deletion and, finally, if it is a group of phones, it represents a phone insertion perhaps combined with substitution.
4. The pronunciation variations are clustered using a decision tree (Breiman et al. 1993) that uses the phonetic features of the contextual information of the left and right contexts. "The tree building algorithms partition the set of pronunciation rules by making a series binary splits, selected from a set of around 70 questions for each of the left and right contexts." The binary splits are selected to minimise the misclassification error rate of the tree. Finally a tree is obtained for each base phone.
   Given that the phone error rate of the speech recognisers is high, a confidence measure based on the number of competing models is used to filter out poorly transcribed data. The confidence measure is given by the following equation:

$$C = \frac{S_m}{S_m^T}$$

   where S_m is the number of active search paths for phone 'm' between the starting and ending frames of the phone and S_m^T is the total number of active search paths between the starting and ending frames of phone 'm'.
5. Finally, a new accent-specific pronunciation dictionary is generated from the original dictionary: The tree provides a list of phone replacements

for each phone within a given context. Each replacement carries a probability, thus the probability of each new pronunciation can be calculated as the product of each of the individual phone replacement probabilities. Later these probabilities are used as weights during the recognition process.

The number of alternative pronunciations is limited using a pronunciation probability threshold to avoid the increase of the word rate due to an excessive increase of the confusability between words in the dictionary.

This block supervised adaptation technique was tested on American English in the Wall Street Journal (WSJ0) task when the system had been trained using the WSJCAM0 British English database (Humphries & Woodland 1997). The training database is composed of 7,861 utterances from 92 speakers. The experiments showed that the word error rate dropped from 30.9% to 24.8%.

3.4 Speaker normalisation techniques

In this section we present two examples of different approaches to Vocal Tract Length Normalisation (VTLN) even though there are many others that have been reported (Ono et al. 1993, Ariki et al. 1996, Eide & Gish 1996, Lee & Rose 1996, Ishii & Tonomura 1997, Westphal et al. 1998). The first one is based on a linear transformation of the frequency axis (Wegmann et al. 1996) and the second one is based on the estimation of the first and second formants position to obtain the conditional probability density functions of the position of the formants. The normalisation factor is later computed to maximise the joint probability density function (Lincoln et al. 1997).

There are other normalisation techniques that are not based on vocal tract normalisation, like the one proposed by Acero & Huang (1996). This technique is an extension of CMN, which was proposed to normalise differences in the channel (Liu, Stern, Huang & Acero 1993, Sankar & Lee 1995). Later CMN was improved to eliminate the dependency of the cepstral mean on the amount of noise, which was achieved by computing two means: one for the noise and other for the speech (Huang, Acero, Alleva, Hwang, Jiang & Mahajan 1995, Sankar & Lee 1995). Finally, Acero & Huang (1996) extended this method to a larger number of acoustic classes.

Additionally, speaker normalisation techniques are increasingly being used in combination with speaker adaptation techniques. For example, Hariharan & Viikki (1999) combine VTLN and MAP, Zhao (1994) combines a speaker normalisation based on estimating a spectral bias term and phone model adaptation. Her main assumption is that the spectral variation sources can be divided into two categories: one acoustic and the other phone-specific. This method is later improved in Zhao (1996). Additionally, Uebel & Woodland (1999) report an interesting study that compares different approaches to VTLN

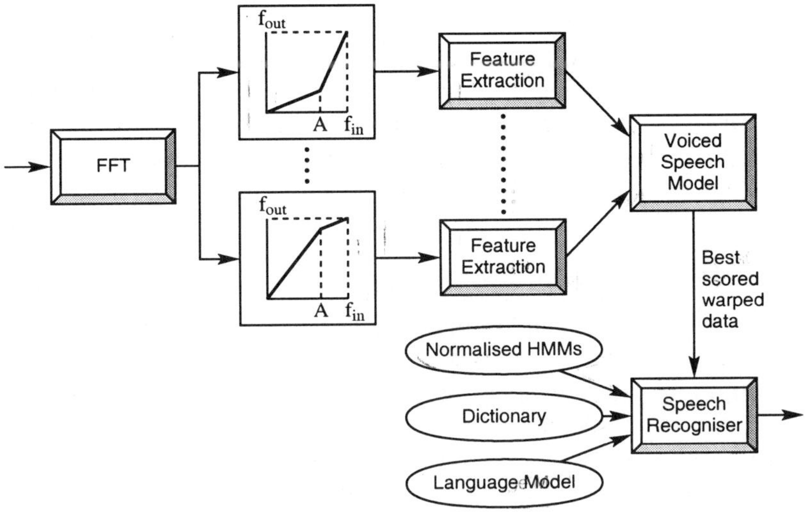

FIGURE 3.12: Block diagram of the frequency warping linear transformation based recogniser.

in combination with MLLR.

3.4.1 Frequency warping linear transformation

The idea of this method (Wegmann et al. 1996) is to re-scale or warp the frequency axis during the feature extraction step, so that speech from all the speakers seem to be produced by a vocal tract of the same standard length. The frequency axis is warped by means of a linear transformation that have fixed points at 0 and 4 KHz. The transformation is composed of two lines: one that goes from the origin to a point 'A' and another from 'A' to 4 KHz. The point 'A' is chosen below 4 KHz, so that the map from 0 KHz to 'A' is a line through the origin with a variable slope (from 0.8 to 1.2) and the map from 'A' to 4 KHz starts at the intersection of the previous line at 'A' and ends at 4KHz (see figure 3.12).

There are ten different transformations. Five of them have slopes greater than one, four have slopes smaller than one and one has just slope one.

The normalisation procedure consists of estimating the parameters of the voiced speech model that will be used to select the best warp scale and train gender independent speaker normalised models. The voiced speech model is a 256-gaussian mixture that is trained using the voiced frames of the incoming utterances. The whole normalisation procedure can be summarised as follows:

1. The training utterances are normalised using the ten previously described warp scales. Later the feature extraction process is carried out for each of the warped utterances.
2. The initial gaussian mixture parameters of the voiced speech model are

estimated using the voiced frames of the unwarped training data.

3. The voiced speech model is used to determine the best warp scale for each of the training speakers. This is done using one utterance per speaker, picking the voiced frames of each of the warped versions of the utterance and scoring them with the voiced speech model. The best scoring warp scale is finally selected.
4. The voiced speech model is re-estimated using the selected warping data in step 3.
5. The steps 3 and 4 are repeated until a convergence criterion is reached.
6. Finally, a set of gender independent speaker normalised models is trained using the normalised speech.

At recognition time, the incoming utterances are speaker normalised using the ten warp scales. The voiced speech model scores their voiced frames and selects the best scoring scale. Next, the recogniser uses the best scored warped data and the normalised models to do the recognition process.

This is a static supervised normalisation method that was tested on the Switchboard CAIP set (Wegmann et al. 1996), whose vocabulary is composed of about 10,000 words. The system was trained using 65 hours of the Switchboard database from 160 speakers (80 female and 80 male) and the experimental results showed an 11.8% word error rate reduction (from 49.8% to 43.9%).

3.4.2 Formant estimation normalisation

This method (Lincoln et al. 1997) scales the frequency axis using the estimates of the first and second formant frequencies of the new speaker. The estimates of the formant frequencies are obtained from a probabilistic model of the formant distributions. The scaling of the frequency axis is then used to normalise the available data of the new speaker. As in the previous method, the frequency axis is usually compressed for the female speakers and expanded for the males. This is what would intuitively be expected since, in general, the length of the vocal tract is shorter for women than for men.

The normalisation procedure consists of computing the first and second formant position for each vowel class and then use both estimates to estimate a unidimensional gaussian distribution for each formant and vowel class. In particular:

1. Formant estimation, which is carried out in three steps:
   - Linear prediction coefficient computation of each frame belonging to a vowel.

$$H(z) = \frac{1}{A(z)} = \frac{1}{\sum_{i=1}^{m} a_i \cdot z^{-i}}$$

Where 'm' is the analysis order and a_i are the predictor coefficients.

- Computation of the predictor polynomial roots to obtain the candidate frequencies and bandwidths of the speech formants:

$$F_i = \frac{\vartheta_i \cdot f_s}{2\pi}$$

$$B_i = \frac{-\ln\left[|r_i| \cdot f_s\right]}{\pi}$$

where f_s is the sampling frequency, ϑ_i is the angle of root r_i, and $|r_i|$ is the magnitude of root r_i.
- Root selection: The roots are sorted into the order of ascending frequency and then the two lowest roots such that $|r_i| > 0.9$, are selected. These roots will be the first (F_1) and the second (F_2) formants for the current frame.

2. Conditional distributions estimation: The parameters of the gaussian distributions for each formant and vowel class are estimated using the F_1 and F_2 estimates for each vowel class and frame, obtaining both $p(f^1|F_1)$ and $p(f^2|F_2)$:

$$p(f^n|F_n) = \frac{1}{\sqrt{2\pi} \cdot \sigma_n} \cdot e^{-\frac{(f^n-\mu_n)^2}{2\sigma_n^2}}$$

3. Normalisation factor estimation, that is carried out in two parts:
   - Computation of the normalisation factor, a_i^S, for every vowel frame 'i' of speaker 'S':

$$a_i^S = \underset{a}{\arg max}\, p\left(a \cdot f_i^1|F_1\right) \cdot p\left(a \cdot f_i^2|F_2\right) \Rightarrow a_i^S = \frac{\frac{f_i^1 \cdot \mu_1}{(\sigma_1)^2} + \frac{f_i^2 \cdot \mu_2}{(\sigma_2)^2}}{\left(\frac{f_i^1}{\sigma_1}\right)^2 + \left(\frac{f_i^2}{\sigma_2}\right)^2}$$

   - Combination of the normalisation factors to obtain a single normalisation factor $a(S)$ for speaker S:

$$a(S) = \frac{\sum_{i=1}^{N} a_i^S \cdot p\left(a_i^S \cdot f_i^1|F_1\right) \cdot p\left(a_i^S \cdot f_i^2|F_2\right)}{\sum_{k=1}^{N} p\left(a_k^S \cdot f_k^1|F_1\right) \cdot p\left(a_k^S \cdot f_k^2|F_2\right)}$$

   where N is the total number of vowel frames for speaker S.
4. Normalisation of the filter bank by scaling the centre frequencies of the mel-scaled filter bank according to the previously computed $a(S)$.
5. Normalisation of the training data.
6. Estimation of the HMM parameters using the normalised training data.

At recognition time, for every new speaker, it is necessary to compute the normalisation factor following steps 1, 2 and 3. The normalisation factor is later used to normalise all the incoming utterances of the new speaker and so that the normalised models can be applied.

This is a block supervised normalisation method that was tested on the TIMIT database (Lincoln et al. 1997). The speech recogniser was also trained in the TIMIT database, that contains utterances from 426 speakers. The experiments, which were carried out in a phone recognition task, showed a 3% relative word error rate reduction.

3.5 Speaker adaptive training (SAT)

Speaker Adaptive training was first proposed by Anastasakos, McDonough, Schwartz & Makhoul (1996), Anastasakos, McDonough & Makhoul (1997), and McDonough, Anastasakos, Zavaliagkos & Gish (1997) and consists of integrating the normalisation in the acoustic model estimation process. It is based on a maximum likelihood formulation which goal is to reduce the effect of speaker variation in the training data. In this way, the variance of the models is reduced and consequently the overlap of the acoustic models. The resulting acoustic models are called *compact models* and later, at recognition time, a MLLR transformation is applied to adapt them to the new speaker.

The goal of SAT is therefore to reduce the overlap of the acoustic models due to interspeaker variation and improve the modelling of the phonetic relevant information. This is achieved by jointly estimating the HMM parameters and affine transformations of the gaussian means in a unified maximum likelihood framework. Each affine transformation, G^r, is applied to its corresponding speaker 'r' of the training database and maps the compact model, λ_c, to a speaker dependent model. In this way, each training observation sequence X^r from a particular speaker 'r' is generated by its corresponding speaker dependent model $G^r(\lambda_c)$. Hence, if the training database is composed of 'R' speakers, the optimum set of parameters of the compact model λ_c and the 'R' speaker transformations $G = \{G^1, \ldots, G^R\}$ have to be estimated. The formulae for estimating the means and the variances of the compact model are developed in Anastasakos et al. (1996) and Anastasakos et al. (1997).

This speaker compensation technique can therefore be summarised in two steps: first a compact model and 'R' transformations are estimated from the training data and then, at recognition time, a new transformation is applied to adapt the compact models to the new speaker. All the transformations are performed using the MLLR speaker adaptation technique (section 3.3.2).

The experimental results (Anastasakos et al. 1996) were obtained with the BYBLOS large vocabulary speech recogniser. The training data was composed of 62 hours of speech from 284 speakers from the SI-284 Wall Street Corpus. The testing data was drawn from the 1994 H1 and 1994 S0 development tasks, which are 20,000 and 5,000 words tasks respectively. For each of the 20 test speakers there were 40 utterances of transcribed data that were used for block supervised adaptation of the acoustic models. In these conditions, the word error rate was reduced 19% and 26% respectively.

This method has been later used in combination with allophone modelling by Huo & Ma (1999) to compensate for the effects of training data in the state tying process, when the training data is very diversified. The basic idea consists of integrating the linear transformation based normalisation technique used in SAT into the decision tree based state tying process. In this way, state tying is able to more accurately model the variability caused by co-articulation of phones in different contexts.

3.6 Conclusions

In this chapter, we address the problem of compensating for the effects of inter/intraspeaker variability in automatic speech recognition. This problem is becoming more and more important as speech technology is being increasingly used in real applications.

Currently, most of the speaker compensation techniques are applied either to the acoustic models or to the features of the speech signal, even though we have pointed out that there are many other areas like pronunciation modelling, language modelling, dialogue strategies design, etc. where compensation might be applied to reduce the mismatch between the speech technology based application and the user. We have also shown that the speech signal has 'additional information' that is not being used in most current state-of-the-art speech recognisers. This additional, non-orthographic information is used by human beings to improve the communication process, which suggests that speech recognisers may increase their word accuracies if they used more knowledge and adaptation techniques.

An important issue is that there is a trend to integrate speaker normalisation techniques in the speech recogniser, so that speaker normalisation techniques are becoming part of the speech recognition process in many state-of-the-art systems. In the same way, there are some adaptation techniques like MLLR and MAP that have become a reference and are commonly used together with other compensation methods.

Finally, we have also shown the trend of the current research in this area and have outlined interesting directions for future research and for selecting the best compensation technique for a particular problem. The fast evolution of these technologies both in improved performance and velocity of compensation make us feel optimistic concerning the use of speech recognition in real services and applications. We are confident that in the near future, we will have 'real' speaker independent speech recognisers, since they will be able to instantaneously adapt to new speakers and provide word accuracies similar to the ones provided by current speaker dependent recognisers.

Acknowledgments

I would like to thank Alejandro Acero (Microsoft) and Pedro Moreno (Compaq) for their valuable comments and advice with the preliminary version of this chapter. I am also grateful to Jean-Claude Junqua (Panasonic) for his advice and for encouraging me to contribute to this book.

I also want to thank my colleagues Luis Hernandez (E.T.S.I. de Telecomunicación de Madrid) for his advice, help and for reviewing this chapter, Paloma León (Telefónica I+D) for her help in phonetics and Miguel Angel Rodriguez (Telefónica I+D) and Gregorio Escalada for their support on speech synthesis methods for artificial and semiartificial database generation. Finally, I also would like to thank Ismael Cortazar (Telefónica I+D) for reviewing the appendix, Joaquin Gonzalez (E.U.I. de Telecomunicación de Madrid) for the new and interesting functionalities he has added to the MATLAB programs supplied at http://www.elsnet.org/ and Frank Healy for reviewing the style and syntax of this chapter.

References

Acero, A. & Huang, X. (1996). Speaker and gender normalisation for continuous density hidden Markov models, *International Conference on Acoustic, Speech and Signal Processing (ICASSP'96)*, Atlanta, Georgia, pp. 342–345.

Ahadi, S. M. & Woodland, P. C. (1995). Rapid speaker adaptation using model prediction, *International Conference on Acoustic, Speech and Signal Processing (ICASSP'95)*, Vol. 1, Detroit, pp. 684–687.

Álvarez, J. (1995). *Modelado Acústico y Teorías de Adaptación MAP para Reconocimiento de Habla Continua en Castellano*, Tesis doctoral, Departamento de Señales, Sistemas y Radiocomunicaciones, E. T. S. I. de Telecomunicación de Madrid.

Álvarez, J., Tapias, D., Crespo, C., Cortazar, I. & Martínez, F. (1997). Development and evaluation of the ATOS conversational system, *International Conference on Acoustic, Speech and Signal Processing (ICASSP'97)*, Vol. 2, Munich, pp. 1139–1142.

Anastasakos, T., McDonough, J. & Makhoul, J. (1997). Speaker adaptive training: A maximum likelihood approach to speaker normalisation, *International Conference on Acoustic, Speech and Signal Processing (ICASSP'97)*, Munich, pp. 1043–1046.

Anastasakos, T., McDonough, J., Schwartz, R. & Makhoul, J. (1996). A compact model for speaker adaptive training, *ICSLP'96*, Philadelphia, PA, pp. 1137–1140.

Ariki, Y., Tagashira, S. & Nishijima, M. (1996). Speaker recognition and speaker normalisation by projection to speaker subspace, *International Conference on Acoustic, Speech and Signal Processing (ICASSP'96)*, Atlanta, Georgia, pp. 319–322.

Bellegarda, J. R., De Souza, P. V., Nadas, A., Nahamoo, D., Picheny, M. A. & Bahl, L. R. (1994). The metamorphic algorithm: A speaker mapping approach to data augmentation, *IEEE Transactions on Speech and Audio Processing* **2**(3): 413–420.

Breiman, L., Friedman, J., Olshen, R. A. & Stone, C. J. (1993). *Classification and Regression Trees*, Chapman & Hall.

Brown, P. F., Lee, C.-H. & Spohrer, J. C. (1983). Bayesian adaptation in speech recognition, *International Conference on Acoustic, Speech and Signal Processing (ICASSP'83)*, Vol. 2, pp. 761–764.

Chesta, C., Siohan, O. & Lee, C.-H. (1999). Maximum a posteriori linear regression for hidden Markov model adaptation, *Proceedings of the 6th European Conference on Speech Communication and Technology (Eurospeech '99)*, Budapest, pp. 211–214.

Chou, W. (1999). Maximum a posteriori linear regression with elliptically symmetric matrix variate priors, *Proceedings of the 6th European Conference on Speech Communication and Technology (Eurospeech '99)*, Vol. 1, Budapest, pp. 1–4.

Choukri, K., Chollet, G. & Grenier, Y. (1986). Spectral transformations through canonical correlation analysis for speaker adaptation in ASR, *International Conference on Acoustic, Speech and Signal Processing (ICASSP'86)*, Vol. 4, Tokyo, pp. 2659–2662.

Collado, J. A., Concejero, P., Tapias, D., Castellanos, M. A. & Hernández, L. A. (1999). Medida de la inteligibilidad del habla: Comparación entre humanos y sistemas automáticos de reconocimiento, *Boletín de Factores Humanos* **21**. http://boletin-fh.tid.es/bole21/art005.htm.

Cox, S. (1993). Speaker adaptation using a predictive model, *Proceedings of the 3rd European Conference on Speech Communication and Technology (Eurospeech '93)*, Vol. 3, Berlin, pp. 2283–2286.

Cox, S. (1995). Predictive speaker adaptation in speech recognition, *Computer Speech and Language* **9**: 1–17.

Cox, S. J., Linford, P. W., Hill, W. B. & Johnston, R. D. (1998). Towards speech recogniser assessment using a human reference standard, *Computer Speech and Language* **12**: 375–391.

DeGroot, M. H. (1970). *Optimal Statistical Decisions*, McGraw-Hill.

Diakoloukas, V. & Digalakis, V. (1997). Adaptation of hidden Markov models using multiple stochastic transformations, *Proceedings of the 5th European Conference on Speech Communication and Technology (Eurospeech '97)*, Vol. 4, Rhodes, Greece, pp. 2063–2066.

Digalakis, V. V., Rtischev, D. & Neumeyer, L. G. (1995). Speaker adaptation using constrained estimation of Gaussian mixtures, *IEEE Transactions on Speech and Audio Processing* **3**(5): 357–365.

Duda, R. O. & Hart, P. E. (eds) (1973). *Pattern Classification and Scene Analysis*, Wiley, New York.

Eide, E. & Gish, H. (1996). A parametric approach to vocal tract length normalization, *International Conference on Acoustic, Speech and Signal Processing (ICASSP'96)*, Atlanta, Georgia, pp. 346–348.

Eskenazi, M. (1993). Trends in speaking styles research, *Proceedings of the 3rd European Conference on Speech Communication and Technology (Eurospeech '93)*, Vol. 1, Berlin, pp. 501–509.

Finke, M. & Waibel, A. (1997). Speaking mode dependent pronunciation modelling in large vocabulary conversational speech recognition, *Proceedings of the 5th European Conference on Speech Communication and Technology (Eurospeech '97)*, Rhodes, Greece, pp. 2379–2382.

Gales, M. J. F. & Woodland, P. C. (1996a). Mean and variance adaptation within the MLLR framework, *Computer Speech and Language* **10**(4): 250–264.

Gales, M. J. F. & Woodland, P. C. (1996b). Variance compensation within the MLLR framework, *Technical Report CUED/F-INFENG/TR242*, Cambridge University. Available via anonymous ftp from: svr-ftp.eng.cam.ac.uk.

Gao, Y., Padmanabhan, M. & Picheny, M. (1997). Speaker adaptation based on preclustering training speakers, *Proceedings of the 5th European Conference on Speech Communication and Technology (Eurospeech '97)*, Vol. 4, Rhodes, Greece, pp. 2091–2094.

García, C. & Tapias, D. (2000). La frecuencia fundamental de la voz y sus efectos en reconocimiento de habla continua, *Sociedad Española para el Procesamiento del Lenguaje Natural, SEPLN '00*, Vigo, Spain, pp. 163–168.

Gauvain, J.-L. & Lee, C.-H. (1994). Maximum a posteriori estimation for multivariate Gaussian mixture observations of Markov chains, *IEEE Transactions on Speech and Audio Processing* **2**(2): 291–298.

Hansen, J. (1988). *Analysis and Compensation of Stressed and Noisy Speech with Application to Robust Automatic Recognition*, PhD thesis, Georgia Institute of Technology.

Hariharan, R. & Viikki, O. (1999). On combining vocal tract length normalisation and speaker adaptation for noise robust speech recognition, *Proceedings of the 6th European Conference on Speech Communication and Technology (Eurospeech '99)*, Vol. 1, Budapest, pp. 215–218.

Huang, J. & Padmanabhan, M. (1999). A study of adaptation techniques on a voicemail transcription tool, *Proceedings of the 6th European Conference on Speech Communication and Technology (Eurospeech '99)*, Vol. 1, Budapest, pp. 13–16.

Huang, X., Acero, A., Alleva, A., Hwang, M. Y., Jiang, L. & Mahajan, M. (1995). Microsoft Windows highly intelligent speech recogniser: Whisper, *International Conference on Acoustic, Speech and Signal Processing (ICASSP'95)*, Vol. 1, Detroit, pp. 93–96.

Humphries, J. J. & Woodland, P. C. (1997). Using accent specific pronunciation modelling for improved large vocabulary continuous speech recognition, *Proceedings of the 5th European Conference on Speech Communication and Technology (Eurospeech '97)*, Vol. 5, Rhodes, Greece, pp. 2367–2370.

Huo, Q. & Lee, C.-H. (1997). On-line adaptive learning of the continuous density hidden Markov model based on approximate recursive Bayes estimate, *IEEE Transactions on Speech and Audio Processing* **5**(2): 161–172.

Huo, Q. & Ma, B. (1999). Irrelevant variability normalization in learning HMM state tying from data based on phonetic decision-tree, *International Conference on Acoustic, Speech and Signal Processing (ICASSP'99)*, Vol. 2, Phoenix, AZ, pp. 577–580.

Imamura, A. (1991). Speaker-adaptive HMM-based speech recognition with a stochastic speaker classifier, *International Conference on Acoustic, Speech and Signal Processing (ICASSP'91)*, Vol. 2, Toronto, pp. 841–844.

Ishii, J. & Tonomura, M. (1997). Speaker normalisation and adaptation based on linear transformation, *International Conference on Acoustic, Speech and Signal Processing (ICASSP'97)*, Vol. 2, Munich, pp. 1055–1058.

Junqua, J.-C. (1993). The Lombard reflex and its role on human listeners and automatic speech recognisers, *Journal of the Acoustical Society of America* **93**(1): 510–524.

Kirby, M. & Sirovich, L. (1990). Application of the Kartunen-Loeve procedure for the characterization of human faces, *IEEE Transactions on Pattern Analysis and Machine Intelligence* **12**(1): 103–108.

Komori, Y., Kosaka, T., Yamada, M. & Yamamoto, H. (1997). Speaker adaptation for context-dependent HMM using spatial relation of both phoneme context hierarchy and speakers, *Proceedings of the 5th European Conference on Speech Communication and Technology (Eurospeech '97)*, Vol. 4, Rhodes, Greece, pp. 2039–2042.

Kosaka, T., Takami, J. & Sagayama, S. (1993). Rapid speaker adaptation using speaker-mixture allophone models applied to speaker independent speech recognition, *International Conference on Acoustic, Speech and Signal Processing (ICASSP'93)*, Vol. 2, Minneapolis, pp. 570–573.

Kuhn, R., Junqua, J.-C., Nguyen, P. & Niedzielski, N. (1999). Rapid speaker adaptation in eigenvoice space. Submitted for publication to IEEE Transactions on Speech and Audio Processing.

Kuhn, R., Nguyen, P., Junqua, J.-C., Boman, R., Niedzielski, N., Fincke, S., Field, K. & Contolini, M. (1999). Fast speaker adaptation using a priori knowledge, *International Conference on Acoustic, Speech and Signal Processing (ICASSP'99)*, Vol. 2, Phoenix, Arizona, pp. 749–752.

Kuhn, R., Nguyen, P., Junqua, J.-C., Goldwasser, L., Niedzielski, N., Fincke, S., Field, K. & Contolini, M. (1998). Eigenvoices for speaker adaptation, *ICSLP'98*, Vol. 5, Sydney, pp. 1771–1774.

Lasry, M. & Stern, R. (1984). A posteriori estimation of correlated jointly Gaussian mean vectors, *IEEE Transactions on Pattern Analysis and Machine Intelligence* **6**(4): 530–535.

Lee, C. H., Lin, C.-H. & Juang, B. H. (1990). A study on speaker adaptation of continuous density HMM parameters, *International Conference on Acoustic, Speech and Signal Processing (ICASSP'90)*, Vol. 1, Alburquerque, pp. 145–148.

Lee, L. & Rose, R. C. (1996). Speaker normalization using efficient frequency warping procedures, *International Conference on Acoustic, Speech and Signal Processing (ICASSP'96)*, Atlanta, Georgia, pp. 353–356.

Leggetter, C. J. (1995). *Improved Acoustic Modelling for HMMs Using Linear Transformations*, PhD thesis, Cambridge University.

Leggetter, C. J. & Woodland, P. C. (1994a). Speaker adaptation of HMMs using linear regression, *Technical Report CUED/F-INFENG/TR.181*, Cambridge University.

Leggetter, C. J. & Woodland, P. C. (1995). Maximum likelihood linear regression for speaker adaptation of continuous density hidden Markov models, *Computer Speech and Language* **9**: 171–185.

Leggetter, C. & Woodland, P. C. (1994b). Speaker adaptation of continuous density HMMs using linear regression, *ICSLP'94*, Vol. 2, Yokohama, pp. 451–454.

León, P. (1996). Informe de resultados de la captura y revisión de la base de datos VESLIM (región central), *Technical report*, Telefónica Investigación y Desarrollo.

Lincoln, M., Cox, S. & Ringland, S. (1997). A fast method of speaker normalisation on using formant estimation, *Proceedings of the 5th European Conference on Speech Communication and Technology (Eurospeech '97)*, Vol. 4, Rhodes, Greece, pp. 2095–2098.

Liu, F., Stern, R. M., Huang, X. & Acero, A. (1993). Efficient cepstral normalisation for robust speech recognition, *Proceedings of ARPA Human Language Technology Workshop*, Princeton, NJ, pp. 69–74.

Llisterri, J. (1992). Speaking styles in speech research, *ELSNET/ESCA/SALT Workshop on Integrating Speech and Natural Language*, Dublin.

Martínez Celdrán, E. (ed.) (1984). *Fonetica*, Teide.

Martínez, F., Tapias, D. & Álvarez, J. (1998). Towards speech rate independence in large vocabulary continuous speech recognition, *International Conference on Acoustic, Speech and Signal Processing (ICASSP'98)*, Vol. 2, Seattle, Washington, pp. 725–728.

Martínez, F., Tapias, D., Álvarez, J. & León, P. (1997). Characteristics of slow, average and fast speech and their effects in large vocabulary continuous speech recognition, *Proceedings of the 5th European Conference on Speech Communication and Technology (Eurospeech '97)*, Vol. 1, Rhodes, Greece, pp. 469–472.

Matsui, T., Matsuoka, T. & Furui, S. (1997). Smoothed n-best based speaker adaptation for speech recognition, *International Conference on Acoustic, Speech and Signal Processing (ICASSP'97)*, Vol. 2, Munich, pp. 1015–1018.

Matsuoka, T. & Lee, C.-H. (1993). A study of on-line Bayesian adaptation for HMM-based speech recognition, *Proceedings of the 3rd European Conference on Speech Communication and Technology (Eurospeech '93)*, Berlin, pp. 815–818.

McDonough, J., Anastasakos, T., Zavaliagkos, G. & Gish, H. (1997). Speaker-adapted training on the switchboard corpus, *International Conference on Acoustic, Speech and Signal Processing (ICASSP'97)*, Vol. 2, Munich, pp. 1059–1062.

Mirghafori, N., Fosler, E. & Morgan, N. (1995). Fast speakers in large vocabulary continuous speech recognition: Analysis & antidotes, *Proceedings of the 4th European Conference on Speech Communication and Technology (Eurospeech '95)*, Madrid, pp. 491–494.

Mirghafori, N., Fosler, E. & Morgan, N. (1996). Towards robustness to fast speech in ASR, *International Conference on Acoustic, Speech and Signal Processing (ICASSP'96)*, Vol. 1, Atlanta, pp. 335–338.

Ohkura, K., Sugiyama, M. & Sagayama, S. (1992). Speaker adaptation based on transfer vector field smoothing with continuous density HMMs, *ICSLP'92*, Alberta, pp. 369–372.

Olive, J. P., Greenwood, A. & Coleman, J. (eds) (1993). *Acoustics of American English Speech*, Springer-Verlag.

Ono, Y., Wakita, H. & Zhao, Y. (1993). Speaker normalization using constrained spectral shifts in auditory filter domain, *Proceedings of the 3rd European Conference on Speech Communication and Technology (Eurospeech '93)*, Berlin, pp. 355–358.

Padmanabhan, M., Bahl, L. R., Namahoo, D. & Picheny, M. A. (1996). Speaker clustering and transformation for speaker adaptation in large-vocabulary speech recognition systems, *International Conference on Acoustic, Speech and Signal Processing (ICASSP'96)*, Vol. 2, Atlanta, Georgia, pp. 701–704.

Padmanabhan, M., Bahl, L. R., Namahoo, D. & Picheny, M. A. (1998). Speaker clustering and transformation for speaker adaptation in speech recognition systems, *IEEE Transactions on Speech and Audio Processing* **6**(1).

Pisoni, D., Bernacki, R., Nusbaum, H. & Yuchtman, M. (1985). Some acoustic-phonetic correlates of speech produced in noise, *International Conference on Acoustic, Speech and Signal Processing (ICASSP'85)*, pp. 1581–1584.

Quilis, A. (ed.) (1988). *Fonetica Acústica de la Lengua Española*, Gredos.

Ravishankar, M. K. (1996). *Efficient Algorithms for Speech Recognition*, Doctoral thesis, Carnegie Mellon University, Pittsburgh, PA.

Richardson, M., Hwang, M., Acero, A. & Huang, X. D. (1999). Improvements on speech recognition for fast talkers, *Proceedings of the 6th European Conference on Speech Communication and Technology (Eurospeech '99)*, Budapest, pp. 411–414.

Rozzi, W. A. & Stern, R. M. (1991). Speaker adaptation in continuous speech recognition via estimation of correlated mean vectors, *International Conference on Acoustic, Speech and Signal Processing (ICASSP'91)*, Toronto, pp. 865–868.

Sankar, A. & Lee, C. H. (1995). Robust speech recognition based on stochastic matching, *International Conference on Acoustic, Speech and Signal Processing (ICASSP'95)*, Vol. 1, Detroit, pp. 121–124.

Siegler, M. A. & Stern, R. M. (1995). On the effects of speech rate in large vocabulary speech recognition systems, *International Conference on Acoustic, Speech and Signal Processing (ICASSP'95)*, Detroit, pp. 612–615.

Stanton, B., Jamieson, L. & Allen, G. (1988). Acoustic-phonetic analysis of loud and lombard speech in simulated cockpit conditions, *International Conference on Acoustic, Speech and Signal Processing (ICASSP'88)*, pp. 331–334.

Stern, R. & Lasry, M. (1987). Dynamic speaker adaptation for feature-based isolated word recognition, *IEEE Transactions on Acoustics, Speech, and Signal Processing* **ASSP-35**(6): 751–763.

Summers, W., Pisoni, D., Bernacki, R., Pedlow, R. & Stockes, M. (1984). Effects of noise in speech production: Acoustic and perceptual analyses, *Journal of the Acoustical Society of America* **84**(3): 917–928.

Tapias, D., García, C. & Cazassus, C. (1999). On the characteristics and effects of loudness during utterance production in continuous speech recognition, *International Conference on Acoustic, Speech and Signal Processing (ICASSP'99)*, Phoenix, AZ, pp. 89–92.

Thomas, T. J., Peckham, J. & Frangoulis, E. (1989). A determination of the sensitivity of speech recognisers to speaker variability, *International Conference on Acoustic, Speech and Signal Processing (ICASSP'89)*, Glasgow, pp. 544–547.

Thomas, T. J., Peckham, J., Frangoulis, E. & Cove, J. (1989). The sensitivity of speech recognisers to speaker variability and speaker variation, *Proceedings of the 1st European Conference on Speech Communication and Technology (Eurospeech '89)*, Paris, pp. 408–411.

Uebel, L. F. & Woodland, P. C. (1999). An investigation into vocal tract length normalisation, *Proceedings of the 6th European Conference on Speech Communication and Technology (Eurospeech '99)*, Vol. 6, Budapest, pp. 2527–2530.

Vroomen, J., Collier, R. & Mozziconacci, S. (1993). Duration and intonation in emotional speech, *Proceedings of the 3rd European Conference on Speech Communication and Technology (Eurospeech '93)*, Vol. 1, Berlin, pp. 570–580.

Wegmann, S., McAllaster, D., Orloff, J. & Peskin, B. (1996). Speaker normalization on conversational telephone speech, *International Conference on Acoustic, Speech and Signal Processing (ICASSP'96)*, Vol. 1, Atlanta, Georgia, pp. 339–341.

Westphal, M., Schultz, T. & Waibel, A. (1998). Linear discriminant — A new criterion for speaker normalization, *ICSLP'98*, Sydney, pp. 827–830.

Womack, B. D. & Hansen, J. H. L. (1996). Improved speech recognition via speaker stress directed classification, *International Conference on Acoustic, Speech and Signal Processing (ICASSP'96)*, Atlanta, Georgia, pp. 53–56.

Zavaliagkos, G., Schwartz, R., McDonough, J. & Makhoul, J. (1995). Adaptation algorithms for large scale HMM recognisers, *Proceedings of the 4th European Conference on Speech Communication and Technology (Eurospeech '95)*, Madrid, pp. 1131–1134.

Zhao, Y. (1994). An acoustic-phonetic-based speaker adaptation technique for improving speaker independent continuous speech recognition, *IEEE Transactions on Speech and Audio Processing* **2**(3): 380–394.

Zhao, Y. (1996). Self-learning speaker and channel adaptation based on spectral variation source decomposition, *Speech Communication* **18**(1): 65–77.

Chapter 4

Robustness in Statistical Language Modeling: Review and Perspectives

Jerome R. Bellegarda
Apple Computer
Two Infinite Loop
Cupertino, California 95014
jerome@apple.com

Abstract Robustness in statistical language modeling refers to the need to maintain adequate speech recognition accuracy as fewer and fewer constraints are placed on the spoken utterances, or more generally when the lexical, syntactic, or semantic characteristics of the discourse in the training and testing tasks differ. Obstacles to robustness involve the dual issues of model coverage and parameter reliability, which are intricately related to the quality and quantity of training data, as well as the estimation paradigm selected. Domain-to-domain differences impose further variations in vocabulary, context, grammar, and style. This chapter reviews a selected subset of recent approaches proposed to deal with some of these issues, and discusses possible future directions of improvement.

4.1 Introduction

4.1.1 Background

Language modeling plays a pivotal role in automatic speech recognition (ASR). It is variously used to constrain the acoustic analysis, guide the search through various (partial) text hypotheses, and/or contribute to the determination of the final transcription (Bahl, Jelinek & Mercer 1983, Jelinek 1985, Rabiner, Juang & Lee 1996). Fundamentally, its function is to encapsulate as much as possible of the syntactic, semantic, and pragmatic characteristics for the task considered. In the search, the successful capture of this information is critical to help determine the most likely sequence of words spoken, because it quantifies which word sequences are acceptable in a given language for a given task, and which are not. In that sense, language modeling can be thought of as a

J.-C. Junqua and G. van Noord (eds.), Robustness in Languages and Speech Technology, 101–121.

way to impose a collection of constraints on words sequences. Two distinct paradigms have traditionally been exploited to impose such constraints: finite-state grammars and stochastic n-grams.

Historically, (context-free) finite-state grammars were investigated first. Because of their conceptual simplicity, they were found to be convenient for self-contained, small vocabulary tasks. For more general use, however, they suffer from two shortcomings. First, they become extremely tedious to deploy in large vocabulary environments. Second, their coverage typically degrades with the size and complexity of the task, which is of course related to the difficulty of characterizing natural language with formal models (Church 1987). Thus, their use has been largely limited to applications such as command and control, where the number of things which can be said is typically restricted, as are the scope and complexity of the formulation.

In dictation and other large vocabulary applications, on the other hand, the spoken utterances can assume an arbitrary degree of complexity. In such situations, stochastic n-grams, such as bigrams and trigrams, have been used extensively (Jelinek 1990). Because they rely on the statistical patterns present in the language, as extracted from a large quantity of relevant training text, they have also come to be referred to as statistical language models (SLMs) (Rosenfeld 1994). This is a bit of a misnomer, since non n-gram language models can certainly also be statistical in nature. Stochastic finite state automata, for example, typically use domain-specific, hand-crafted non-terminals, but their rule probabilities are in fact estimated automatically from a training corpus (Jelinek & Lafferty 1991). Yet, in large vocabulary applications, they are usually applied in conjunction with n-grams, rather than by themselves (see, for example Jurafsky, Wooters, Segal, Stolcke, Fosler, Tajchman & Morgan 1995, Stolcke & Segal 1994).

Thus, in the following, we will focus exclusively on the type of data-driven approaches to ASR language modeling involving some dependence on the n-gram paradigm. In this context, the terminology SLM will refer to an n-gram language model, or a closely related approach based on an n-gram.

4.1.2 SLM Robustness

The success of an SLM in capturing relevant syntactic, semantic, and pragmatic information from the training data is directly related to its ability to suitably discriminate between different strings of n words. This ability is in turn heavily influenced by two related issues, coverage and estimation. Coverage hinges on the selection of the underlying vocabulary, while estimation centers around the choice of n. Both are critically dependent on the quality and quantity of training data available. Coverage-related trade-offs include incurring more errors due to unknown words (low coverage) vs. losing accuracy from increased acoustic confusability (very large vocabulary). Estimation-related trade-offs include settling for weaker predictive power (low n) vs. suffering from more unreliable parameter estimates (higher n). The aim of this chapter is to review some of the solutions available to make such trade-offs less severe.

Clearly, the more constraints can be placed on the spoken utterances, the more manageable coverage and estimation become. Since constraints restrict what can be said and how it can be said, they help both in the prediction of the kind of text that can be generated, which increases coverage, and in the specification and collection of the corresponding training corpus, which makes estimation more reliable. Conversely, as fewer and fewer constraints are considered, system performance tends to deteriorate.

On the other hand, the more constraints are placed on the spoken utterances, the more likely it is that the test utterances might not exactly conform to these constraints. Generally speaking, we can expect a degradation in performance when the lexical, syntactic, or semantic characteristics of the discourse in the training and testing tasks differ. As a result, placing too many constraints on the training environment might very well turn out to be counter-productive in the long run if the task generalizes poorly.

There is therefore an inherent trade-off between coverage/estimation reliability and generalization properties. Robustness in statistical language modeling refers to the need to maintain adequate speech recognition accuracy in the face of this trade-off. Basically, the issue is to arrive at a "good" estimate of the conditional probability of a word given some context, despite potential weaknesses in the training data, the limitations of the n-gram paradigm, and a possible mismatch between training and testing conditions. This goal can be attained through five general strategies: (i) carefully optimize coverage, (ii) use robust estimation techniques, (iii) aggregate information, (iv) extend the effective span of the model, and/or (v) dynamically adapt to the testing conditions. Clearly, (i) and (ii) arise naturally from the above discussion. The other three strategies are discussed below.

In the information aggregation approach (iii), the idea is to take advantage of observations of other words that behave "like" this word in the particular context considered. It has worked quite well as a way of taking into account lexical and syntactic information, perhaps due to the natural fit with syntactic classes. It has not been so fruitful as a method for capturing semantic information. One possible explanation is that the semantic fabric relevant to a task tends to be distributed over a much larger span than can be addressed with the n-gram paradigm.

This has motivated the span extension approach (iv), where the idea is to extend and/or complement the n-gram paradigm with information extracted from large-span units (i.e., comprising a large number of words). Structured language models, for instance, rely on syntactic units at the sentence level, in the form of parse sub-trees. Alternatively, the units can also be semantically motivated. For example, in dictation applications, the semantic span of interest is the entire document being dictated. Trigger pairs and latent semantic analysis, among other methods, have been proposed to extract this kind of information.

Finally, in the language model adaptation approach (v), the basic premise is to exploit specific knowledge about the recognition task, as accumulated during the recognition process itself. The rationale is that the performance of any statistical model tends to worsen on tasks whose characteristics differ substantially from those encountered during the training phase. Here the crux of the matter is the integration of dynamic information about the peculiarities of the testing task.

4.1.3 Organization

The above strategies lead to the following organization of the material in this chapter.

In the next section, we address the issue of coverage optimization, at both lexical and n-gram levels. Section 4.3 is devoted to robust estimation techniques, including classical interpolation and back-off, as well as minimum discrimination information approaches. Robustness based on information aggregation is covered in Section 4.4, which examines both class-based and mixture-based language models. In Section 4.5, we turn to the data-driven extraction of large-span information, as well as its incorporation into the n-gram framework, as relevant to the issue of span extension. Section 4.6 discusses adaptation techniques, as exemplified by dynamic language models like cache and adaptive mixture n-grams. Finally, in Section 4.7 we offer some concluding remarks and perspectives for future improvement.

4.2 Coverage Optimization

4.2.1 Lexical vs. N-Gram

Coverage can be broken down into two parts: lexical and n-gram coverage. Lexical coverage is important because current systems cannot recognize any "unknown" word. Hence, the word error rate of a speech recognition system is at least equal to the percentage of spoken words not present in the underlying vocabulary (Jelinek 1990). Selecting the vocabulary to have maximal coverage, so that the expected text has as few unknown words as possible, remains a fairly intensive effort (Rosenfeld 1995). Yet, the accuracy of the system may deteriorate if the vocabulary is not tailored properly.

N-gram coverage is also of interest to maximize system performance. Since the SLM estimates its n-gram parameters during training, it cannot directly estimate probabilities for n-grams which do not occur in the training corpus. In such situations, the language model usually relies on lower order n-grams, which may be less pertinent to the task at hand. It is no coincidence that recognition errors are much more likely to occur within n-grams which have not been observed in the training data (Chase, Rosenfeld & Ward 1994). It it therefore highly probable that n-gram coverage translates directly into improved recognition accuracy.

4.2.2 Lexical Coverage

The motivation for maximizing lexical coverage is twofold. First, every unknown—also called out-of-vocabulary (OOV)—word is essentially guaranteed to generate one substitution error, no matter how sophisticated the language model might be. Second, this error might also cause the next word to be wrongly recognized, because the language model no longer operates on the correct context. This is referred to as the "ripple effect" of OOV words. To avoid such errors, it is imperative to optimize the underlying vocabulary so as to reduce the number of OOV words present in the expected text during recognition.

Vocabulary optimization is inherently task-dependent, so it would seem prudent to refrain from making sweeping generalizations. Unfortunately, lexical coverage analysis results have only been reported for a small number of large vocabulary tasks. The corpus which has been the most closely examined, by far, is the North American Business (NAB) collection of news articles from the Wall Street Journal and other business publications (Kubala, Bellegarda, Cohen, Pallett, Paul, Phillips, Rajasekaran, Richardson, Riley, Rosenfeld, Roth & Weintraub 1994). Thus, results have to be interpreted in the context of this specific task.

These remarks notwithstanding, some general principles seem to emerge (Rosenfeld 1995). First, coverage is strongly affected by the amount of training data used to construct the lexicon, but this effect becomes much less pronounced as the size of the training corpus increases past 30 to 50 million words. Second, the source and the recency of the data are important: a properly tailored training set can reduce the size required to achieve a given OOV rate by as much as a factor of 5. Finally, for a given task the optimal vocabulary size depends on the trade-off between OOV rate and acoustic confusability: for the NAB task, the optimal size was found to be between 55,000 and 80,000 words (Rosenfeld 1995).

The latter point follows from the fact that, as the vocabulary grows, fewer errors are caused by OOV words, while more errors are introduced due to increased acoustic confusability between words. Thus, the exact trade-off depends, to a large extent, on the magnitude of the ripple effect mentioned above. For the NAB corpus, it has been estimated that each OOV word results in an average of 1.2 word recognition errors (Rosenfeld 1995). Note that the problem may be more severe for languages which are more inflectional than English. French and German, for example, typically require larger vocabulary sizes to achieve coverage similar to English (Roukos 1997).

4.2.3 N-Gram Coverage

Lexical coverage analysis is but a subset of the larger problem of n-gram coverage (corresponding to $n = 1$). For $n > 1$, however, the analysis is complicated by the much lower frequency of the n-grams (as compared to the least frequent words in the vocabulary). For example, in the official ARPA bigram used in NAB evaluations, the least frequent bigrams have a frequency of about 1 in 10 million. In optimizing bigram coverage, we are thus trying to distinguish events which occur once every 10 million words from those which occur slightly more rarely. To differentiate somewhat reliably between such events, we need to observe them enough times for the difference in their counts to be statistically meaningful, i.e., we must have at least 100 to 200 million words available. For most tasks, it is not easy to accumulate corpora of that size, and the problem worsens considerably for increasing values of n.

To complicate things further, there is a potential interplay between n-gram coverage and the evolution of the language. Accumulating millions (billions?) of words of data from a relevant source obviously takes time, during which the language patterns may shift, degrading the relevance of the older data. Viewing language as a non-stationary stochastic source, Rosenfeld (1995) postulated the following principle: *One can never determine accurately both the extent and the time frame of a linguistic phenomenon.* As a corollary, he concluded that we are inherently incapable of detecting transient and rare linguistic events.

In addition, languages other than English may have specific characteristics which act as yet another source of complication. For example, in German and other languages, aggregated (compound) words can occur abundantly and often spontaneously (Spies 1995). Similarly, tense, gender, and number agreements greatly multiply the size of the potential n-gram space for French (Brousseau, Drouin, Foster, Isabelle, Kuhn, Normandin & Plamondon 1995). Such issues all have unfortunate consequences for n-gram coverage.

4.3 Robust Estimation

4.3.1 Classical Smoothing

A direct consequence of such less than perfect n-gram coverage is that most strings of n words (with $n > 1$) are never observed. In addition, a large number of the strings that actually occur may do so very infrequently. This phenomenon, sometimes referred to as "lopsided sparsity," makes it difficult to provide reliable probability estimates for most n-grams. This has motivated the investigation of a number of smoothing techniques over the years to increase the reliability of these estimates.

Two widely used approaches are interpolation (Jelinek & Mercer 1980) and back-off (Katz 1987). Both of them rely on the discounting and redistribution paradigm, where a portion of the probability mass corresponding to frequent events is re-distributed across infrequent and never observed events (Kneser & Ney 1995, Witten & Bell 1991). To illustrate, let w_q denote the current word (i.e., about to be predicted), and $H_{q-1}^{n} = w_{q-1}w_{q-2}\dots w_{q-n+1}$ and $H_{q-1}^{n-1} = w_{q-1}w_{q-2}\dots w_{q-n+2}$ the n-gram and $(n-1)$-gram history (context) for this particular word, respectively. Also denote by $0 \leq \tilde{\Pr}(w_q|H^n_{q-1}) \leq \Pr(w_q|H^n_{q-1})$ the discounted language model "probability," i.e., what remains of $\Pr(w_q|H^n_{q-1})$ after a portion is set aside for re-distribution.

With this notation, the interpolated language model probability is given by:

$$\Pr(w_q|H^n_{q-1}) = \tilde{\Pr}(w_q|H^n_{q-1}) + \lambda(H^n_{q-1})\Pr(w_q|H^{n-1}_{q-1}), \tag{4.1}$$

where:

$$\lambda(H^n_{q-1}) \quad + \sum_{\{w:\tilde{\Pr}(w|H^n_{q-1})>0\}} \tilde{\Pr}(w|H^n_{q-1}) = 1. \tag{4.2}$$

Similarly, the back-off language model probability is given by:

$$\Pr(w_q|H^n_{q-1}) = \begin{cases} \tilde{\Pr}(w_q|H^n_{q-1}), & \text{if } \tilde{\Pr}(w_q|H^n_{q-1}) > 0; \\ \beta(H^n_{q-1})\Pr(w_q|H^{n-1}_{q-1}) & \text{otherwise}, \end{cases} \tag{4.3}$$

with:

$$\sum_{\{w:\tilde{\Pr}(w|H^n_{q-1})>0\}} \tilde{\Pr}(w|H^n_{q-1}) \quad + \quad \beta(H^n_{q-1}) \sum_{\{w:\tilde{\Pr}(w|H^n_{q-1})=0\}} \Pr(w|H^{n-1}_{q-1}) = 1. \tag{4.4}$$

The main issue becomes the specification of $\tilde{\Pr}(w_q|H^n_{q-1})$, which governs how much of the probability mass is set aside for re-distribution. Clearly, this step has a plurality of solutions. Among the many well-known approaches which have been proposed, let us cite linear discounting (Jelinek 1990), absolute discounting (Ney, Essen & Kneser 1994), floor discounting (Witten & Bell 1991), and Good-Turing discounting (Katz 1987). An excellent overview can be found in Federico & de Mori (1998); see also Chen (1996) for a more exhaustive analysis.

4.3.2 Maximum Entropy

The above classical smoothing approaches attempt to combine knowledge from a hierarchy of information sources (such as n-gram and $(n-1)$-gram), in order to arrive at a more robust estimate of the n-gram probability sought. A different way to combine knowledge is provided by the maximum entropy criterion, which leads to minimum discrimination information (MDI) estimation (Della Pietra, Della Pietra, Mercer & Roukos 1992).

In this approach, knowledge sources are introduced in terms of constraints that the underlying distribution should satisfy. These constraints are usually set on single events of the joint probability distribution (such as, for example, a word and a history) in such a way that the constraint functions obey the marginal probabilities observed in the training data. (Additional constraints for the never observed words may then be added for each history, as necessary to satisfy optimization conditions.)

To illustrate, take as before the problem of deriving $\Pr(w_q|H_{q-1}^n)$, and consider the associated event of the joint probability distribution, corresponding to the particular word $w = w_q$ and the history $H = H_{q-1}^n$. In the maximum entropy framework, the resulting joint n-gram distribution $\Pr(H, w)$ is estimated subject to the linear constraints:

$$\sum_{\{(H,w)\}} \delta_k(H,w)\,\Pr(H,w) = \gamma_k\,, \qquad 1 \leq k \leq K\,, \tag{4.5}$$

where δ_k is the characteristic function of an appropriate subset of the sample space, γ_k denotes the relevant empirical marginal probability, and K is the number of linearly independent constraints considered. Constraints like (4.5) are usually set only where marginal probabilities can be reliably estimated, i.e., for well-observed events (H, w). Similar constraints can be placed using all lower order n-gram histories $H = H_{q-1}^{n-i}$, $1 \leq i \leq n-2$.

It can be shown (Darroch & Ratcliff 1972) that the joint distribution $\Pr(H, w)$ satisfying the constraints (4.5) belongs to the exponential family (Mood, Graybill & Boes 1974). It has the parametric form:

$$\Pr(H,w) = \frac{1}{Z(H,w)} \prod_{k=1}^{K} \exp\{\lambda_k\, \delta_k(H,w)\}\,, \tag{4.6}$$

where λ_k is the MDI parameter associated with the kth linear constraint in (4.5), and $Z(H, w)$ is a suitable normalization factor. The λ parameters are typically trained using the generalized iterative scaling (GIS) algorithm (Darroch & Ratcliff 1972). This algorithm converges to a unique solution, provided the constraints (4.5) are consistent. Recent improvements on the GIS algorithm have been reported in Della Pietra, Della Pietra & Lafferty (1997). The reader is referred to Lafferty & Suhm (1995) and Rosenfeld (1996) for further details on MDI estimation.

4.4 Information Aggregation

4.4.1 Class Models

For a given estimation paradigm, robustness can also be achieved by aggregating information from similar, rare events. Class models take advantage of observations of other words that behave "like" the word under study in this particular context. Clearly, grouping words into classes and accumulating statistics at the class level rather than the word level makes the frequency counts more reliable and thereby improves the robustness of the estimation (e.g., see Essen & Steinbiss 1992).

In general, the functional form of the language model probability becomes (with the usual standard assumptions):

$$\Pr(w_q|H_{q-1}^n) = \sum_{\{C_q\}} \sum_{\{C_{q-1}^n\}} \Pr(w_q|C_q)\Pr(C_q|C_{q-1}^n)\Pr(C_{q-1}^n|H_{q-1}^n), \tag{4.7}$$

where $\{C_q\}$ is the set of the possible classes for word w_q, and $\{C_{q-1}^n\}$ is the set of possible class histories associated with the word history H_{q-1}^n.

In practice, it is common to reduce the space of possible class histories to just the most likely one to appear given H_{q-1}^n, as observed in the training data. In that case, (4.7) can be simplified to:

$$\Pr(w_q|H_{q-1}^n) = \sum_{\{C_q\}} \Pr(w_q|C_q)\Pr(C_q|C_{q-1}^n). \tag{4.8}$$

The language model probability thus comprises a class n-gram component—$\Pr(C_q|C_{q-1}^n)$—and a class assignment component—$\Pr(w_q|C_q)$.

A number of variants have been developed on this theme, using grammatical constraints such as part-of-speech, or morphological units such as lemma, or both (Maltese & Mancini 1992). More recently, algorithms have evolved to automatically determine word classes without explicit syntactic or semantic knowledge: see, for example, Jardino & Adda (1993) and Tamoto & Kawabata (1995). In Jardino & Adda (1993), all words are gathered into a single class at the beginning of the procedure, and are successively split to maximize the average mutual information of adjacent classes. In Tamoto & Kawabata (1995), a similar divisive clustering is proposed, based on binomial *a posteriori* distributions on word co-occurrences. A number of other papers have described related approaches, with slight variations in the optimization criterion or distance metric used for clustering (Farhat, Isabelle & O'Shaughnessy 1996, Isotani & Matsunaga 1994, Jardino 1996, Niesler & Woodland 1996).

4.4.2 Mixture Models

Another way to aggregate information is across several domains. Mixture n-gram language models do so by interpolating several n-grams trained on different corpora. Although the original motivation was language model adaptation (Kneser & Steinbiss 1993), this approach can also be seen to increase the reliability of the probability estimates (Bellegarda & Nahamoo 1990, Iyer & Ostendorf 1999).

For mixture models, the functional form of the SLM probability is:

$$\Pr(w_q|H_{q-1}^n) = \sum_{k=1}^{K} \lambda_k(H_{q-1}^n)\Pr_k(w_q|H_{q-1}^n), \tag{4.9}$$

where K represents the number of different corpora used in training, and the interpolation coefficients $\lambda_k(H_{q-1}^n)$ are typically estimated using the EM algorithm on a text passage closely related to the task under consideration.

Thus, the probability of a word n-gram depends on the frequency of this n-gram in each of several corpora, which on the average tends to improve the overall coverage. Since the parameters λ_k can be reliably estimated using only a comparatively small amount of data, this approach enables a better estimation of the language model probabilities. This can be construed as another form of count pooling, qualitatively similar to the information aggregation occurring in class models. This particular usage is further discussed in, e.g., Adda, Jardino & Gauvain (1999). We will come back to mixture models later in this chapter regarding their use in adaptation.

4.5 Span Extension

4.5.1 Variable Length Models

Span extension seeks to address a different facet of SLM robustness, having to do with the inherent local horizon of n-grams (Niesler & Woodland 1996). Consider, for instance, predicting the word *fell* from the word *stocks* in the two equivalent phrases:

$$\textit{stocks fell sharply as a result of the announcement}\,, \tag{4.10}$$

and

$$\textit{stocks, as a result of the announcement, sharply fell}\,. \tag{4.11}$$

While a simple bigram is adequate in (4.10), a 9-gram would be necessary in (4.11) to correctly predict *fell*. This is currently unrealistic, however, as the estimation issues discussed earlier typically restrict n to the values $n \leq 3$ (Woodland, Odell, Valtchev & Young 1994).

Variable length models seek to avoid the intrinsic limit on predictive power resulting from such artificially low values of n. The basic idea is to include frequent word compounds in the n-gram count accumulation, so as to take into account larger spans when the associated probability mass is large enough (Kneser 1996). How the admissible word compounds are determined varies according to the particular approach selected. Phrase n-grams (cf. Kenne, O'Kane & Pearcy 1995) iteratively join adjacent word pairs with high mutual information until perplexity no longer improves. Decision tree language models (Bahl, Brown, de Souza & Mercer 1989) use classification trees (see, for example Bellegarda 1996, Chou 1988) to determine equivalence classes on the history. Many other implementations based on the same philosophy have been investigated.

These techniques have been shown to extend the effective context of the basic n-gram language model. Still, going from n to, say, $2n$, does not really solve the dilemma exemplified in (4.10) and (4.11). The fundamental problem is that the n-gram paradigm is inherently unable to capture high level, global knowledge about the syntactic and semantic characteristics of the task at hand.

4.5.2 Use of Structure

As mentioned in the introduction, in the present state of affairs, high level syntactic knowledge usually implies some variant of context-free grammar. Assuming a suitable parser is available for the domain considered, syntactic information can thus be used in conjunction with the n-gram paradigm to incorporate large-span constraints into the recognition. How these constraints are incorporated varies from estimating n-gram probabilities from grammar-generated data (Zue, Glass, Goodine, Leung, Phillips, Polifroni & Seneff 1991) to computing a linear interpolation of the two models (Jurafsky et al. 1995). When the syntactic information is used specifically to determine equivalence classes on the n-gram history, so-called structured SLMs are generated.

An early attempt along those lines was the prediction suffix tree n-gram (Pereira, Singer & Tishby 1995), which derives equivalence classes using a special form of probabilistic finite state automata. A more general framework for structured language modeling evolved to take into account the hierarchical nature of natural language (Chelba, Engle, Jelinek, Jimenez, Khudanpur, Mangu, Printz, Ristad, Rosenfeld, Stolcke & Wu 1997). In this approach, the model assigns a probability $\Pr(W, T)$ to every sentence W and its every possible binary parse T. The terminals of T are the words of W with part-of-speech tags, and non-terminal nodes are annotated with the headword of the phrase spanned by the associated parse sub-tree. The standard n-gram language model is then modified to operate given the last $(n-1)$ headwords as opposed to the last $(n-1)$ words. In the example above, the top two headwords would be *stocks* and *fell* in both cases, thereby solving the problem.

The main caveat in structured language modeling is the reliance on the parser, and particularly the implicit assumption that the correct parse will in fact be assigned a high probability. The basic framework was recently extended to operate efficiently in a left-to-right manner (Chelba & Jelinek 1999, Jelinek & Chelba 1999), through careful optimization of both chart parsing (Younger 1967) and search modules. Also noteworthy is a somewhat complementary line of research (Zhang, Black & Finch 1999), which exploits the syntactic structure contained in the sentences prior to the one featuring the word being predicted.

4.5.3 Topics and Triggers

Handling high level semantic information is perhaps even more elusive, as it is likely to be diffused across several paragraphs. One way to address this issue is to take into account a (possibly large) set of topics, usually pre-defined from a hand-labelled hierarchy, which covers the relevant semantic space of the current document. The associated language model probability can then be expressed as (cf. Gildea & Hoffman 1999, Schwartz, Imai, Kubala, Nguyen & Makhoul 1997):

$$\Pr(w_q|H_{q-1}^n) = \sum_{k=1}^{K} \Pr(w_q|T_k)\Pr(T_k|H_{q-1}^n)\,, \tag{4.12}$$

where T_k is one of K topics in the set considered. (Note that (4.12) does not capture the local structure of the language, so it must be combined with the standard n-gram in some way: see, for example, Martin, Liermann & Ney (1997).) The main uncertainty in this approach is the granularity required in the topic clustering procedure. To illustrate, in (4.10) and (4.11), even perfect knowledge of the general topic (most likely, "stock market trends") does not help much.

An alternative solution is to use long distance dependencies between word pairs which show significant correlation in the training corpus. In the above example, suppose that the training data reveals a significant correlation between *stocks* and *fell*. Then the presence of *stocks* in the document could automatically trigger *fell*, causing its probability estimate to change. Because this behavior would occur in both (4.10) and (4.11), proximity being irrelevant in this kind of model, the two phrases would lead to the same result.

In this approach, the pair (*stocks*, *fell*) is said to form a word trigger pair (Lau, Rosenfeld & Roukos 1993). In practice, word pairs with high mutual information are searched for inside a window of fixed duration. Unfortunately, trigger pair selection is a complex issue: different pairs display markedly different behavior, which limits the potential of low frequency word triggers (Rosenfeld 1996). Still, self-triggers have been shown to be particularly powerful and robust (Lau et al. 1993), which underscores the desirability of exploiting correlations between the current word and features of the document history.

4.5.4 Latent Semantic Analysis

Recent work has sought to extend the word trigger concept by using a more systematic framework to handle the trigger pair selection (Bellegarda 1997, Bellegarda, Butzberger, Chow, Coccaro & Naik 1996). This is based on a paradigm originally formulated in the context of information retrieval, called latent semantic analysis (LSA) (Berry, Dumais & O'Brien 1995, Deerwester, Dumais, Furnas, Landauer & Harshman 1990, Dumais 1994, Foltz & Dumais 1992, Landauer & Dumais 1997, Landauer, Laham, Rehder & Schreiner 1998, Story 1996). The LSA approach defines a mapping between the underlying vocabulary $\mathcal{V}$, the training text corpus $\mathcal{T}$, and a vector space $\mathcal{S}$, whereby each word w_i in $\mathcal{V}$ is represented by a vector u_i in $\mathcal{S}$ and each document d_j in $\mathcal{T}$ is represented by a vector v_j in $\mathcal{S}$ (Bellegarda 1998b).

The first step is the construction of a matrix (W) of co-occurrences between words and documents. In marked contrast with n-gram modeling, word order is ignored: the matrix W is accumulated from the available training data by simply keeping track of which word is found in what document. Among other possibilities, a suitable expression for the $(i,j)^{\text{th}}$ element of W is given by (cf. Bellegarda et al. 1996, Dumais 1991):

$$w_{i,j} = (1 - \varepsilon_i)\,\frac{c_{i,j}}{n_j}\,, \tag{4.13}$$

where $c_{i,j}$ is the number of times w_i occurs in d_j, n_j is the total number of words present in d_j, and ε_i is the normalized entropy of w_i in the corpus $\mathcal{T}$, given by $\varepsilon_i = -\,(1/\log N)\sum (c_{i,j}/t_i)\,\log(c_{i,j}/t_i)$, with $t_i = \sum c_{i,j}$.

The second step is to compute the singular value decomposition (SVD) of W as (Berry 1992, Berry & Sameh 1989, Cullum & Willoughby 1985):

$$W \approx \hat{W} = U\,S\,V^T\,, \tag{4.14}$$

where U is the $(M \times R)$ matrix of left singular vectors u_i $(1 \leq i \leq M)$, S is the $(R \times R)$ diagonal matrix of singular values, V is the $(N \times R)$ matrix of right singular vectors v_j $(1 \leq j \leq N)$, $R \ll M (\ll N)$ is the order of the decomposition, and ^{T} denotes matrix transposition. The left singular vectors represent the words in the given vocabulary, and the right singular vectors represent the documents in the given corpus. Thus, the space $\mathcal{S}$ sought is the one spanned by U and V. An important property of this space is that two words whose representations are "close" (in some suitable metric) tend to appear in the same kind of documents, whether or not they actually occur within identical word contexts in those documents. Conversely, two documents whose representations are "close" tend to convey the same semantic meaning, whether or not they contain the same word constructs. Thus, we can expect that the respective representations of words and documents that are semantically linked would also be "close" in the LSA space $\mathcal{S}$.

The third step is to leverage this property for language modeling purposes. Let $H_{q-1}^{(lsa)}$ denote the admissible LSA history (context) for this particular word, i.e., the current document up to word w_{q-1}, denoted by $\tilde{d}_{q-1}$. Then the associated LSA language model probability is given by:

$$\Pr(w_q | H_{q-1}^{(lsa)}, \mathcal{S}) = \Pr(w_q | \tilde{d}_{q-1})\,, \tag{4.15}$$

where the conditioning on $\mathcal{S}$ reflects the fact that the probability depends on the particular vector space arising from the SVD representation, and $\tilde{d}_{q-1}$ has a representation in the space $\mathcal{S}$ given by:

$$\tilde{v}_{q-1} = \tilde{d}_{q-1}^T\,U\,S^{-1}\,, \tag{4.16}$$

through a straightforward extension of (4.14). The expression (4.15) is referred to as the direct LSA model. In Bellegarda (1998a), we have also introduced a number of clustered models with attractive smoothing properties. For instance, if we assume that a set of word clusters C_k, $1 \leq k \leq K$, has been produced in $\mathcal{S}$, then we can expand (4.15) as:

$$\Pr(w_q|\tilde{d}_{q-1}) = \sum_{k=1}^{K} \Pr(w_q|C_k) \Pr(C_k|\tilde{d}_{q-1}), \tag{4.17}$$

which is referred to as the word-clustered LSA model. (Note, in passing, the data-driven equivalence to (4.12), if each word cluster is assumed to define a topic.) This model has been shown to result in even better performance (Bellegarda 1998a, Bellegarda 1999). Document-clustered LSA models can be derived in a similar way (Bellegarda 1998b, Gotoh & Renals 1997).

Finally, the fourth step is to integrate the above with the conventional n-gram formalism. This integration can occur in a number of ways, such as straightforward interpolation, or within the maximum entropy framework (Rosenfeld 1996). Alternatively, if we denote by $\bar{H}_{q-1}$ the overall available history (comprising an n-gram component as well as the LSA component mentioned above), then a suitable expression for the integrated probability is given by (Bellegarda 1998b):

$$\Pr(w_q|\bar{H}_{q-1}) = \frac{\Pr(w_q|w_{q-1}w_{q-2}\ldots w_{q-n+1}) \Pr(\tilde{d}_{q-1}|w_q)}{\sum_{w_i \in \mathcal{V}} \Pr(w_i|w_{q-1}w_{q-2}\ldots w_{q-n+1}) \Pr(\tilde{d}_{q-1}|w_i)}. \tag{4.18}$$

Note that, if $\Pr(\tilde{d}_{q-1}|w_q)$ is viewed as a prior probability on the current document history, then (4.18) simply translates the classical Bayesian estimation of the n-gram (local) probability using a prior distribution obtained from (global) LSA. The end result, in effect, is a modified n-gram language model incorporating large-span semantic information.

4.6 Language Model Adaptation

4.6.1 Cache Models

The last facet of SLM robustness is to address the degradation in performance which occurs when there is a mismatch between the lexical, syntactic, or semantic characteristics of the training task and those of the testing task. This requires dynamically modifying the language model statistics according to the current history, as hypothesized given the text corpus.

The most widely used approach is to use a cache memory (Kuhn & de Mori 1990). Cache models exploit self-triggering words inside the text corpus to capture short-term (dynamic) shifts in word-use frequencies which cannot be captured by the underlying (static) n-gram model. Due to practical limitations on the size of the cache window, coupled with reliability considerations mentioned earlier, this strategy is only feasible for $n = 1$. In an effort to propagate the power of the method to higher order cases, the cache component has been extensively applied in conjunction with a class model of the form (4.8). The idea is to replace the probability $\Pr(w_q|C_q)$ in (4.8) with:

$$\tilde{\Pr}(w_q|C_q) = \lambda \Pr_{\text{training}}(w_q|C_q) + (1-\lambda)\Pr_{\text{cache}}(w_q|C_q)\,, \qquad (4.19)$$

where the interpolation parameter $0 \leq \lambda \leq 1$ is typically estimated with the EM algorithm, or empirically. Furthermore, this parameter can also be made a function of recency (Clarkson & Robinson 1997).

4.6.2 Adaptive Mixture Models

As mentioned earlier, mixture models can also be used for adaptation, as long as the interpolation coefficients $\lambda_k(H_{q-1}^n)$ in (4.9) are estimated on the hypothesized history for the word under consideration. In that case they are referred to as adaptive mixture SLMs. In essence, the mixture approach provides the framework necessary to generalize the cache concept to $n > 1$. Several variations have been developed along these lines, including performing interpolation at the sentence rather than the word level (Iyer, Ostendorf & Rohlicek 1994). Adaptive mixtures have also been used in addition to the original cache framework (Clarkson & Robinson 1997). For an insightful review of mixture and cache language models, see Iyer & Ostendorf (1999).

A noteworthy situation arises when the mixture involves only two components, one trained on a very large corpus (resulting in the so-called "static" SLM component), and another trained on a presumably much smaller corpus specific to the current task (resulting in the "dynamic" component). In that case the adaptive mixture problem becomes somewhat analogous to the speaker adaptation problem in acoustic modeling. The robust estimation techniques reviewed in Section 4.3 can be applied to achieve an appropriate level of smoothing between static and dynamic components.

4.7 Conclusion

4.7.1 Summary

Obstacles to language model robustness are intricately related to the quality and quantity of training data, as well as the estimation paradigm selected. Domain-to-domain differences impose further variations in vocabulary, context, grammar, and style. The crux of the problem is therefore to derive a good estimate of the conditional probability of a word given some context, despite potential weaknesses in the training data, the limitations of the n-gram paradigm, and a possible mismatch between training and testing conditions.

In this chapter, we have distinguished between five major approaches for achieving SLM robustness: coverage optimization, robust estimation, information aggregation, span extension, and language model adaptation.

The need for coverage optimization and robust estimation comes from the fact that, for a given large vocabulary application, enough training material of the quality required by commmon estimation techniques will probably not be readily available. Among the typical ways to cope with the estimation of a large number of parameters using sparse data are interpolation, back-off, and maximum entropy.

Information aggregation seeks to define equivalence classes on elements of the context to reduce their cardinality, and therefore the number of parameters to estimate in the associated models. This strategy encompasses methods based on, for instance, parts of speech, morphological analysis, word classes, and compound phrases.

Span extension aims at complementing standard n-grams with more global information gathered at the sentence or document level. This seeks to encapsulate higher level knowledge into the SLM, making it more responsive to the idiosyncrasies of natural language. Falling in this category are structured language models, as well as methods based on word triggers and latent semantic analysis.

Finally, language model adaptation seeks to exploit specific knowledge about the recognition task, as accumulated during the recognition process itself, to update the SLM appropriately. This strategy comprises methods, such as cache memory language models, which directly address the mismatch between training and testing conditions.

4.7.2 Perspectives

Given this inventory of somewhat disparate techniques, it is legitimate to ask where the field is headed. Arguably, the past decade has mostly brought comparatively small, incremental improvements over the basic trigram language model. Are order of magnitude breakthroughs still possible, and if so, what is the most promising direction for attaining them?

This question opens up a new perspective on the five approaches discussed in this chapter. While they each tackle a different aspect of language modeling robustness, the underlying philosophy falls into two categories. In the case of coverage optimization, robust estimation, and information aggregation, the fundamental *raison d'être* is to alleviate a lack of data problem. Span extension and language model adaptation, on the other hand, make use of additional information, be it syntactic structure, semantic fabric, or task specificity. As a result, these latter two approaches appear more likely to harbor a potential breakthrough.

In fact, current research using structured SLMs, hybrid latent semantic SLMs, and advanced adaptive SLMs is still in its infancy. A concerted effort to explore the use of deep structure (such as predicate argument structure) has yet to emerge, partly because of a relative dearth of parsed data. There is no clear consensus regarding the optimal integration of latent semantic information into the n-gram paradigm. And adaptive frameworks going beyond the original unigram cache concept have been difficult to evaluate, due to the challenge of conducting meaningful comparisons across different tasks.

Still, in the author's opinion, the near future is likely to bring significant improvements in language modeling performance, mostly as a result of cumulative advances made in both span extension and SLM adaptation. The most probable scenario is one where large span syntactic knowledge, global semantic analysis, and pragmatic task information each play a role in making the prediction of the current word given the observed context more accurate and more robust.

More generally, none of the five major strategies summarized above are mutually exclusive. As a result, they could all conceivably be used in combination to achieve the desired level of robustness required by a given application. The challenge, of course, is first in integrating these various knowledge sources into an efficient language model component, and second in integrating this language model with the acoustic component of the speech recognition system, all within the resource constraints of the application.

References

Adda, G., Jardino, M. & Gauvain, J. L. (1999). Language modeling for broadcast news transcription, *Proceedings of the Sixth European Conference Speech Communication and Technology*, Vol. 4, Budapest, Hungary, pp. 1759–1762.

Bahl, L. R., Brown, P. F., de Souza, P. V. & Mercer, R. L. (1989). A tree-based statistical language model for natural language speech recognition, *IEEE Transactions on Acoustics, Speech, and Signal Processing* **ASSP-37**(7): 1001–1008.

Bahl, L. R., Jelinek, F. & Mercer, R. L. (1983). A maximum likelihood approach to continuous speech recognition, *IEEE Transactions on Pattern Analysis and Machine Intelligence* **PAMI-5**(2): 179–190.

Bellegarda, J. R. (1996). Context-dependent vector clustering for speech recognition, *in* C.-H. Lee, F. K. Soong & K. K. Paliwal (eds), *Automatic Speech and Speaker Recognition: Advanced Topics*, Kluwer Academic Publishers, New York, chapter 6, pp. 133–157.

Bellegarda, J. R. (1997). A latent semantic analysis framework for large-span language modeling, *Proceedings of the Fifth European Conference Speech Communication and Technology*, Vol. 3, Rhodes, Greece, pp. 1451–1454.

Bellegarda, J. R. (1998a). Exploiting both local and global constraints for multi-span statistical language modeling, *Proceedings of the 1998 IEEE International Conference on Acoustics, Speech, and Signal Processing*, Vol. 2, Seattle, WA, pp. 677–680.

Bellegarda, J. R. (1998b). A multi-span language modeling framework for large vocabulary speech recognition, *IEEE Transactions on Speech and Audio Processing* **6**(5): 456–467.

Bellegarda, J. R. (1999). Speech recognition experiments using multi-span statistical language modeling, *Proceedings of the 1999 IEEE International Conference on Acoustics, Speech, and Signal Processing*, Vol. II, Phoenix, AZ, pp. 717–720.

Bellegarda, J. R., Butzberger, J. W., Chow, Y.-L., Coccaro, N. B. & Naik, D. (1996). A novel word clustering algorithm based on latent semantic analysis, *Proceedings of the 1996 IEEE International Conference on Acoustics, Speech, and Signal Processing*, Vol. I, Atlanta, GA, pp. 172–175.

Bellegarda, J. R. & Nahamoo, D. (1990). Tied mixture continuous parameter modeling for speech recognition, *IEEE Transactions on Acoustics, Speech, and Signal Processing* **ASSP-38**(12): 2033–2045.

Berry, M. & Sameh, A. (1989). An overview of parallel algorithms for the singular value and dense symmetric eigenvalue problems, *Journal of Computational Applied Mathematics* **27**: 191–213.

Berry, M. W. (1992). Large-scale sparse singular value computations, *International Journal for Supercomputer Applications* **6**(1): 13–49.

Berry, M. W., Dumais, S. T. & O'Brien, G. W. (1995). Using linear algebra for intelligent information retrieval, *SIAM Review* **37**(4): 573–595.

Brousseau, J., Drouin, C., Foster, G., Isabelle, P., Kuhn, R., Normandin, Y. & Plamondon, P. (1995). French speech recognition in an automatic dictation system for translators: The TransTalk project, *Proceedings of the Fourth European Conference Speech Communication and Technology*, Vol. 1, Madrid, pp. 193–196.

Chase, L., Rosenfeld, R. & Ward, W. (1994). Error-responsive modifications to speech recognizers: Negative n-grams, *Proceedings of the 1994 International Conference Spoken Language Processing*, Yokohama.

Chelba, C., Engle, D., Jelinek, F., Jimenez, V., Khudanpur, S., Mangu, L., Printz, H., Ristad, E. S., Rosenfeld, R., Stolcke, A. & Wu, D. (1997). Structure and performance of a dependency language model, *Proceedings of the Fifth European Conference Speech Communication and Technology*, Vol. 5, Rhodes, Greece, pp. 2775–2778.

Chelba, C. & Jelinek, F. (1999). Recognition performance of a structured language model, *Proceedings of the Sixth European Conference Speech Communication and Technology*, Vol. 4, Budapest, pp. 1567–1570.

Chen, S. (1996). *Building Probabilistic Models for Natural Language*, PhD thesis, Harvard University, Cambridge, MA.

Chou, P. A. (1988). *Applications of Information Theory to Pattern Recognition and the Design of Decision Trees and Trellises*, PhD thesis, Stanford University, Stanford, CA.

Church, K. W. (1987). *Phonological Parsing in Speech Recognition*, Kluwer Academic Publishers, New York.

Clarkson, P. R. & Robinson, A. J. (1997). Language model adaptation using mixtures and an exponentially decaying cache, *Proceedings of the 1997 IEEE International Conference on Acoustics, Speech, and Signal Processing*, Vol. 1, Munich, pp. 799–802.

Cullum, J. K. & Willoughby, R. A. (1985). Real rectangular matrices, *Lanczos Algorithms for Large Symmetric Eigenvalue Computations*, Vol. 1 Theory, Brickhauser, Boston, chapter 5.

Darroch, J. N. & Ratcliff, D. (1972). Generalized iterative scaling for log-linear models, *Annals of Mathematical Statistics* **43**(5): 1470–1480.

Deerwester, S., Dumais, S. T., Furnas, G. W., Landauer, T. K. & Harshman, R. (1990). Indexing by latent semantic analysis, *Journal of the American Society for Information Science* **41**: 391–407.

Della Pietra, S., Della Pietra, V. & Lafferty, J. (1997). Inducing features of random fields, *IEEE Transactions on Pattern Analysis and Machine Intelligence* **PAMI-19**(1): 1–13.

Della Pietra, S., Della Pietra, V., Mercer, R. & Roukos, S. (1992). Adaptive language model estimation using minimum discrimination estimation, *Proceedings of the 1992 IEEE International Conference on Acoustics, Speech, and Signal Processing*, Vol. I, San Francisco, CA, pp. 633–636.

Dumais, S. T. (1991). Improving the retrieval of information from external sources, *Behavior Research on Methods, Instrumentation, and Computers* **23**(2): 229–236.

Dumais, S. T. (1994). Latent semantic indexing (LSI) and TREC-2, *in* D. Harman (ed.), *Second Text REtrieval Conference (TREC-2)*, NIST Publication 500-215, pp. 105–116.

Essen, U. & Steinbiss, V. (1992). Co-occurrence smoothing for stochastic language modeling, *Proceedings of the 1992 IEEE International Conference on Acoustics, Speech, and Signal Processing*, San Francisco, CA, pp. 161–164.

Farhat, A., Isabelle, J. & O'Shaughnessy, D. (1996). Clustering words for statistical language models based on contextual word similarity, *Proceedings of the 1996 IEEE International Conference on Acoustics, Speech, and Signal Processing*, Vol. I, Atlanta, GA, pp. 180–183.

Federico, M. & de Mori, R. (1998). Language modeling, *in* R. de Mori (ed.), *Spoken Dialogues with Computers*, Academic Press, London, chapter 7, pp. 199–230.

Foltz, P. W. & Dumais, S. T. (1992). Personalized information delivery: An analysis of information filtering methods, *Communications of the ACM* **35**(12): 51–60.

Gildea, D. & Hoffman, T. (1999). Topic-based language modeling using EM, *Proceedings of the Sixth European Conference Speech Communication and Technology*, Vol. 5, Budapest, pp. 2167–2170.

Gotoh, Y. & Renals, S. (1997). Document space models using latent semantic analysis, *Proceedings of the Fifth European Conference Speech Communication and Technology*, Vol. 3, Rhodes, Greece, pp. 1443–1448.

Isotani, R. & Matsunaga, S. (1994). A stochastic language model for speech recognition integrating local and global constraints, *Proceedings of the 1994 IEEE International Conference on Acoustics, Speech, and Signal Processing*, Vol. II, Adelaide, Australia, pp. 5–8.

Iyer, R. & Ostendorf, M. (1999). Modeling long distance dependencies in language: Topic mixtures versus dynamic cache models, *IEEE Transactions on Speech and Audio Processing* **7**(1): 30–39.

Iyer, R., Ostendorf, M. & Rohlicek, J. R. (1994). Language modeling with sentence-level mixtures, *Proceedings of the ARPA Speech and Natural Language Workshop*, Morgan Kaufmann Publishers, pp. 82–86.

Jardino, M. (1996). Multilingual stochastic n-gram class language models, *Proceedings of the 1996 IEEE International Conference on Acoustics, Speech, and Signal Processing*, Vol. I, Atlanta, GA, pp. 161–163.

Jardino, M. & Adda, G. (1993). Automatic word classification using simulated annealing, *Proceedings of the 1993 IEEE International Conference on Acoustics, Speech, and Signal Processing*, Minneapolis, MN, pp. 41–44.

Jelinek, F. (1985). The development of an experimental discrete dictation recognizer, *Proceedings of the IEEE* **73**(11): 1616–1624.

Jelinek, F. (1990). Self-organized language modeling for speech recognition, *in* A. Waibel & K.-F. Lee (eds), *Readings in Speech Recognition*, Morgan Kaufmann Publishers, pp. 450–506.

Jelinek, F. & Chelba, C. (1999). Putting language into language modeling, *Proceedings of the Sixth European Conference Speech Communication and Technology*, Vol. 1, Budapest, pp. KN1–KN5.

Jelinek, F. & Lafferty, J. D. (1991). Computation of the probability of initial substring generation by stochastic context-free grammars, *Computational Linguistics* **17**: 315–323.

Jelinek, F. & Mercer, R. L. (1980). Interpolated estimation of Markov source parameters from sparse data, *Pattern Recognition in Practice*, Amsterdam, pp. 381–397.

Jurafsky, D., Wooters, C., Segal, J., Stolcke, A., Fosler, E., Tajchman, G. & Morgan, N. (1995). Using a stochastic context-free grammar as a language model for speech recognition, *Proceedings of the 1995 IEEE International Conference on Acoustics, Speech, and Signal Processing*, Vol. I, Detroit, MI, pp. 189–192.

Katz, S. M. (1987). Estimation of probabilities from sparse data for the language model component of a speech recognizer, *IEEE Transactions on Acoustics, Speech, and Signal Processing* **ASSP-35**: 400–401.

Kenne, P. E., O'Kane, M. & Pearcy, H. G. (1995). Language modeling of spontaneous speech in a court context, *Proceedings of the Fourth European Conference Speech Communication and Technology*, Vol. 3, Madrid, pp. 1801–1804.

Kneser, R. (1996). Statistical language modeling using a variable context, *Proceedings of the 1996 International Conference on Spoken Language Processing*, Philadelphia, PA, pp. 494–497.

Kneser, R. & Ney, H. (1995). Improved backing-off for n-gram language modeling, *Proceedings of the 1995 IEEE International Conference on Acoustics, Speech, and Signal Processing*, Vol. I, Detroit, MI, pp. 181–184.

Kneser, R. & Steinbiss, V. (1993). On the dynamic adaptation of stochastic language models, *Proceedings of the 1993 IEEE International Conference on Acoustics, Speech, and Signal Processing*, Vol. II, Minneapolis, MN, pp. 586–588.

Kubala, F., Bellegarda, J. R., Cohen, J. R., Pallett, D., Paul, D. B., Phillips, M., Rajasekaran, R., Richardson, F., Riley, M., Rosenfeld, R., Roth, R. & Weintraub, M. (1994). The hub and spoke paradigm for CSR evaluation, *Proceedings of the ARPA Speech and Natural Language Workshop*, Morgan Kaufmann Publishers, pp. 40–44.

Kuhn, R. & de Mori, R. (1990). A cache-based natural language method for speech recognition, *IEEE Transactions on Pattern Analysis and Machine Intelligence* **PAMI-12**(6): 570–582.

Lafferty, J. D. & Suhm, B. (1995). Cluster expansion and iterative scaling for maximum entropy language models, *in* K. Hanson & R. Silver (eds), *Maximum Entropy and Bayesian Methods*, Kluwer Academic Publishers, Norwell, MA.

Landauer, T. K. & Dumais, S. T. (1997). Solution to Plato's problem: The latent semantic analysis theory of acquisition, induction, and representation of knowledge, *Psychological Review* **104**(2): 211–240.

Landauer, T. K., Laham, D., Rehder, B. & Schreiner, M. E. (1998). How well can passage meaning be derived without using word order: A comparison of latent semantic analysis and humans, *Proceedings of the Cognitive Science Society.*

Lau, R., Rosenfeld, R. & Roukos, S. (1993). Trigger-based language models: A maximum entropy approach, *Proceedings of the 1993 IEEE International Conference on Acoustics, Speech, and Signal Processing*, Vol. II, Minneapolis, MN, pp. 45–48.

Maltese, G. & Mancini, F. (1992). An automatic technique to include grammatical and morphological information in a trigram-based statistical language model, *Proceedings of the 1992 IEEE International Conference on Acoustics, Speech, and Signal Processing*, San Francisco, CA, pp. 157–160.

Martin, S. C., Liermann, J. & Ney, H. (1997). Adaptive topic-dependent language modelling using word-based varigrams, *Proceedings of the Fifth European Conference Speech Communication and Technology*, Vol. 3, Rhodes, Greece, pp. 1447–1450.

Mood, A., Graybill, F. & Boes, D. (1974). *Introduction to the Theory of Statistics*, McGraw-Hill, New York.

Ney, H., Essen, U. & Kneser, R. (1994). On structuring probabilistic dependences in stochastic language modeling, *Computer, Speech, and Language* **8**: 1–38.

Niesler, T. & Woodland, P. (1996). A variable-length category-based n-gram language model, *Proceedings of the 1996 IEEE International Conference on Acoustics, Speech, and Signal Processing*, Vol. I, Atlanta, GA, pp. 164–167.

Pereira, F. C., Singer, Y. & Tishby, N. (1995). Beyond word n-grams, *in* D. Yarowsky & K. Church (eds), *Proceedings of the Third Workshop on Very Large Corpora*, Massachusetts Institute of Technology, Cambridge, MA, pp. 95–106.

Rabiner, L. R., Juang, B.-H. & Lee, C.-H. (1996). An overview of automatic speech recognition, *in* C.-H. Lee, F. K. Soong & K. K. Paliwal (eds), *Automatic Speech and Speaker Recognition: Advanced Topics*, Kluwer Academic Publishers, Boston, MA, chapter 1, pp. 1–30.

Rosenfeld, R. (1994). The CMU statistical language modeling toolkit and its use in the 1994 ARPA CSR evaluation, *Proceedings of the ARPA Speech and Natural Language Workshop*, Morgan Kaufmann Publishers.

Rosenfeld, R. (1995). Optimizing lexical and n-gram coverage via judicious use of linguistic data, *Proceedings of the Fourth European Conference on Speech Communication and Technology*, Madrid, pp. 1763–1766.

Rosenfeld, R. (1996). A maximum entropy approach to adaptive statistical language modeling, *Computer Speech and Language* **10**: 187–228.

Roukos, S. (1997). Language representation, *in* R. Cole (ed.), *Survey of the State of the Art in Human Language Technology*, Cambridge University Press, chapter 6.

Schwartz, R., Imai, T., Kubala, F., Nguyen, L. & Makhoul, J. (1997). A maximum likelihood model for topic classification of broadcast news, *Proceedings of the Fifth European Conference Speech Communication and Technology*, Vol. 3, Rhodes, Greece, pp. 1455–1458.

Spies, M. (1995). A language model for compound words in speech recognition, *Proceedings of the Fourth European Conference on Speech Communication and Technology*, Madrid, pp. 1767–1770.

Stolcke, A. & Segal, J. (1994). Precise n-gram probabilities from stochastic context-free grammars, *Proceedings of the 32nd Meeting of the Association for Computational Linguistics*, Las Cruces, NM, pp. 74–79.

Story, R. E. (1996). An explanation of the effectiveness of latent semantic indexing by means of a bayesian regression model, *Information Processing & Management* **32**(3): 329–344.

Tamoto, M. & Kawabata, T. (1995). Clustering word category based on binomial posteriori co-occurrence distribution, *Proceedings of the 1995 IEEE International Conference on Acoustics, Speech, and Signal Processing*, Vol. I, Detroit, MI, pp. 165–168.

Witten, I. H. & Bell, T. C. (1991). The zero-frequency problem: Estimating the probability of novel events in adaptive text compression, *IEEE Transactions on Information Theory* **37**(4): 1085–1094.

Woodland, P. C., Odell, J. J., Valtchev, V. & Young, S. J. (1994). Large vocabulary continuous speech recognition using HTK, *Proceedings of the 1994 IEEE International Conference on Acoustics, Speech, and Signal Processing*, Adelaide, Australia, pp. 125–128.

Younger, D. H. (1967). Recognition and parsing of context-free languages in time N^3, *Information & Control* **10**: 198–208.

Zhang, R., Black, E. & Finch, A. (1999). Using detailed linguistic structure in language modeling, *Proceedings of the Sixth European Conference Speech Communication and Technology*, Vol. 4, Budapest, pp. 1815–1818.

Zue, V., Glass, J., Goodine, D., Leung, H., Phillips, M., Polifroni, J. & Seneff, S. (1991). Integration of speech recognition and natural language processing in the MIT voyager system, *Proceedings of the 1991 IEEE International Conference on Acoustics, Speech, and Signal Processing*, Toronto, pp. 713–716.

Chapter 5

IMPROVING ROBUSTNESS BY MODELING SPONTANEOUS SPEECH EVENTS

Peter A. Heeman
Oregon Graduate Institute
heeman@cse.ogi.edu

James F. Allen
University of Rochester
james@cs.rochester.edu

Abstract In spontaneous speech, speakers segment their speech into intonational phrases, and make repairs to what they are saying. However, techniques for understanding spontaneous speech tend to treat these events as noise, in the same manner as they handle out-of-grammar constructions and misrecognitions. In our approach, we advocate that these events should be explicitly modeled, and that they must be resolved early in the processing stream. We put forward a statistical language model, which can be used during speech recognition, that models these events. This not only improves speech recognition perplexity and POS tagging, but also results in much richer output from the recognizer, with speech repairs resolved and intonational phrase boundaries identified. Syntactic and semantic processing can thus focus on dealing with out-of-grammar constructions and misrecognitions.

5.1 Introduction

To enable spoken dialogue systems to advance towards more collaborative interaction between humans and computers, we need to deal with language as it is actually spoken. In natural speech, speakers group words into intonational phrases and make repairs to what they are saying. Consider the following speaker's turn from the Trains corpus (Heeman & Allen 1995).

J.-C. Junqua and G. van Noord (eds.), Robustness in Languages and Speech Technology, 123–152.

***Example* 1 (d93-13.3 utt63)**
um it'll be there it'll get to Dansville at three a.m. and then you wanna do you take tho- want to take those back to Elmira so engine E two with three boxcars will be back in Elmira at six a.m. is that what you wanna do

From reading the word transcription, the reader should immediately notice the prevalence of *speech repairs*, where speakers go back and change or repeat something they just said. Fortunately for hearers, speech repairs tend to have a standard form. The *reparandum* is the stretch of speech that the speaker is replacing; it might end in the middle of a word, resulting in a word fragment. The end of the reparandum is called the *interruption point*. There can also be an editing term, consisting of filler words, such as 'uh' and 'um', or cue phrases, such as 'let's see', 'well', and 'okay'. This is then followed by the *alteration*, which is the replacement for the reparandum. Below, we illustrate this analysis on the first repair from the above speaker turn.

***Example* 2 (Repair)**
um $\underbrace{\text{it'll be there}}_{\textit{reparandum}}\underset{\textit{ip}}{\uparrow}$ $\underbrace{\text{it'll get to Dansville}}_{\textit{alteration}}$ at three a.m.

Speech repairs are very prevalent in spontaneous speech. In the Trains corpus, 10% of all words are part of the editing term or reparandum of a speech repair, and 54% of all speaker turns with at least 10 words have at least one repair. To determine the speaker's intended message, speech repairs need to be *resolved*: they need to be *detected*, by finding their interruption point, and *corrected*, by determing the extents of the reparanda and editing terms.[1]

In addition to making repairs, speakers also break their turn of speaking into intonational phrases, which are signaled through variations in the pitch contour, segmental lengthening and pauses. Previous research has shown that intonational information can reduce syntactic ambiguity for humans (Beach 1991) and in computer parsers (Bear & Price 1990, Ostendorf, Wightman & Veilleux 1993). Other researchers have proposed segmenting speech into speech acts (i.e. Mast, Kompe, Harbeck, Kießling, Niemann, Nöth, Schukat-Talamazzini & Warnke 1996) or linguistically, based on having a single clause (i.e. Meteer & Iyer 1996). However, there is no clear consensus as to the right approach. Although intonational phrases might not be the ideal unit for modeling interaction in dialogue, it definitely captures the speaker's intention and will undoubtedly be a major component of any definition (Traum & Heeman 1997).

1. The reparanda and editing terms, however, should not simply be expunged from the speech, as it might contain valuable information, such as the identify of an anaphoric reference: "Peter was . . . well he was fired".

Now that we have introduced the spontaneous speech events, we show our example annotated in terms of them. Repair reparanda are indicated in italic, with the alteration starting on a new line indented to start at the reparandum onset. Intonational phrase boundaries are marked with '%'.

***Example* 3 (d93-13.3 utt63)**

um *it'll be there*
        it'll get to Dansville at three a.m. %
and then *you wanna*
                do you *take tho-*
                            want to take those back to Elmira %
so engine E two with three boxcars will be back in Elmira at six a.m. %
is that what you wanna do %

Although the spontaneous speech events of speech repairs and intonational phrasing are much more common in human-human speech than in human-computer speech (Oviatt 1995), this will change as people become increasingly more comfortable with human-computer interaction and start focusing on the task before them, rather than on the form of their interaction (Price 1997). Hence, spoken language systems will increasingly need to deal with these events.

5.1.1 Robust Parsing Approach

One line of research that has become popular for dealing with speech repairs is to use robust parsing techniques. For understanding spontaneous speech, speech repairs are not the only phenomena that create problems; one also needs to deal with word misrecognitions and out-of-grammar constructions. All three of these problems tend to be lumped together and given to a robust parser. Ward (1991) used a robust semantic parser to look for sequences of words that matched grammar fragments associated with slots of case frames. The parser would try to fill as many slots as possible. If a slot is only partially filled, it is abandoned. If a slot is filled more than once, the latter value is taken (Young & Matessa 1991). In this volume, Rosé & Lavie (2000) describe using a robust parser, which incorporated a skipping mechanism, with a feature unification grammar; and van Noord (2000) describes using a skipping mechanism in parsing word graphs.

5.1.2 Modeling Spontaneous Speech Events

Rather than view spontaneous speech events as noise in the input to a robust parser, we advocate that speech repairs and intonational phrasing should be explicitly modeled. There are local cues, such as editing terms, word correspondences, pauses, that give evidence for these events. Hence, we should be able to automatically identify the intonational phrases and resolve the speech repairs.

By modeling these events, we will have a richer understanding of the speech. This will simplify later syntactic and semantic processing, since such processing can start from enriched output rather than trying to cope with the apparent ill-formedness that spontaneous speech events cause. This will also make it easier for these processes to deal with the other problems of understanding spontaneous speech: namely misrecognitions and out-of-grammar constructions.

Speech repairs and intonatinal phrasing are intertwined with the speech recognition problem of predicting the next word given the previous context (Heeman & Allen 1999). Hence, our approach is to redefine the speech recognition problem so that it includes the resolution of speech repairs and identification of intonational phrases. We also include the tasks of part-of-speech (POS) tagging and discourse marker identification, since these tasks are also intertwined with resolving speech repairs and identifying intonational phrasing. Since all tasks are being resolved in the same model, we can account for the interactions between the tasks in a framework that can compare alternative hypotheses for the speakers' turn. Not only does this allow us to model the spontaneous speech events, but it also results in an improved language model, evidenced by both improved POS tagging and better probability estimates of the next word. Furthermore, speech repairs and phrase boundaries have acoustic correlates, such as pauses between words. By resolving speech repairs and identifying intonational phrases during speech recognition, these acoustic cues, which otherwise would be treated as noise, can give evidence as to the occurrence of these events, and further improve speech recognition results.

5.1.3 Overview of the Chapter

We next describe the Trains corpus and the annotation of speech repairs and intonational phrases. We then introduce our baseline language model, which incorporates POS tagging and discourse marker identification, and we introduce the machine learning techniques we use for for estimating the probability distrbutions. We then augment our baseline model with speech repair and intonational phrase detection and speech repair correction, and give a sample run of the model. We then give the results of running our model on the Trains corpus, and compare our work with previous work in modeling speech repairs and intonational phrasing. Finally, we present the conclusions and future work.

5.2 The Trains Corpus

For our research work, we used the Trains corpus, a corpus of human-human task-oriented dialogs available from the Linguistics Data Consortium. The corpus consists of six and a half hours of speech produced by 34 different speakers solving 20 different problems. Each word was transcribed using its orthographic spelling, unless it was mispronounced and the speaker subsequently repairs the mispronunciation. Contractions, including words such as 'wanna', were transcribed as single words. Word fragments were annotated with a trailing dash. Our POS tagset is based on the Penn tagset but modified to include tags for discourse markers and end-of-turns, and to provide richer syntactic information (Heeman 1997). Contractions are annotated using '∧' to conjoin the tag for each part; for instance, 'wanna' is annotated as **'VBP∧TO'**.

For annotating the intonational phrase boundaries, we used the ToBI (Tones and Break Indices) definition (Silverman, Beckman, Pitrelli, Ostendorf, Wightman, Price, Pierrehumbert & Hirschberg 1992) of intonational phrases. Intonational phrases are determined by both the pitch contour and the perceived juncture between each pair of words, where the perceived juncture takes into account both interword pauses and preboundary lengthening (normalized duration of the final consonants). In the corpus, there were 5535 intonational phrase boundaries, not included those at the end of speakers' turns. Speakers' turns almost always ended with an intonational phrase boundary.

For annotating speech repairs, we developed a scheme based on the one proposed by Bear, Dowding, Shriberg & Price (1993). The scheme allows the extent of the reparanda and editing terms to be annotated, along with the type of the repair. Following loosely in the footsteps of the work of Hindle (1983), we divide them into the following categories: *fresh starts*, *modification repairs*, and *abridged repairs*.

Fresh starts occur where the speaker abandons the current utterance and starts again, where the abandonment seems to be acoustically signaled either in the editing term or at the onset of the alteration. Example 4 illustrates a fresh start where the speaker abandons the partial utterance 'I need to send,' and replaces it by the question 'how many boxcars can one engine take.'

***Example* 4 (d93-14.3 utt2)**

I need to send ↑ let's see how many boxcars can one engine take

reparandum ip editing term alteration

The second type are modification repairs, which comprise the remainder of repairs with a non-empty reparandum. The example below illustrates this type of repair.

***Example* 5 (d92a-1.2 utt40)**

you can carry them both on ↑ tow both on the same engine

reparandum ip alteration

Modification repairs tend to have strong word correspondences between the reparandum and alteration, which can help the hearer determine the reparandum onset as well as signal that a repair occurred. In the example above, there are word matches on the instances of 'both' and 'on', and a replacement of the verb 'carry' by 'tow'. Modification repairs can in fact consist solely of the reparandum being repeated by the alteration.

The third type are the abridged repairs. These repairs consist of an editing term, but with no reparandum, as the following example illustrates.

***Example* 6 (d93-14.3 utt42)**

we need to $\underset{ip}{\uparrow}$ $\underbrace{\text{um}}_{\textit{editing term}}$ manage to get the bananas to Dansville more quickly

For these repairs, the hearer has to determine that an editing term occurred, which can be difficult for phrases such as 'let's see' or 'well' since they can also have a sentential interpretation. The hearer also has to determine that the reparandum is empty. As the example above illustrates, this is not necessarily a trivial task because of the spurious word correspondences between 'need to' and 'manage to'. In the Trains corpus, there are 423 abridged repairs, 1302 modification repairs, 671 fresh starts.

5.3 POS-Based Language Model

In this section, we present a speech recognition language model that incorporates POS tagging. Here, POS tags are viewed as part of the output of the speech recognizer rather than as intermediate objects. Not only is this syntactic information needed for modeling the occurrence of speech repairs and intonational phrases, but it will also be useful for higher level syntactic and semantic processes. Incorporating POS tagging can also be seen as a first step in tightening the coupling between speech recognition and natural language processing so as to be able to make use of richer knowledge of natural language than simple word-based language models provide.

5.3.1 Word-based Language Models

The goal of speech recognition is to find the most probable sequence of words $\hat{W}$ given the acoustic signal A (Jelinek 1985).

$$\begin{aligned} \hat{W} &= \arg\max_{W} \Pr(W|A) \\ &= \arg\max_{W} \Pr(A|W)\Pr(W) \end{aligned} \tag{5.1}$$

The first term, $\Pr(A|W)$, is the *acoustic model* and the second term, $\Pr(W)$, is the *language model*. We rewrite W explicitly as the sequence of words $W_1W_2W_3\ldots W_N$, where N is the number of words in the sequence. For expository ease, we use $W_{i,j}$ to refer to $W_i \ldots W_j$. We now rewrite $\Pr(W_{1,N})$ as follows.

$$\Pr(W_{1,N}) = \prod_{i=1}^{N} \Pr(W_i|W_{1,i\text{-}1}) \tag{5.2}$$

The above equation gives us the probability of the word sequence as the product of the probability of each word given its previous lexical context.

5.3.2 Incorporating POS Tags

To incorporate POS tags into the language model, we redefine the speech recognition problem so as to include finding the best POS and discourse marker sequence along with the best word sequence. For the word sequence W, let P be a POS sequence. The goal of the speech recognition process is to now solve the following.

$$\begin{aligned} \hat{W}\hat{P} &= \arg\max_{WP} \Pr(WP|A) \\ &= \arg\max_{WP} \Pr(A|WP)\Pr(WP) \end{aligned} \tag{5.3}$$

The first term $\Pr(A|WP)$ is the acoustic model, which can be approximated by $\Pr(A|W)$. The second term $\Pr(WP)$ is the POS-based language model and accounts for both the sequence of words and their POS assignment. We rewrite this term as follows.

$$\begin{aligned} \Pr(W_{1,N}P_{1,N}) &= \prod_{i=1}^{N} \Pr(W_iP_i|W_{1,i\text{-}1}P_{1,i\text{-}1}) \\ &= \prod_{i=1}^{N} \Pr(W_i|W_{1,i\text{-}1}P_{1,i})\Pr(P_i|W_{1,i\text{-}1}P_{1,i\text{-}1}) \end{aligned} \tag{5.4}$$

Equation 5.4 involves two probability distributions that need to be estimated. To successfully use POS tags in a language model, we need to estimate these probability distributions as best possible.

5.3.3 Estimating the Probabilities

To estimate the probability distributions, we follow the approach of Bahl, Brown, de Souza & Mercer (1989) and use a decision tree learning algorithm (Breiman, Friedman, Olshen & Stone 1984) to partition the context into equivalence classes. The algorithm starts with a single node. It then finds a question to ask about the node in order to partition the node into two *leaves*, each more informative as to which event occurred than the parent node. Information theoretic metrics, such as minimizing entropy, are used to decide which question to propose. The proposed question is then verified using heldout data: if the split does not lead to a decrease in entropy according to the heldout data, the split is rejected and the node is not further explored. This process continues with the new leaves and results in a hierarchical partitioning of the context. After the tree is grown, relative frequencies are calculated for each node, and these probabilities are then interpolated with their parent node's probabilities using a second heldout dataset.

Using the decision tree algorithm to estimate probabilities is attractive since the algorithm can choose which parts of the context are relevant, and in what order. Hence, this approach lends itself more readily to allowing extra contextual information to be included, such as both the word identities and POS tags, and even hierarchical clusterings of them. If the extra information is not relevant, it will not be used. The approach of using decision trees will become even more critical in the next two sections where the probability distributions will be conditioned on an even richer context.

Questions about POS Tags

The context that we use for estimating the probabilities includes both word identities and POS tags. To make effective use of this information, we allow the decision tree algorithm to generalize between words and POS tags that behave similarly. To learn which ones behave similarly, Black, Jelinek, Lafferty, Magerman, Mercer & Roukos (1992) used the clustering algorithm of Brown, Della Pietra, deSouza, Lai & Mercer (1992) to build a hierarchical classification tree. The algorithm starts with each POS tag in a separate class and iteratively finds two classes to merge that results in the smallest loss of information about POS adjacency. This continues until only a single class remains. The order in which classes were merged, however, gives a binary tree with the root corresponding to the entire tagset, each leaf to a single POS tag, and intermediate nodes to groupings of the tags that are statistically similar. The path from the root to a tag gives the binary encoding for the tag. The decision tree algorithm can ask which partition a tag belongs to by asking questions about its binary encoding.

Questions about Word Identities

For handling word identities, one could follow the approach used for handling the POS tags and view the POS tags and word identities as two separate sources of information. Instead, we view the word identities as a further refinement of the POS tags (Heeman 1997). We start the clustering algorithm with a separate class for each combination of word and POS tag that exists in the training data. Classes are only merged if the POS tags are the same. The result is a word classification tree for each tag. This approach means that the trees will not be polluted by words that are ambiguous as to their tag, as exemplified by the word 'loads', which is used in the corpus as a third-person present tensed verb and as a plural noun. Furthermore, this approach simplifies the clustering task because the hand annotations of the POS tags resolve a lot of the difficulty that the algorithm would otherwise have to learn.

Other Questions

We allow two other types of information to be used as part of the context: numeric and categorical information. Although this type of information is not used in this section, they will be used in the next two sections. For a numeric variable N, the decision tree searches for questions of the form 'is $N >= n$', where n is a numeric constant. For a categorical variable C, it searches over questions of the form 'is $C \in S$' where S is a subset of the possible values of C. We also allow composite questions (Bahl et al. 1989), which are boolean combinations of elementary questions.

5.4 Identifying Speech Repairs and Intonational Phrases

In the previous section, we presented a POS-based language model. This model did not account for the occurrence of speech repairs nor intonational phrases. Ignoring these events when building a statistical language model leads to probabilistic estimates for the words and POS tags that are less precise, since they mix contexts that cross intonational boundaries and interruption points of speech repairs with *fluent* stretches of speech.

The problem with incorporating speech repair and intonational phrase detection into a language model is that there is not a reliable signal for detecting repairs (Bear, Dowding & Shriberg 1992) nor intonational phrases. Rather, there are a number of sources of information that give evidence as to the occurrence of these events. These sources include the presence of pauses, fillers, cue phrases, discourse markers, word fragments, word correspondences and syntactic anomalies. In this section, we augment our POS-based language model so that it also detects intonational phrases and speech repairs, along with their editing terms. To model the occurrence of intonational boundaries and speech repairs, we introduce three extra variables into the language model: the *repair* variable R_i, the *editing term* variable E_i and the *intonation* variable I_i. The probability distributions of the resulting model take into account most of the sources of evidence that signal spontaneous speech events.

5.4.1 Speech Repairs

The repair variable indicates the occurrence of speech repairs and its type: whether it is a modification repair, fresh start, or an abridged repair. The type of repair is important since the strategy that a hearer uses to correct a repair depends on the type of repair. For fresh starts, the hearer must determine the beginning of the current utterance. For modification repairs, the hearer can make use of the correspondences between the reparandum and alteration to determine the reparandum onset. For abridged repairs, there is no reparandum, and so simply knowing that it is abridged gives the correction.

For repairs that do not have an editing term, the interruption point is where the local context is disrupted, and hence is the logical place to tag such repairs. For repairs with an editing term, we tag the repair at the end of the editing term. This leads to the following definition of the repair variable R_i for the transition between word $W_{i\text{-}1}$ and W_i.

$$R_i = \begin{cases} \textbf{Mod} & \text{if } W_i \text{ is the alteration onset of a modification repair} \\ \textbf{Can} & \text{if } W_i \text{ is the alteration onset of a fresh start (or } \textit{cancel}\text{)} \\ \textbf{Abr} & \text{if } W_i \text{ is the alteration onset of an abridged repair} \\ \textbf{null} & \text{otherwise} \end{cases}$$

5.4.2 Editing Terms

Speech repairs often have an editing term, which follows the interruption point. Whether a word is being used as an editing term is not easy to determine. Phrases such as 'let me see' can be used as part of the sentential content of a sentence or as and editing term. Even fillers, such as 'um' and 'uh', only count as part of an editing term when they are not utterance initial. Hence, we need to model the occurrence of editing terms along with the occurrence of speech repairs. In our model, the variable E_i indicates the type of editing term transition between word $W_{i\text{-}1}$ and W_i.

$$E_i = \begin{cases} \textbf{Push} & \text{if } W_{i\text{-}1} \text{ is not part of an editing term but } W_i \text{ is} \\ \textbf{ET} & \text{if } W_{i\text{-}1} \text{ and } W_i \text{ are both part of an editing term} \\ \textbf{Pop} & \text{if } W_{i\text{-}1} \text{ is part of an editing term but } W_i \text{ is not} \\ \textbf{null} & \text{otherwise} \end{cases}$$

Below, we give an example and show all non-null editing term and repair tags.

Example 7 **(d93-10.4 utt30)**
that'll get there at four a.m. **Push** oh **ET** sorry **Pop Mod** at eleven a.m.

5.4.3 Intonational Phrases

The final variable is I_i, which marks the occurrence of intonational phrase boundaries.

$$I_i = \begin{cases} \textbf{\%} & \text{if } W_{i\text{-}1} \text{ ends an intonational phrase} \\ \textbf{null} & \text{otherwise} \end{cases}$$

The intonation variable is separate from the editing term and repair variables since it is not restricted by the value of the other two. For instance, an editing term could end an intonational phrase, especially on the end of a cue phrase such as 'let's see', as can the reparandum, as Example 8 below demonstrates.

Example 8 **(d92a-2.1 utt29)**
that's the one with the bananas **% Push** I **ET** mean **Pop Mod** that's taking the bananas

5.4.4 Redefining the Speech Recognition Problem

We now redefine the speech recognition problem so that its goal is to find the sequence of words and the corresponding POS, intonation, editing term and repair tags that is most probable given the acoustic signal.

$$\begin{aligned}\hat{W}\hat{P}\hat{R}\hat{E}\hat{I} &= \arg\max_{WPREI} \Pr(WPREI|A) \\ &= \arg\max_{WPREI} \Pr(A|WPREI)\Pr(WPREI) \end{aligned} \tag{5.5}$$

The second term is the language model probability, and can be rewritten as follows.

$$\begin{aligned}\Pr(W_{1,N}P_{1,N}R_{1,N}E_{1,N}I_{1,N}) &= \prod_{i=1}^{N} \Pr(W_iP_iR_iE_iI_i|W_{1,i\text{-}1}P_{1,i\text{-}1}R_{1,i\text{-}1}E_{1,i\text{-}1}I_{1,i\text{-}1}) \\ &= \prod_{i=1}^{N} \Pr(I_i|W_{1,i\text{-}1}P_{1,i\text{-}1}R_{1,i\text{-}1}E_{1,i\text{-}1}I_{1,i\text{-}1}) \\ &\quad \Pr(E_i|W_{1,i\text{-}1}P_{1,i\text{-}1}R_{1,i\text{-}1}E_{1,i\text{-}1}I_{1,i}) \\ &\quad \Pr(R_i|W_{1,i\text{-}1}P_{1,i\text{-}1}R_{1,i\text{-}1}E_{1,i}I_{1,i}) \\ &\quad \Pr(P_i|W_{1,i\text{-}1}P_{1,i\text{-}1}R_{1,i}E_{1,i}I_{1,i}) \\ &\quad \Pr(W_i|W_{1,i\text{-}1}P_{1,i}R_{1,i}E_{1,i}I_{1,i}) \end{aligned} \tag{5.6}$$

5.4.5 Representing the Context

Equation 5.6 requires five probability distributions to be estimated. The context for each includes all of the words, POS, intonation, repair and editing term tags that have been hypothesized, each as a separate piece of information. In principal, we could give all of this information to the decision tree algorithm and let it decide what information to use in constructing equivalence classes. However, repairs, editing terms and even intonation phrases do not occur in the same abundance as fluent speech and are not as constrained. Hence, it will be difficult to properly estimate the probability distributions.

Consider the following example of a speech repair without an editing term.

***Example* 9 (d92-1 utt53)**
engine E two picks **Mod** takes the two boxcars

When predicting the first word of the alteration 'takes', it is inappropriate to ask about the preceding words, such as 'picks', without realizing that there is a modification repair in between. The same also holds for intonational boundaries and editing term pushes and pops. In the example below, a question should only be asked about 'is' in the realization that it ends an intonational phrase.

***Example* 10 (d92a-1.2 utt3)**
you'll have to tell me what the problem is **%** I don't have their labels

Although the intonation, repair and editing term tags are part of the context and so can be used in partitioning it, the problem is that null intonation, repair and editing term tags dominate the training examples. So, we are bound to run into contexts in which there are not enough intonational phrases and repairs for the decision tree algorithm to learn the importance of using this information, and instead it might blindly subdivide the context based on some subdivision of the POS tags. The solution is analogous to what is done in POS tagging of written text: we give a view of the words and POS tags with the non-null repair, non-null intonation and editing term push and pop tags inserted. By inserting these tags into the word and POS sequence, it will be more difficult for the learning algorithm to ignore them.

Now consider the following examples, which both start with 'so we need to'.

***Example* 11 (d92a-2.2 utt6)**
so we need to **Push** um **Pop Abr** get a tanker of OJ to Avon

***Example* 12 (d93-11.1 utt46)**
so we need to get the three tankers

This is then followed by the verb 'get', except the first has an editing term in between. In predicting this word, the editing term hinders the decision tree algorithm from generalizing with non-abridged examples. The same thing happens with fresh starts and modification repairs. To allow generalizations between repairs with an editing term and those without, we use a view of the context with completed editing terms removed (cf. Stolcke & Shriberg 1996b).

To illustrate the augmented word and POS contexts given to the decision tree, consider the following example.

***Example* 13 (d93-18.1 utt47)**
it takes one **Push** you **ET** know **Pop Mod** two hours **%**

For the correct interpretation for the POS tag of 'you', the context includes the previous words together with the tag that indicates we are starting an editing term: 'it/**PRP** takes/**VBP** one/**CD Push**.' The context for the editing term **Pop** is 'it/**PRP** takes/**VBP** one/**CD Push** you/**PRP** know/**VBP**.' The repair tag is predicted after the editing term is completed, and hence has the editing term cleaned up: 'it/**PRP** takes/**VBP** one/**CD**' (we also give it the context with the editing term not cleaned up). The context for the POS tag of 'two' is 'it/**PRP** takes/**VBP** one/**CD Mod**.'

We also include two variables that indicate whether we are processing an editing term without forcing it to look for an editing term **Push** in the context: **ET-state** indicates whether we are processing an editing term and whether a cue phrase was seen; and **ET-prev** indicates the number of editing term words seen so far.

The contexts given to the decision tree algorithm encode basic knowledge about the effects of speech repairs, editing terms and intonational phrase boundaries. This allows the limited amount of training data to be used more effectively in estimating the probability districtions.

5.5 Correcting Speech Repairs

The previous section focused on the detection of speech repairs, editing terms and intonational phrases. But for repairs, we have only addressed half of the problem; the other half is determining the extent of the reparandum. Hindle (1983) and Kikui & Morimoto (1994) focused on correcting speech repairs, assuming the interruption point would be already detected. Although the model of the previous section detects repairs, this model is not effective enough. One of its crucial shortcomings is that it does not use as evidence whether there is a suitable correction (Heeman, Loken-Kim & Allen 1996). Since hearers are often unaware of speech repairs (Martin & Strange 1968), they must be able to correct them as the utterance is unfolding and as an indistinguishable event from detecting them and recognizing the words involved.

Recently, Stolcke & Shriberg (1996b) presented a word-based model for speech recognition that models simple word deletion and repetition patterns. They used the prediction of the repair to clean up the context and help predict what word will occur next. Although their model is limited to simple types of repairs, it provides a starting point for incorporating speech repair correction into a statistical language model.

5.5.1 Our Approach

There are several sources of information that give evidence as to the extent of the reparandum. Probably the most widely used is the presence of word correspondences between the reparandum and alteration, both at the word level and at the level of syntactic constituents (Levelt 1983, Hindle 1983, Bear et al. 1992, Heeman & Allen 1994, Kikui & Morimoto 1994). Second, there tends to be a fluent transition from the speech that precedes the onset of the reparandum to the alteration (Kikui & Morimoto 1994). This source is very important for repairs that do not have initial retracing, and is the mainstay of the 'parser-first' approach (e.g. Dowding, Gawron, Appelt, Bear, Cherny, Moore & Moran 1993)—keep trying alternative corrections until one of them parses. Third, there are certain regularities for where speakers restart. Reparandum onsets tend to be at constituent boundaries (Nooteboom 1980), and in particular, at boundaries where a co-ordinated constituent can be placed (Levelt 1983). Hence, reparandum onsets can be partially predicted without even looking at the alteration.

To model the correction of speech repairs, we add three more variables to our language model that enables us to make use of the above sources of evidence. For each non-abridged repair, we hypothesize the reparandum onset, and as we process the subsequent words, we hypothesize to which word in the reparandum it corresponds (or licences it), and the correspondence type. With this expanded model, the words of the alteration should be better predicted by the proper hypothesis of the correction variables then by some other interpretation. Consider the following example with strong word correspondences.

***Example* 14 ((d93-3.2 utt45)**

which engine $\underbrace{\text{are we}}_{\textit{reparandum}}\ \underset{\textit{ip}}{\uparrow}$ are we taking

If we predicted that a modification repair occurred and that the reparandum consists of 'are we', then the probability of 'are' being the first word of the alteration would be very high since it matches the first word of the reparandum. Conversely, if we are not predicting a modification repair with reparandum 'are we', then the probability of seeing 'are' would be much lower. The same holds for predicting the next word, 'we': it is more likely under the repair interpretation. As we process the words of the alteration, the repair interpretation will better account for the words that follow it, strengthening the interpretation.

When predicting the first word of the alteration, we can also make use of the second source of evidence identified above: the context provided by the words that precede the reparandum. Consider the following repair in which the first two words of the alteration are inserted.

***Example* 15 (d93-16.2 utt66)**

and two tankers $\underbrace{\text{to}}_{\textit{reparandum}}\ \underset{\textit{ip}}{\uparrow}$ of OJ to Dansville

Here, if we know the reparandum is 'to', then we know that the first word of the reparandum must be a fluent continuation of the speech before the onset of the reparandum. In fact, we see that the repair interpretation provides better context for predicting the first word of the alteration than a hypothesis that predicts either the wrong reparandum onset or predicts no repair at all.

We also make use of the third source of information. When we initially hypothesize the reparandum onset, we can take into account the a priori probability that it will occur at that point. In the following example, the words 'should' and 'the' are preferred by Levelt's co-ordinated constituent rule (Levelt 1983), and hence should have a higher score. Exceptions to the rule, such as this one, should have a lower score.

***Example* 16 (d93-10.4 utt30)**
the two boxcars $\underbrace{\text{of orange juice should}}_{\textit{reparandum}}\uparrow_{\textit{ip}}$ er of oranges should be made into ...

5.5.2 Reparandum Onset

After we predict a modification repair or a fresh start, we need to predict the reparandum onset. Consider the following two examples of modification repairs.

***Example* 17 (d93-16.3 utt9)**
to fill $\underbrace{\text{the engine}}_{\textit{reparandum}}\uparrow_{\textit{ip}}$ the boxcars with bananas

***Example* 18 (d93-25.6 utt31)**
drop off $\underbrace{\text{the one tanker}}_{\textit{reparandum}}\uparrow_{\textit{ip}}$ the two tankers

Although the examples differ in the length of the reparandum, their reparanda both start at the onset of a noun phrase. This same phenomena also exists for fresh starts where reparandum onsets are likely to follow an intonational boundary, the beginning of the turn, or a discourse marker. In order to allow generalizations across different reparandum lengths, we query each potential onset to see how likely it is as the onset. For $R_i \in \{\mathbf{Mod}, \mathbf{Can}\}$ and $j < i$, we define O_{ij} as follows.

$$O_{ij} = \begin{cases} \mathbf{Onset} & W_j \text{ is the reparandum onset of repair } R_i \\ \mathbf{null} & \text{otherwise} \end{cases}$$

We normalize the probabilities to ensure that $\sum_j (O_{ij} = \mathbf{Onset}) = 1$.

In the previous section, the context given to the decision tree algorithm included non-null repair tags, non-null intonation tags, and and editing pushes and excluded completed editing terms. Now that we are also hypothesizing reparanda onsets, we can exclude the reparanda of previous repairs. Consider the following example of *overlapping* repairs, repairs in which we can not separate their reparanda and alterations.

***Example* 19 (d93-16.3 utt4)**

what's the shortest route from engine $\underset{ip}{\uparrow}$ from $\underset{ip}{\uparrow}$ for engine two at Elmira

The reparandum of the first repair is 'from engine'. In predicting the reparandum of the second, we work from the cleaned up context: 'what's the shortest route from.'

The context used in estimating how likely a word is as the reparandum onset also includes the word we are querying. We also include the words and POS tags that precede the proposed reparandum onset, thus allowing the decision tree to check if the onset is at a suitable constituent boundary. Since reparanda rarely extend over more than one utterance, we include three variables that help indicate whether an utterance boundary is being crossed. The first indicates the number of intonational phrase boundaries embedded in the proposed reparandum. The second indicates the number of discourse markers in the reparandum. Discourse markers at the beginning of the reparandum are not included, and if discourse markers appear consecutively, the group is only counted once. The third indicates the number of fillers in the reparandum.

Another source of information is the presence of other repairs in the turn. In the Trains corpus, 35.6% of non-abridged repairs overlap. If a repair overlaps a previous one then its reparandum onset is likely to co-occur with the alteration onset of the previous repair. Hence we include a variable that indicates whether there is a previous repair, and if there is, whether the proposed onset coincides with, precedes, or follows the alteration onset of the preceding repair.

5.5.3 The Active Repair

Determining word correspondences is complicated by the occurrence of overlapping repairs. To keep our approach simple, we allow at most one previous word to license the correspondence. Consider again Example 19. Here, one could argue that the word 'for' corresponds to the word 'from' from either the reparandum of the first or second repair. In either case, the correspondence to the word 'engine' is from the reparandum of the first repair. Our approach is to first decide which repair the correspondence will be to and then decide which word of that repair's reparandum will license the current word. We always choose the most recent repair that has words in its reparandum that have not yet licensed a correspondence (other than a word fragment). Hence, the active repair for predicting the word 'for' is the second repair, while the active repair for predicting 'engine' is the first repair. For predicting the word 'two', neither the first nor second repair has any unlicensed words in their reparandum, and hence it will not have an active repair.

5.5.4 Licensing a Correspondence

If we are in the midst of processing a repair, we can use the reparandum to help predict the current word W_i and its POS tag D_i. In order to do this, we need to determine which word in the reparandum of the active repair will *license* the current word. As illustrated in Figure 5.1, word correspondences for speech

FIGURE 5.1: Cross Serial Correspondences between Reparandum and Alteration

repairs tend to exhibit a cross serial dependency (Heeman & Allen 1994); in other words, if we have a correspondence between w_j in the reparandum and w_k in the alteration, any correspondence with a word in the alteration after w_k will be to a word that is after w_j. Hence, if there is already a correspondence for the repair then the licensing word will follow the last correspondence in the reparandum.

The licensing word might need to skip over words due to deleted words in the reparandum or inserted words in the alteration. In the example below, the word 'tow' is licensed by 'carry', but the word 'them' must be skipped over before processing the licensing between the two instances of 'both'.

***Example* 20 (d92a-1.2 utt40)**

you can $\underbrace{\text{carry them both on}}_{\textit{reparandum}}\uparrow_{\textit{ip}}$ tow both on the same engine

The next example illustrates the opposite problem: the word 'two' has no correspondence with any word in the reparandum.

***Example* 21 (d93-15.4 utt45)**

and fill $\underbrace{\text{my boxcars full of oranges}}_{\textit{reparandum}}\uparrow_{\textit{ip}}$ my two boxcars full of oranges

For words that have no correspondence, we define the licensing word as the first available word in the alternation, in this case 'boxcars'. We leave it to the correspondence variable to encode that there is no correspondence. This gives us the following definition for the correspondence licensor, L_{ij} where i is the current word and j runs over all words in the reparandum of the active repair that come after the last word in the reparandum with a correspondence.

$$L_{ij} = \begin{cases} \textbf{Corr} & W_j \text{ licenses the current word} \\ \textbf{Corr} & W_i \text{ is inserted and } W_j \text{ is first available word in reparandum} \\ \textbf{null} & \text{otherwise} \end{cases}$$

Just as with the reparandum onset, we estimate the probability by querying each eligible word. The context for this query includes information about the proposed word, namely its POS tag. We also include information about the repair structure that has been found so far. If the previous word was a word match, there is a good chance that the current word will involve a word match to the next word. We include variables that indicates the number of words skipped in the reparandum and alteration since the last correspondence, the number of words since the onset of the reparandum and alteration, the number of words to the end of the reparandum, the type of repair and the reparandum length. We also include information about the POS and word context prior to the current word. This allows the decision tree to judge whether the proposed word is syntactically able to license the next word.

5.5.5 The Word Correspondence

Now that we have decided which word in the reparandum will potentially license the current word, we need to predict the type of correspondence. We focus on correspondences involving exact word match (identical POS tag and word), word replacements (same POS tag), or no such correspondence.

$$C_i = \begin{cases} \textbf{m} & W_i \text{ is a word match of the word indicated by } L_i \\ \textbf{r} & W_i \text{ is a word replacement of the word indicated by } L_i \\ \textbf{x} & W_i \text{ has no correspondence (inserted word)} \\ \textbf{null} & \text{No active repair} \end{cases}$$

The context used for estimating the correspondence variable is exactly the same as that used for estimating the licensor.

5.5.6 Redefining the Speech Recognition Problem

Now that we have introduced the correction tags, we redefine the speech recognition problem so that it includes finding the most probable corrections tags.

$$\begin{aligned} \hat{W}\hat{P}\hat{C}\hat{L}\hat{O}\hat{R}\hat{E}\hat{I} &= \arg\max_{WPCLOREI} \Pr(WPCLOREI|A) \\ &= \arg\max_{WPCLOREI} \Pr(A|WPCLOREI)\Pr(WPCLOREI) \end{aligned} \quad (5.7)$$

The second term is the language model and can be rewritten as we did for Equation 5.5.

In Section 5.5.2, we discussed that the word and POS context for the probability distributions can now exclude the reparanda of previous repairs. This not only applies to the three new probability distributions, but to the other five as well. Consider the following example.

***Example* 22 (d93-13.1 utt64)**
pick up and load two (*reparandum*) ↑*ip* um the two boxcars on engine two

In processing the word 'the', if we hypothesized that it follows a modification repair with editing term 'um' and reparandum 'two', then we can now generalize with fluent examples, such as the following, in hypothesizing its POS tag and the word identity.

***Example* 23 (d93-12.4 utt97)**
and to make the orange juice and load the tankers

For predicting the word and POS tags, we have an additional source of information, namely the values of the correspondence licensor and the correspondence type. Rather than use these two variables as part of the context that we give the decision tree algorithm, we use these tags to override the decision tree probability. If a word replacement or word match was hypothesized, we assign all of the POS probability to the appropriate POS tag. If a word match was hypothesized, we assign all of the word probability to the appropriate word.

5.6 Example

This section illustrates the workings of the algorithm. We illustrate the algorithm where it is constrained to the actual word transcription.[2] The algorithm incrementally considers all possible interpretations proceeding one word at a time. Low scoring paths are pruned so as to keep the search space tractible. Consider the following example.

***Example* 24 (d92a-2.1 utt95)**
okay % uh and that will take a total of (*reparandum*) ↑*ip* um let's see (*et*) total of s- (*reparandum*) ↑*ip* of seven hours

Rather than try to show all of the competing hypotheses, we focus on the correct interpretation, which, for this example, happens to be the winning interpretation. We contrast the probabilities of the correct tags with those of its competitors. For reference, we give a simplified view of the context that is used for each probability. Full results of the algorithm will be given in the next section.

2. In other work, we have used the language model to rescore word graphs (Heeman 1999).

5.6.1 Predicting 'um' as the Onset of an Editing Term

Below, we give the probabilities involved in the correct interpretation of the word 'um' given the correct interpretation of the words 'okay uh and that will take a total of'. We start with the intonation variable. The correct tag of **null** is significantly preferred over the alternative, mainly because intonational boundaries rarely follow prepositions.

$$\Pr(I_{10}{=}\textbf{null} \mid \text{a total of}) = 0.9997$$
$$\Pr(I_{10}{=}\% \mid \text{a total of}) = 0.0003$$

For I_{10} = **null**, we give the alternatives for the editing term tag. Since an editing term is not in progress, the only possible values are **Push** and **null**.

$$\Pr(E_{10}{=}\textbf{Push} \mid \text{a total of}) = 0.242$$
$$\Pr(E_{10}{=}\textbf{null} \mid \text{a total of}) = 0.758$$

With E_{10} = **Push**, the only allowable repair tag is **null**. Since no repair has been started, the reparandum onset O_{10} must be **null**. Similarly, since no repair is in progress, L_{10}, the correspondence licensor, and C_{10}, the correspondence type, must both be null.

We next hypothesize the POS tag. Below we list all of the tags that have a probability greater than 1%. Since we are starting an editing term, we see that POS tags associated with the first word of an editing term have a high probability, such as **UH_FP** for 'um', **AC** for 'okay', **CC_D** for 'or', **UH_D** for 'well', and **VB** for the 'let' in 'let's see'.

$$\Pr(D_{10}{=}\textbf{UH\_FP} \mid \text{a total of } \textbf{Push}) = 0.731$$
$$\Pr(D_{10}{=}\textbf{AC} \mid \text{a total of } \textbf{Push}) = 0.177$$
$$\Pr(D_{10}{=}\textbf{CC\_D} \mid \text{a total of } \textbf{Push}) = 0.026$$
$$\Pr(D_{10}{=}\textbf{UH\_D} \mid \text{a total of } \textbf{Push}) = 0.020$$
$$\Pr(D_{10}{=}\textbf{VB} \mid \text{a total of } \textbf{Push}) = 0.026$$

For D_{10} set to **UH_FP**, the word choices are 'um', 'uh', and 'er'.

$$\Pr(W_{10}{=}\text{um} \mid \text{a total of } \textbf{Push UH\_FP}) = 0.508$$
$$\Pr(W_{10}{=}\text{uh} \mid \text{a total of } \textbf{Push UH\_FP}) = 0.488$$
$$\Pr(W_{10}{=}\text{er} \mid \text{a total of } \textbf{Push UH\_FP}) = 0.004$$

Given the correct interpretation of the previous words, the probability of the filler 'um' along with the correct tags is 0.090.

5.6.2 Predicting 'total' as the Alteration Onset

We now give the probabilities involved in the second instance of 'total', which is the alteration onset of the first repair, whose editing term 'um let's see', which ends an intonational phrase, has just finished. Again we start with the intonation variable.

$$\Pr(I_{14}{=}\% \mid \text{a total of } \textbf{Push} \text{ um let's see}) = 0.902$$
$$\Pr(I_{14}{=}\textbf{null} \mid \text{a total of } \textbf{Push} \text{ um let's see}) = 0.098$$

For I_{14} = %, the editing term probabilities are given below. Since an editing term is in progress, the only possibilities are that it is continued or that it has ended.

Pr(E_{14}=**Pop** | a total of **Push** um let's see %) = 0.830

Pr(E_{14}=**ET** | a total of **Push** um let's see %) = 0.170

For E_{14} = **Pop**, we give the probabilities for the repair variable. Since an editing term has just ended, the null tag for the repair variable is ruled out. Note the modification interpretation receives a score approximately one third of that of a fresh start. However, the repair interpretation catches up after the alteration is processed.

Pr(R_{14}=**Mod** | a total of **Push** um let's see % **Pop**) = 0.228

Pr(R_{14}=**Can** | a total of **Push** um let's see % **Pop**) = 0.644

Pr(R_{14}=**Abr** | a total of **Push** um let's see % **Pop**) = 0.128

For R_{14} = **Mod**, we give the probabilities assigned to the possible reparandum onsets. For each, we give the proposed reparandum onset, X, and the words that precede it.

Pr($O_{14,X}$=**Onset** |W=take a total X=of R=**Mod**) = 0.589

Pr($O_{14,X}$=**Onset** |W=will take a X=total R=**Mod**) = 0.126

Pr($O_{14,X}$=**Onset** |W=that will take X=a R=**Mod**) = 0.145

Pr($O_{14,X}$=**Onset** |W=and that will X=take R=**Mod**) = 0.023

Pr($O_{14,X}$=**Onset** |W=uh and that X=will R=**Mod**) = 0.016

Pr($O_{14,X}$=**Onset** |W=% uh and X=that R=**Mod**) = 0.047

Pr($O_{14,X}$=**Onset** |W=okay % uh X=and R=**Mod**) = 0.047

Pr($O_{14,X}$=**Onset** |W=<turn> okay % X=uh R=**Mod**) = 0.003

Pr($O_{14,X}$=**Onset** |W=<turn> X=okay R=**Mod**) = 0.003

With 'total' as the reparandum onset, there are two possibilities for which word of the reparandum will license the current word—either the word 'total' or 'of'.

Pr($L_{10,X}$=**Corr** |W=will take a X=total R=**Mod**) = 0.973

Pr($L_{10,X}$=**Corr** |W=will take a X=of R=**Mod**) = 0.027

With 'total' as the correspondence licensor, we need to decide the type of correspondence: whether it is a word match, word replacement or otherwise.

Pr(C_{14}=**m** |W=will take a L=total R=**Mod**) = 0.5882

Pr(C_{14}=**r** |W=will take a L=total R=**Mod**) = 0.1790

Pr(C_{14}=**x** |W=will take a L=total R=**Mod**) = 0.2328

For the correct interpretation, the word correspondence is a word match with the word 'total' and POS tag **NN**. Hence, the POS tag and identity of the current word are both fixed and hence have a probability of 1. Given the correct interpretation of the previous words, the probability of the word 'total' along with the correct tags is 0.0111.

5.7 Results and Comparison

In this section, we present the results of running our model on the Trains corpus. We first explain the methodology that we use throughout the experiments, we then give results that indicate modeling speech repairs and intonational phrasing improves language modeling and POS tagging. We then give results for the tasks of identifying intonational phrase endings, detecting speech repairs and correcting them. We also compare our results with those reported by other researchers. This comparison is not exact because other researchers used different corpora, and employed different inputs. Also, our approach is the only one that has combined the detection and correction of speech repairs, and identifion of intonational phrase boundaries, POS tags, and discourse markers, in a speech recognition model. Hence our comparison is with systems that only address part of the problem.

5.7.1 Experimental Setup

We tested our model on the hand-collect transcripts of the Trains Corpus in order to determine how well it could detect and correct speech repairs, and identify intonational phrases. We used a six-fold cross-validation procedure. The dialogs were divided into six partitions and each was tested using a model built from the other five. Changes in speaker are marked in the word transcription with the special token <**turn**>. We treat contractions, such as 'that'll' and 'gonna', as separate words, treating them as 'that' and ''ll' for the first example, and 'going' and 'ta' for the second. We also changed all word fragments into a common token <**fragment**>. In searching for the best sequence of POS tags for the transcribed words, we follow the technique proposed by Chow & Schwartz (1989) and only keep a small number of alternative paths by pruning the low probability paths after processing each word.

5.7.2 Perplexity, Recall and Precision

A way to measure the effectiveness of the language model is to measure the *perplexity* that it assigns to a test corpus (Bahl, Baker, Jelinek & Mercer 1977). Perplexity is an estimate of how well the language model is able to predict the next word of a test corpus in terms of the number of alternatives that need to be considered at each point. For word-based language models, with estimated probability distribution of $\hat{\Pr}(w_i|w_{1,i\text{-}1})$, the perplexity of a test set $w_{1,N}$ is calculated as 2^H, where H is the entropy, which is defined as $H = -\frac{1}{N}\sum_{i=1}^{N}\log_2 \hat{\Pr}(w_i|w_{1,i\text{-}1})$.

We report results on identifying intonational phrase boundaries and speech repairs in terms of *recall, precision* and *error rate*. The recall rate is the number of times that the algorithm correctly identifies an event over the total number of times that it actually occurred. The precision rate is the number of times the algorithm correctly identifies it over the total number of times it identifies it. The error rate is the number of errors in identifying an event over the number of times that the event occurred.

5.7.3 POS Tagging and Perplexity

Table 5.1 shows that POS tagging and word perplexity benefit from modeling intonational phrases and speech repairs. The second column gives the results of the POS-based language model of Section 5.3. Column three adds

TABLE 5.1: Comparison of POS tagging, discourse marker identification and perplexity rates

	WP	WPCLOREI	WPCLOREIS
POS Errors	1711	1652	1563
POS Error Rate	2.93	2.83	2.68
Word Perplexity	24.04	22.96	22.35

speech repair and intonational phrase modeling and results in an improvement to word perplexity and POS tagging. Column four adds silence information. The amount of silence that followed a word was used to adjust the probability distributions for the repair, editing term and intonation variables (Heeman & Allen 1999). Silence information was not used to adjust the POS nor word probability distributions. We see that modeling speech repairs and intonational phrases results in an overall perplexity reduction of 7.0% over the POS-based model. We also see a significant improvement in POS tagging with an error rate reduction of 8.6% over the POS-based model. As we further improve the modeling of the user's utterance, we should expect to see further improvements in the language model.

5.7.4 Intonational Phrases

In Table 5.2, we give the results of our full model in detecting intonational phrase boundaries. We purposely divide intonational phrase boundaries that occur within a speaker's turn from those that occur at the end. The reason for this is that our model uses the end-of-turn information as part of its input, and since almost all turns end with an intonational phrase boundary, it easily learns this regularity. As for the within turn boundaries, the model achieves a recall rate of 71.8% and a precison of 70.8%.

TABLE 5.2: Intonational phrase results for full model

	Recall	Precision	Error Rate
Within Turn	71.76	70.82	57.79
End of Turn	98.05	94.17	8.00
All Boundaries	84.76	82.53	33.17

In comparison to our work, Wightman & Ostendorf (1994) made much fuller use of acoustical information: using preboundary lengthening, pausal durations, as well as other acoustic cues to automatically label intonational phrases and word accents. They trained a decision tree to estimate the probability of a phrase boundary given the acoustic context. These probabilities were fed into a Markov model whose state is the boundary type of the previous word. For training and testing their algorithm, they used a single-speaker corpus of news stories read by a public radio announcer. With this speaker-dependent model, they achieved a recall rate of 78.1% and a precision of 76.8%. However, it is unclear how well this will adapt to spontaneous speech, where repairs might interfere with the cues that they use, and to speaker independent testing.

Wang & Hirschberg (1992) also looked at detecting intonational phrases. Using automatically-labeled features, including POS tag of the current word, category of the constituent being built, distance from last boundary, and word accent, they built decision trees to classify each word as to whether it has an intonational boundary. With this approach, they achieved a recall rate of 79.5% and a precision rate of 82.7% on a subset of the ATIS corpus. Excluding end-of-turn data gives a recall rate of 72.2% and a precision of 76.2%. These results group speech repairs with intonational boundaries. In their corpus, there were 424 disfluencies and 405 turn-internal boundaries. The performance of the decision tree that does not classify disfluencies as intonational boundaries is significantly worse. However, these results were achieved with one-tenth the data of the Trains corpus.

Kompe, Kießling, Niemann, Nöth, Schukat-Talamazzini, Zottmann & Batliner (1995) combined acoustic cues with a statistical language model to find intonational phrases. They combined normalized syllable duration, length of pauses, pitch contour and energy using a multi-layered perceptron that estimates the probability $\Pr(v_i|c_i)$, where v_i indicates if there is a boundary after the current word and c_i is the acoustic features of the neighboring six syllables. This score is combined with the score from a statistical language model, which determines the probability of the word sequence with the hypothesized phrase boundary inserted using a backoff strategy. Building on this work, Mast et al. (1996) segmented speech into speech acts as the first step in automatically classifying them and achieved a recognition accuracy of 92.5% on turn internal boundaries using Verbmobil dialogs. This translates into a recall rate of 85.0%, a precision of 53.1% and an error rate of 90.1%. Their model, which employs rich acoustic modeling, does not account for interactions with speech repairs, POS tags, nor does it redefine the speech recognition language model.

Meteer & Iyer (1996) investigated whether modeling linguistic segments, segments with a single independent clause, improves language modeling. They computed the probability of the sequence of words with the hypothesized segment boundaries inserted into the sequence. Working on the Switchboard corpus, they found that predicting linguistic boundaries improved perplexity from 130 to 127. Similar to this work, Stolcke & Shriberg (1996a) investigated how the language-model can find the boundaries. Their best results were obtained by using POS tags as part of the input, as well as the word identities of certain word classes, in particular fillers, conjunctions, and certain discourse markers. However, this work does not incorporate the automatic POS tagging and discourse marker identification.

5.7.5 Speech Repairs

Table 5.3 gives the results of detecting speech repairs using our full model. The first row gives the recall and precision rates using the measure we refer to as *All Repairs*, in which we ignore errors that result from improperly identifying the type of repair, and hence scores a repair as correctly detected as long as it was identified as either an abridged repair, modification repair or fresh start. Furthermore, when multiple repairs have contiguous reparanda, we count all repairs involved (of the hand-annotations) as correct as long as the combined reparandum is correctly identified. Hence, for Example 19 given earlier, as long as the overall reparandum was identified as 'from engine from', both of the hand-annotated repairs are counted as correct. We see that we are able to

TABLE 5.3: Speech repair detection for full model

	Recall	Precision	Error Rate
All Repairs	76.79	86.66	35.01
Abridged	75.88	82.51	40.18
Modification	80.87	83.37	35.25
Fresh Starts	48.58	69.21	73.02
Modification & Fresh Starts	73.69	83.85	40.49

detect speech repairs with a recall of 76.8% and a precision of 86.7%. The next three rows of Table 5.3 give the rates at which each individual type of repair was correctly identified. The measure we used here is slightly different from the *All Repairs* measure: misclassications of a repair type were counted as wrong if the extent of the repair was not correctly identified. Here we see that fresh starts are the most difficult type of repair to detect. The fifth column combines the results of the modification repairs and fresh starts and does not count misclassifications between these two types of repairs.

Table 5.4 gives the results for correcting speech repairs. For all of the measures, a repair is counted as correctly corrected if it was identified and the extent of the reparadum was correctly determined. Our overall recall rate is 65.9% with a precision of 74.3%.

A number of other researchers have addressed the issue of detecting and correcting speech repairs. Bear et al. (1992) investigated the use of pattern matching of the word correspondences, global and local syntactic and semantic ill-formedness, and acoustic cues as evidence for detecting speech repairs. They tested their pattern matcher on a subset of the ATIS corpus from which they removed all *trivial* repairs, repairs that involve only the removal of a word fragment or a filler. For their pattern matching results, they achieved a detection recall rate of 76% with a precision of 62%, and a correction recall rate

TABLE 5.4: Speech repair correction for full model

	Recall	Precision	Error Rate
All Repairs	65.85	74.32	56.88
Abridged	75.65	82.26	40.66
Modification	77.95	80.36	41.09
Fresh Starts	36.21	51.59	97.76
Modification & Fresh Starts	63.76	72.54	60.36

of 44% with a precision of 35%. They also combined syntactic and semantic knowledge in a 'parser-first' approach—first try to parse the input and if that fails, invoke repair strategies based on word patterns in the input. In a test set containing 26 repairs (Dowding et al. 1993), they obtained a detection recall rate of 42% with a precision of 85%, and a correction recall rate of 31% with a precision of 62%.

Nakatani & Hirschberg (1994) proposed that speech repairs should be detected in a *speech-first* model using acoustic-prosodic cues, without relying on a word transcription. In order to test their theory, they built a decision tree using a training corpus of 148 turns of speech. They used hand-transcribed prosodic-acoustic features such as silence duration, energy, and pitch, as well as traditional text-first cues such as presence of word fragments, fillers, word matches, word replacements, POS tags, and position of the word in the turn and obtained a detection recall rate of 86.1% with a precision of 91.2%. The cues they found relevant were duration of pauses between words, word fragments, and lexical matching within a window of three words. Note that in their corpus 73.3% of the repairs were accompanied by a word fragment, as opposed to 32% of the modification repairs and fresh starts in the Trains corpus. Hence, word fragments are a stronger indicator of speech repairs in their corpus than in the Trains corpus. Also note that their training and test sets only included turns with speech repairs; hence their "findings should be seen more as indicative of the relative importance of various predictors of [speech repair] location than as a true test of repair site location."

Stolcke & Shriberg (1996b) incorporated repair resolution into a word-based language model. They limited the types of repairs to single and double word repetitions and deletions, deletions from the beginning of the sentence and fillers. In predicting a word, they sum over the probability distributions for each type of repair (including no repair at all). For hypotheses that include a repair, the prediction of the next word is based upon a cleaned-up representation of the context, as well as taking into account if they are predicting a single or double word repetition. Surprisingly, they found that this model actually degrades performance, in terms of perplexity and word error rate. They attributed this to their treatment of fillers: utterance-medial fillers should be cleaned up before predicting the next word, whereas utterance-initial ones should be left intact, a distinction that we make in our model by modeling intonational phrases.

Siu & Ostendorf (1996) extended a language model to account for three roles that words such as fillers can play in an utterance: utterance initial, part of a non-abridged repair, or part of an abridged repair. By using training data with these roles marked and a function-specific variable n-gram model (i.e. use different context for the probability estimates depending on the function of the word), and summing over each possible role, they achieved a perplexity reduction of 82.9 to 81.1.

5.8 Conclusion and Future Work

In this chapter, we redefined the speech recognition language model so that it also identifies intonational phrases and resolves speech repairs. This allows the language model to better account for the words involved in a speaker's turn and allows it to return a more meaningful analysis of the speaker's turn for later processing. The model incorporates identifying intonational phrases, POS tags, and detecting and correcting speech repairs; hence, interactions that exist between these tasks, as well as the task of predicting the next word, can be modeled.

Constraining our model to the hand transcription, it is able to identify 71.8% of all turn-internal intonational boundaries with a precision of 70.8%, and detect and correct 65.9% of all speech repairs with a precision of 74.3%. These results are partially attributable to accounting for the interaction between these tasks (Heeman & Allen 1999). Speech repairs and intonational phrases create discontinuities that traditional speech recognition language models and POS taggers have difficulty modeling. Modeling speech repairs and intonational phrases results in an 8.6% improvement in POS tagging and a 7.0% improvement in perplexity. Part of this improvement is from exploiting silences to give evidence of the speech repairs and intonational phrase boundaries.

More work still needs to be done. First, with the exception of pauses, we have not consider acoustic cues. This is a rich source of information for detecting (and distinguishing between) intonational phrases and interruption points of speech repairs. It would also help in determining the reparandum onset of fresh starts, which tend to occur at intonational boundaries. Acoustic modeling is also needed to identify word fragments. The second area is extending the model to incorporate higher level syntactic and semantic processing. This would not only allow us to give a much richer output from the model, but it would also allow us to account for interactions between this higher level knowledge and modeling speakers' utterances, especially in detecting the ill-formedness that often occur with speech repairs. It would also also aid in finding richer correspondences between the reparandum and alteration, such as between the noun phrase and pronoun in the following example.

***Example* 25 (d93-14.3 utt27)**

$\underbrace{\text{the engine can take as many}}_{\textit{reparandum}}\uparrow_{\textit{ip}}\ \underbrace{\text{um}}_{\textit{et}}\ \underbrace{\text{it can take}}_{\textit{alteration}}$ up to three loaded boxcars

The third and most important area is to incorporate our work into a speech recognizer. We have already used our POS-based model to rescore word-graphs, which results in a one percent absolute reduction in word error rate in comparison to a word-based model Our full model, which accounts for intonational phrases and speech repairs, leads to a further reduction, as well as returns a richer understanding of the speech (Heeman 1999).

Acknowledgments

Funding gratefully received from NSERC Canada, NSF under grant IRI-9623665, DARPA—Rome Laboratory under research contract F30602-95-1-0025, ONR/DARPA under grant N00014-92-J-1512, ONR under grant N0014-95-1-1088, ATR Interpreting Telecommunications Laboratory and CNET, France Télécom.

References

Bahl, L. R., Baker, J. K., Jelinek, F. & Mercer, R. L. (1977). Perplexity—a measure of the difficulty of speech recognition tasks, *Proceedings of the 94th Meeting of the Acoustical Society of America*.

Bahl, L. R., Brown, P. F., de Souza, P. V. & Mercer, R. L. (1989). A tree-based statistical language model for natural language speech recognition, *IEEE Transactions on Acoustics, Speech, and Signal Processing* **36**(7): 1001–1008.

Beach, C. M. (1991). The interpretation of prosodic patterns at points of syntactic structure ambiguity: Evidence for cue trading relations, *Journal of Memory and Language* **30**(6): 644–663.

Bear, J., Dowding, J. & Shriberg, E. (1992). Integrating multiple knowledge sources for detection and correction of repairs in human-computer dialog, *Proceedings of the 30th Annual Meeting of the Association for Computational Linguistics*, pp. 56–63.

Bear, J., Dowding, J., Shriberg, E. & Price, P. (1993). A system for labeling self-repairs in speech, *Technical Note 522*, SRI International.

Bear, J. & Price, P. (1990). Prosody, syntax, and parsing, *Proceedings of the 28th Annual Meeting of the Association for Computational Linguistics*, Pittsburgh, pp. 17–22.

Black, E., Jelinek, F., Lafferty, J., Magerman, D., Mercer, R. & Roukos, S. (1992). Towards history-based grammars: Using richer models for probabilistic parsing, *Proceedings of the DARPA Speech and Natural Language Workshop*, Morgan Kaufman, pp. 134–139.

Breiman, L., Friedman, J. H., Olshen, R. A. & Stone, C. J. (1984). *Classification and Regression Trees*, Wadsworth & Brooks, Monterrey, CA.

Brown, P. F., Della Pietra, V. J., deSouza, P. V., Lai, J. C. & Mercer, R. L. (1992). Class-based n-gram models of natural language, *Computational Linguistics* **18**(4): 467–479.

Chow, Y. S. & Schwartz, R. (1989). The n-best algorithm: An efficient procedure for finding top n sentence hypotheses, *Proceedings of the DARPA Speech and Natural Language Workshop*, Morgan Kaufman, San Mateo, California, pp. 199–202.

Dowding, J., Gawron, J. M., Appelt, D., Bear, J., Cherny, L., Moore, R. & Moran, D. (1993). Gemini: A natural language system for spoken-language understanding, *Proceedings of the 31st Annual Meeting of the Association for Computational Linguistics*, pp. 54–61.

Heeman, P. A. (1997). Speech repairs, intonational boundaries and discourse markers: Modeling speakers' utterances in spoken dialog, *Technical Report 673*, Department of Computer Science, University of Rochester. Doctoral dissertation.

Heeman, P. A. (1999). Modeling speech repairs and intonational phrasing to improve speech recognition, *Automatic Speech Recognition and Understanding Workshop*, Keystone Colorado.

Heeman, P. A. & Allen, J. F. (1995). The Trains spoken dialog corpus, *CD-ROM*, Linguistics Data Consortium.

Heeman, P. A. & Allen, J. F. (1999). Speech repairs, intonational phrases and discourse markers: Modeling speakers' utterances in spoken dialog, *Computational Linguistics* **25**(4): 527–572.

Heeman, P. A., Loken-Kim, K. & Allen, J. F. (1996). Combining the detection and correction of speech repairs, *Proceedings of the 4th International Conference on Spoken Language Processing*, Philadephia, pp. 358–361. Also appears in *International Symposium on Spoken Dialogue*, 1996, pp. 133–136.

Heeman, P. & Allen, J. (1994). Detecting and correcting speech repairs, *Proceedings of the 32nd Annual Meeting of the Association for Computational Linguistics*, Las Cruces, New Mexico, pp. 295–302.

Hindle, D. (1983). Deterministic parsing of syntactic non-fluencies, *Proceedings of the 21st Annual Meeting of the Association for Computational Linguistics*, pp. 123–128.

Jelinek, F. (1985). Self-organized language modeling for speech recognition, *Technical report*, IBM T. J. Watson Research Center, Continuous Speech Recognition Group, Yorktown Heights, NY.

Kikui, G.-i. & Morimoto, T. (1994). Similarity-based identification of repairs in Japanese spoken language, *Proceedings of the 3rd International Conference on Spoken Language Processing*, pp. 915–918.

Kompe, R., Kießling, A., Niemann, H., Nöth, E., Schukat-Talamazzini, E. G., Zottmann, A. & Batliner, A. (1995). Prosodic scoring of word hypotheses graphs, *Proceedings of the 4th European Conference on Speech Communication and Technology (Eurospeech '95)*, Madrid, pp. 1333–1336.

Levelt, W. J. M. (1983). Monitoring and self-repair in speech, *Cognition* **14**: 41–104.

Martin, J. G. & Strange, W. (1968). The perception of hesitation in spontaneous speech, *Perception and Psychophysics* **53**: 1–15.

Mast, M., Kompe, R., Harbeck, S., Kießling, A., Niemann, H., Nöth, E., Schukat-Talamazzini, E. G. & Warnke, V. (1996). Dialog act classification with the help of prosody, *Proceedings of the 4th International Conference on Spoken Language Processing*, Philadelphia, pp. 1728–1731.

Meteer, M. & Iyer, R. (1996). Modeling conversational speech for speech recognition, *Proceedings of the Conference on Empirical Methods in Natural Language Processing*, Philadelphia, pp. 33–47.

Nakatani, C. H. & Hirschberg, J. (1994). A corpus-based study of repair cues in spontaneous speech, *Journal of the Acoustical Society of America* **95**(3): 1603–1616.

Nooteboom, S. G. (1980). Speaking and unspeaking: Detection and correction of phonological and lexical errors, *in* V. A. Fromkin (ed.), *Errors in Linguistic Performance: Slips of the Tongue, Ear, Pen, and Hand*, Academic Press, New York, pp. 87–96.

Ostendorf, M., Wightman, C. & Veilleux, N. (1993). Parse scoring with prosodic information: an analysis/synthesis approach, *Computer Speech and Language* **7**(2): 193–210.

Oviatt, S. (1995). Predicting spoken disfluencies during human-computer interaction, *Computer Speech and Language* **9**: 19–35.

Price, P. (1997). Spoken language understanding, *in* R. Cole, J. Mariani, H. Uszkoreit, A. Zaenen & V. Zue (eds), *Survey of the State of the Art in Human Language Technology*, Cambridge University Press.

Rosé, C. P. & Lavie, A. (2000). Balancing robustness and efficiency in unification-augmented context-free parsers for large practical applications, *Robustness in Language and Speech Technology*, Elsnet. This volume.

Silverman, K., Beckman, M., Pitrelli, J., Ostendorf, M., Wightman, C., Price, P., Pierrehumbert, J. & Hirschberg, J. (1992). ToBI: A standard for labelling English prosody, *Proceedings of the 2nd International Conference on Spoken Language Processing*, pp. 867–870.

Siu, M. & Ostendorf, M. (1996). Modeling disfluencies in conversational speech, *Proceedings of the 4th International Conference on Spoken Language Processing*, pp. 382–391.

Stolcke, A. & Shriberg, E. (1996a). Automatic linguistic segmentation of conversational speech, *Proceedings of the 4th International Conference on Spoken Language Processing*, pp. 1001–1004.

Stolcke, A. & Shriberg, E. (1996b). Statistical language modeling for speech disfluencies, *Proceedings of the International Conference on Audio, Speech and Signal Processing*, pp. 405–408.

Traum, D. R. & Heeman, P. A. (1997). Utterance units in spoken dialogue, *in* E. Maier, M. Mast & S. LuperFoy (eds), *Dialogue Processing in Spoken Language Systems*, Lecture Notes in Artificial Intelligence, Springer-Verlag, Heidelberg, pp. 125–140.

van Noord, G. (2000). Robust parsing of word graphs, *Robustness in Language and Speech Technology*, Elsnet. This volume.

Wang, M. Q. & Hirschberg, J. (1992). Automatic classification of intonational phrase boundaries, *Computer Speech and Language* **6**: 175–196.

Ward, W. (1991). Understanding spontaneous speech: The Phoenix system, *Proceedings of the International Conference on Audio, Speech and Signal Processing*, pp. 365–367.

Wightman, C. W. & Ostendorf, M. (1994). Automatic labeling of prosodic patterns, *IEEE Transactions on Speech and Audio Processing* **2**(4): 469–481.

Young, S. R. & Matessa, M. (1991). Using pragmatic and semantic knowledge to correct parsing of spoken language utterances, *Proceedings of the 2nd European Conference on Speech Communication and Technology (Eurospeech '91)*, Genova, Italy, pp. 223–227.

Chapter 6

Regular Approximation of Context-Free Grammars through Transformation

Mehryar Mohri
AT&T Labs — Research
mohri@research.att.com

Mark-Jan Nederhof
AT&T Labs — Research
nederhof@research.att.com

Abstract We present an algorithm for approximating context-free languages with regular languages. The algorithm is based on a simple transformation that applies to any context-free grammar and guarantees that the result can be compiled into a finite automaton. The resulting grammar contains at most one new nonterminal for any nonterminal symbol of the input grammar. The result thus remains readable and if necessary modifiable. We extend the approximation algorithm to the case of weighted context-free grammars. We also report experiments with several grammars showing that the size of the minimal deterministic automata accepting the resulting approximations is of practical use for applications such as speech recognition.

6.1 Introduction

Despite the availability of extensive literature on the topic of efficient context-free parsing, for large and very ambiguous grammars, context-free parsing poses a serious problem in many practical applications such as real-time speech recognition. For most grammars used in those applications, rules are annotated with weights, and efficient processing of weights forms an additional challenge for the implementation of context-free parsers.

Much more attractive computational properties can be attributed to the use of finite (weighted) automata both in theory and in practice. This commu-

J.-C. Junqua and G. van Noord (eds.), Robustness in Languages and Speech Technology, 153–163.

nication deals with ideas that allow us to make use of these advantageous properties by approximating context-free languages with regular languages. Such (weighted) approximations could play a crucial role in constructing robust speech recognition systems because they are computationally less demanding than context-free grammars and because in general they can give a more accurate model of the syntactic and semantic properties of natural languages than classical n-gram language models.

Several approximations have been described in the existing literature, among which are Pereira & Wright (1997), Grimley Evans (1997), and Johnson (1998); an extensive bibliography of approximations can be found in Nederhof (2000). None of these approximations however provides much insight into how the language is changed during the approximation process, and it is difficult or impossible to influence this process in order to fine-tune the approximating language to the application.

For example, Pereira & Wright (1997) proposes the construction of a specific kind of pushdown automaton from the grammar, which is subsequently approximated to be a finite automaton. Since here the structure of a pushdown automaton is so different from the structure of the context-free language from which it is constructed, it is close to impossible for the grammar writer to predict or influence the approximating language that will result based on his understanding of the grammar.

Our objective is to solve this problem by defining an approximation through a simple transformation of the original grammar. The grammar writer may inspect the transformed grammar, in which the structure of the original grammar is still visible, and change it in a way suitable for the application.

Our approximation algorithm applies to any context-free grammar and guarantees that the result can be compiled into a finite automaton. The resulting grammar contains at most one new nonterminal for any nonterminal symbol of the input grammar, and new rules are formed out of rules from the input grammar by means of a straightforward decomposition. The result thus remains readable and if necessary modifiable. We show that the algorithm can be extended to the case of weighted context-free grammars.

Experiments with several grammars show that the size of the minimal deterministic automata accepting the resulting approximations is of practical use for applications such as speech recognition.

6.2 Preliminaries

A *context-free grammar* G is a 4-tuple (Σ, N, P, S), where Σ and N are two finite disjoint sets of terminals and nonterminals, respectively, $S \in N$ is the start symbol, and P is a finite set of rules. Each rule has the form $A \to \alpha$ with $A \in N$ and $\alpha \in V^*$, where V denotes $N \cup \Sigma$. The relation $\to$ on $N \times V^*$ is extended to a relation on $V^* \times V^*$ in the usual way and the transitive and reflexive closure of $\rightarrow$ is denoted by $\xrightarrow{*}$. The language *generated* by G is given by:

$$L(G) = \{w \in \Sigma^* : S \xrightarrow{*} w\}$$

and is called a *context-free language*. We denote the empty string by ϵ. We let $|G|$ denote the number of rules of a grammar G. We also define the size of G as the total number of occurrences in P of symbols from V, and denote it by $||G||$. We generally use symbols $A, B, C, \ldots$ to range over N, symbols $a, b, c, \ldots$ to range over Σ, symbols X, Y, Z to range over V, symbols $\alpha, \beta, \gamma, \ldots$ to range over V^*, and symbols $v, w, x, \ldots$ to range over Σ^*. We further consider *regular languages*, the class of languages accepted by finite automata. We assume that the reader is familiar with these concepts; for more details we refer to Hopcroft & Ullman (1979).

6.3 Algorithm

6.3.1 Analysis and Compilation

Context-free grammars may generate languages that are not regular. We describe a subclass of grammars, *strongly regular grammars*, that are guaranteed to generate regular languages. This class of grammars coincides with that of grammars without self-embedding (Chomsky 1959). Furthermore, strongly regular grammars can be mapped into equivalent finite automata using an efficient algorithm.

We then present our approximation algorithm, which transforms any grammar into one that is strongly regular. Note that a mapping from an arbitrary CFG generating a regular language into a corresponding finite automaton cannot be realized by any algorithm (Ullian 1967). Therefore, neither our approximation, nor any other, can guarantee that the language is preserved when the grammar already generates a regular language. However, this is guaranteed when the grammar is strongly regular.

Let $\mathcal{R}$ be the relation defined on the set of nonterminals N of G by:

$$A \,\mathcal{R}\, B \;\Leftrightarrow\; (\exists \alpha, \beta \in V^* : A \xrightarrow{*} \alpha B \beta) \wedge (\exists \alpha, \beta \in V^* : B \xrightarrow{*} \alpha A \beta)$$

It is not hard to show that $\mathcal{R}$ defines an equivalence relation. $\mathcal{R}$ partitions N into subsets called sets of *mutually recursive nonterminals*.

For convenience, we refer to the rules with left-hand side $A \in N$ as *the rules of* A, and more generally, to the rules with left-hand side $A \in M$, for some $M \subseteq N$, as *the rules of* M. *Strongly regular grammars* are grammars in which the rules of each set M of mutually recursive nonterminals are either all right-linear or all left-linear.[1]

---

1. Nonterminals that do not belong to M are considered as terminals here, for determining if a rule of M is right-linear or left-linear.

There exist efficient algorithms for computing a finite automaton accepting the language generated by a strongly regular grammar. The finite automaton can be constructed off-line, as shown in Nederhof (1997), but as proposed by Mohri & Pereira (1998) one may also construct an alternative, compact representation of the regular language, from which the finite automaton may be computed; this compact representation may however also be used on-line for processing of input. The steps of this construction and the processing of input are sketched as follows:

1. determine the sets of mutually recursive nonterminals. This can be done in linear time in the size of the input grammar G by computing the strongly connected components of the graph of the grammar.[2]
2. construct a finite-state machine $\mathcal{K}(M)$ for each set of mutually recursive nonterminals M following the classical construction of an automaton from a regular grammar (Aho & Ullman 1973). $\mathcal{K}(M)$ is in effect a finite automaton for which the initial state (in the case of a right-linear set M) or the set of final states (in the case of a left-linear set) have been left unspecified. A finite automaton $\mathcal{N}(A)$ describing the language corresponding to a nonterminal $A \in M$ can be easily derived from $\mathcal{K}(M)$ by specifying the state corresponding to the non-terminal A as an initial or final state depending on the type of the set (right-linear or left-linear case). Thus, $\mathcal{K}(M)$ combined with the start symbol S of the grammar now give a compact representation of the approximating language.
3. for each input string w, we first obtain $\mathcal{N}(S)$ from the $\mathcal{K}(M)$ that satisfies $S \in M$, and this automaton is then expanded in a lazy way by substituting other automata $\mathcal{N}(A)$ for occurrences of A in $\mathcal{N}(S)$ that are encountered while processing w.

Thus, the states and transitions of the finite automaton for the complete language are constructed on demand, when needed for the recognition of an actual input string. Moreover, the construction of the compact representation is optimal in the sense that it requires work linear in the size of the input grammar, or formally $\mathcal{O}(\|G\|)$. The compact representation can be further optimized using ϵ-removal, determinization, and minimization algorithms, possibly in combination with the substitution of subautomata.

6.3.2 Transformation

We now describe a grammar transformation that creates strongly regular grammars. For each set of mutually recursive nonterminals M such that the corresponding rules are not all right-linear or not all left-linear with respect to the nonterminals of M, we apply a grammar transformation defined as follows:[3]

1. For each nonterminal $A \in M$, introduce a new nonterminal $A' \notin N$, and

---

2. The *graph* of a grammar has one node for each nonterminal and has an edge from node A to node B if nonterminal B appears on the right-hand side of a rule with left-hand side A.
3. Other rules of the input grammar are left unchanged.

add the following rule to the grammar:[4]

$$A' \rightarrow \epsilon$$

2. Consider each rule with left-hand side $A \in M$:

$$A \rightarrow \alpha_0 \; B_1 \; \alpha_1 \; B_2 \; \alpha_2 \; \cdots \; B_m \; \alpha_m$$

with $m \geq 0$, $B_1, \ldots, B_m \in M$, $\alpha_0 \ldots \alpha_m \in (\Sigma \cup (N - M))^*$, and replace it by the following set of rules:

$$\begin{array}{rcl} A & \rightarrow & \alpha_0 \; B_1 \\ B_1' & \rightarrow & \alpha_1 \; B_2 \\ B_2' & \rightarrow & \alpha_2 \; B_3 \\ & \ldots & \\ B_{m-1}' & \rightarrow & \alpha_{m-1} \; B_m \\ B_m' & \rightarrow & \alpha_m \; A' \end{array}$$

(In the case where $m = 0$, this set of rules merely contains $A \rightarrow \alpha_0 \; A'$.) Since the rules of M are replaced by right-linear rules, the resulting grammar is strongly regular and can be compiled into a finite automaton as discussed above.

An attractive property of the grammar transformation is that it can be applied to large grammars: at most one new nonterminal is introduced for each nonterminal of the input grammar, and the size of the resulting grammar is at most twice that of the input grammar.

A further convenient property is that the transformed grammar can be used for parsing. The nonterminals of the form X and X' correspond to the beginning and end of recognition of strings generated by X in the original grammar. This fact can be used to compile the transformed grammar into a finite-state transducer that outputs bracketed strings equivalent to parse trees. The resulting parse trees retain much of the structure of the original grammar. See Nederhof (1998) for a related idea.

The language generated by the transformed grammar is a superset of that of the original grammar. Indeed, it is clear from the way a rule in the original grammar is split into several rules, that any string accepted by a series of derivations in the input grammar is also accepted in the resulting grammar.

Since the size of the result of the transformation is comparable to that of the input grammar and since the symbols that are newly introduced still admit a syntactic interpretation, the resulting grammar remains readable, and if necessary can be modified by the user. Finally, we shall see that in several experiments the transformation also leads to automata of reasonable size and therefore that it is of practical interest.

---

4. This can be refined by adding the rule only when A is directly reachable from another strongly connected component.

$$
\begin{array}{lcl}
E & \rightarrow & E + T \\
E & \rightarrow & T \\
T & \rightarrow & T * F \\
T & \rightarrow & F \\
F & \rightarrow & (\ E\ ) \\
F & \rightarrow & a
\end{array}
$$

(A)

$$
\begin{array}{lcl@{\qquad}lcl}
E' & \rightarrow & \epsilon & T & \rightarrow & T \\
T' & \rightarrow & \epsilon & T' & \rightarrow & * \ F \\
F' & \rightarrow & \epsilon & F' & \rightarrow & T' \\
E & \rightarrow & E & T & \rightarrow & F \\
E' & \rightarrow & + \ T & F' & \rightarrow & T' \\
T' & \rightarrow & E' & F & \rightarrow & (\ E \\
E & \rightarrow & T & E' & \rightarrow &)\ F' \\
T' & \rightarrow & E' & F & \rightarrow & a\ F'
\end{array}
$$

(B)

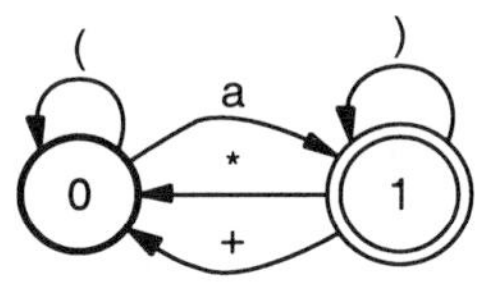

(C)

FIGURE 6.1: Regular approximation of a simple context-free grammar. (A) Grammar of arithmetic expressions. (B) Transformed grammar. (C) Finite automaton accepting the language generated by the transformed grammar.

Our approximation algorithm works in two stages: first the grammar is approximated using the transformation just defined. Then the resulting grammar is compiled into a finite automaton. As mentioned before, the compilation can be performed on demand, in which case the automaton is only constructed and expanded as far as it is required for the recognition of the input string, or the automaton can be completely expanded and optimized. Both options can be useful in practice, and the choice depends on time and space trade-offs. Note that as long as we do not substitute or optimize the subautomata, the total construction comprising the grammar transformation and the compilation is linear in the size of the grammar.

The grammar transformation is illustrated by figures 6.1-(a) and 6.1-(b). Figure 6.1-(a) is a simple grammar for well-formed arithmetic expressions. Figure 6.1-(b) shows the result after transformation. In the original grammar, the nonterminals E, T, and F can be interpreted as corresponding to an expression, a term, or a factor. In the transformed grammar, F for example corresponds to a suffix of an expression starting with a factor, and F' corresponds to a suffix of an expression just after the end of a factor.

The transformed grammar is strongly regular and can be compiled into the finite automaton of figure 6.1-(c).

Our grammar transformation bears some similarity to a transformation from Nederhof (1997) (equivalent to the RTN approximation from Nederhof 2000). In fact, that approximation can be viewed as a more refined variant of our new transformation. A disadvantage of the old approximation method is however that it requires a quadratic number ($\mathcal{O}(|N|^2)$) of nonterminals in the resulting grammar, which may be unacceptable for some large grammars used in practice.

Nevertheless, in light of the approximation method just presented, we can give a simpler formulation of the older approximation as follows. For any set of mutually recursive nonterminals M, the grammar is transformed in the following steps.

1. Let M' be defined by: $A \in M'$ if and only if $A \in M$ and either $A = S$ or $\exists(B \rightarrow \alpha A \beta) \in P : B \notin M$.
2. Introduce two new nonterminals $A^B\uparrow$ and A^B for each $A \in M$ and $B \in M'$.
3. For each nonterminal $A \in M'$ add the following rule to the grammar:

$$A^A\uparrow \;\rightarrow\; \epsilon$$

4. Consider each rule with left-hand side $A \in M$:

$$A \;\rightarrow\; \alpha_0 \; B_1 \; \alpha_1 \; B_2 \; \alpha_2 \; \cdots \; B_m \; \alpha_m$$

with $m \geq 0$, $B_1, \ldots, B_m \in M$, $\alpha_0 \ldots \alpha_m \in (\Sigma \cup (N - M))^*$, and replace it by the following set of rules, for each $C \in M'$:

$$\begin{array}{rcl} A^C & \rightarrow & \alpha_0 \; B_1^C \\ B_1^C\uparrow & \rightarrow & \alpha_1 \; B_2^C \\ B_2^C\uparrow & \rightarrow & \alpha_2 \; B_3^C \\ & \cdots & \\ B_{m-1}^C\uparrow & \rightarrow & \alpha_{m-1} \; B_m^C \\ B_m^C\uparrow & \rightarrow & \alpha_m \; A^C\uparrow \end{array}$$

5. Replace each occurrence of $A \in M'$ in the old rules of the grammar by A^A.

The subset $M' \subseteq M$ represents the set of nonterminals in M that may be reached from parts of the grammar that are not involved in recursion in the set M; also the start symbol S can be in this set. Such nonterminals are at the root of a subtree of the parse tree constructed from nonterminals in M. These nonterminals are maintained in the superscripts of the new nonterminals, which is needed to ensure that the approximating grammar commits itself to a single nonterminal at the root of such a subtree.

Omitting the superscripts in the transformed grammar leads to the transformation presented before: A^B simplifies to A, and $A^B\uparrow$ to A', irrespective of B.

6.4 Weighted grammars

Grammars used in many applications such as those related to speech processing incorporate weights. These weights, which are often interpreted as probabilities, are used to rank different hypotheses for the purpose of disambiguation.

A *weighted context-free grammar* is a context-free grammar in which rules are additionally annotated with weights. The weight set $\mathbb{K}$ has in general the algebraic structure of a semiring $(\mathbb{K}, \oplus, \otimes, \overline{0}, \overline{1})$, with $\overline{0}$ as the identity element of $\oplus$ and $\overline{1}$ the identity element of $\otimes$. Weights from rules used in a single derivation are combined using the $\otimes$-product and the weights from alternative derivations (in the case of ambiguous grammars) are combined using the $\oplus$-sum. We will assume here that the $\oplus$-sum of the weights is well-defined and in $\mathbb{K}$ in all such cases.

The approximation method through transformation of context-free grammars can be extended to the weighted case in at least the following three cases:

- An unweighted grammar and a corpus are given. The grammar is transformed and an equivalent finite automaton is constructed from it. Weights are then assigned to its transitions by applying a learning algorithm based on the corpus.
- A weighted grammar is given. A corpus is constructed from that grammar and the weights are removed from the grammar. The rest of the process is identical to the previous case.
- A weighted grammar is given. The transformation is extended to assign weights to the output rules in such a way that the weight of a rule broken up into m rules by the transformation equals the $\otimes$-product of those m rules.

Here, we consider this last case and thus extend our transformation to weighted grammars as follows. For each set of mutually recursive nonterminals M such that the corresponding rules are not all right-linear and not all left-linear, the following transformation is applied:

1. For each nonterminal symbol $A \in M$, introduce a new nonterminal $A' \notin N$, and add the following rule with weight $\overline{1}$ to the grammar:

$$A' \to \epsilon \;/\overline{1}$$

2. Consider each rule with left-hand side $A \in M$ and weight $x \in \mathbb{K}$:

$$A \to \alpha_0\ B_1\ \alpha_1\ B_2\ \alpha_2\ \cdots\ B_m\ \alpha_m\ /x$$

with $m \geq 0$, $B_1, \ldots, B_m \in M$, $\alpha_0 \ldots \alpha_m \in (\Sigma \cup (N - M))^*$, and replace it by the following set of rules:

$$\begin{array}{rcl} A & \to & \alpha_0\ B_1\ /x_0 \\ B'_1 & \to & \alpha_1\ B_2\ /x_1 \\ B'_2 & \to & \alpha_2\ B_3\ /x_2 \\ & \cdots & \\ B'_{m-1} & \to & \alpha_{m-1}\ B_m\ /x_{m-1} \\ B'_m & \to & \alpha_m\ A'\ /x_m \end{array}$$

with $x_0, x_1, \ldots, x_m \in \mathbb{K}$, and $x_0 \otimes x_1 \otimes \cdots \otimes x_m = x$.

TABLE 6.1: Approximation experiments.

G	$\|G\|$	$\|G_{app}\|$	$\|A_{exp}\|$	$\|A\|$	transf. (s)	compil. (s)	opt. (s)
G_1	19	39	104	31	.14	.11	.40
G_2	47	82	354	95	.15	.12	.37
G_3	516	852	17121	3220	.24	.23	54
G_4	1434	2351	141304	1828	.35	.41	36
G_5	1641	4153	6924	7467	.59	.74	54
G_6	846	1992	55070	97	.38	.53	16

There exists at least one admissible choice for the values of $x_0, x_1, \ldots, x_m$ defined by: $x_0 = x$, and $x_1 = \cdots = x_m = \overline{1}$ in any semiring, but in general the weight x can be distributed over the new rules in many different ways.

Let D be a derivation in the original weighted grammar which derives string w with weight x. Then there will be a derivation D' deriving w with the same weight x in the transformed grammar. However, the transformed grammar may additionally admit derivations of strings w with weights for which no corresponding derivation can be found in original grammar.

As in the unweighted case, the resulting weighted grammar can be compiled into a weighted automaton (Mohri & Pereira 1998).

6.5 Experiments

We have a full implementation of the approximation algorithm presented in the previous section in the general case of weighted context-free grammars. The algorithm and the corresponding utilities have been incorporated in a set of general-purpose grammar tools, the GRM library (Mohri 2000).

Using that implementation, we carried out experiments with grammars of small to large sizes. Grammars G_1, G_2, G_3 and G_4 were used in various experiments by Nederhof (2000). Our approximation results in these experiments exactly coincide with those obtained using the RTN approximation from Nederhof (2000). Grammar G_5 was obtained from Carroll (1993), and grammar G_6 from Schoorl & Belder (1990).

Table 6.1 illustrates our approximation results for each of these grammars. The second column gives the number of rules of each grammar. The next column gives the number of rules of the transformed grammar. Column $|A_{exp}|$ gives the size (in terms of the number of transitions) of the expanded automaton A_{exp} recognizing the transformed grammar, as produced by our algorithm. The next column gives the size of the minimal deterministic automaton A equivalent to A_{exp}. The time required to apply the grammar transformation is indicated in column *transf.*, the time to compile that grammar into A_{exp} is given in column *compil.*, and the time needed to compute the minimal deterministic automaton A, by optimizing the subautomata in combination with expansion, is given in column *opt.*, using an SGI Origin 2000.

The results show that both the size of the transformed grammar and the size of the resulting automata are small enough for practical use for demanding applications such as real-time speech recognition. This contrasts with other existing approximation methods that were shown to produce automata of very large size even for small grammars of less than fifty rules (Nederhof 2000). The approximation time including the two stages of transformation and compilation into a finite automaton is extremely fast and therefore does not present any practical obstacle.

We also applied our approximation algorithm to a weighted grammar of about $25,000$ rules used for translation at AT&T. The transformed grammar had about $36,000$ rules. The whole approximation process including the creation of a finite automaton accepting that grammar took about one minute using the same algorithm on the same machine.

6.6 Conclusion

We presented an algorithm for approximating context-free languages with regular languages. The algorithm was shown to be efficient. Experiments with several grammars showed that the size of the resulting automata is practical for use in many applications.

References

Aho, A. V. & Ullman, J. D. (1973). *The Theory of Parsing, Translation and Compiling*, Prentice-Hall.

Carroll, J. A. (1993). Practical unification-based parsing of natural language, *Technical Report No. 314*, University of Cambridge, Computer Laboratory, England. PhD thesis.

Chomsky, N. (1959). On certain formal properties of grammars, *Information and Control* **2**: 137–167.

Grimley Evans, E. (1997). Approximating context-free grammars with a finite-state calculus, *35th Annual Meeting of the ACL*, pp. 452–459.

Hopcroft, J. E. & Ullman, J. D. (1979). *Introduction to Automata Theory, Languages, and Computation*, Addison-Wesley.

Johnson, M. (1998). Finite-state approximation of constraint-based grammars using left-corner grammar transforms, *36th Annual Meeting of the ACL and 17th International Conference on Computational Linguistics*, Vol. 1, pp. 619–623.

Mohri, M. (2000). Weighted grammar tools: The GRM library, *in* J.-C. Junqua & G. van Noord (eds), *Robustness in Language and Speech Technology*, Kluwer Academic Publishers. This volume.

Mohri, M. & Pereira, F. C. N. (1998). Dynamic compilation of weighted context-free grammars, *36th Annual Meeting of the ACL and 17th International Conference on Computational Linguistics*, Vol. 2, pp. 891–897.

Nederhof, M.-J. (1997). Regular approximations of CFLs: A grammatical view, *International Workshop on Parsing Technologies*, Massachusetts Institute of Technology, pp. 159–170.

Nederhof, M.-J. (1998). Context-free parsing through regular approximation, *Proceedings of the International Workshop on Finite State Methods in Natural Language Processing*, Ankara, Turkey, pp. 13–24.

Nederhof, M.-J. (2000). Practical experiments with regular approximation of context-free languages, *Computational Linguistics* **26**(1): 17–44.

Pereira, F. C. N. & Wright, R. N. (1997). Finite-state approximation of phrase-structure grammars, *in* E. Roche & Y. Schabes (eds), *Finite-State Language Processing*, MIT Press, pp. 149–173.

Schoorl, J. J. & Belder, S. (1990). Computational linguistics at Delft: A status report, *Report WTM/TT 90–09*, Delft University of Technology, Applied Linguistics Unit.

Ullian, J. S. (1967). Partial algorithm problems for context free languages, *Information and Control* **11**: 80–101.

Chapter 7

WEIGHTED GRAMMAR TOOLS: THE GRM LIBRARY

Mehryar Mohri
AT&T Labs — Research
180 Park Avenue, Rm E147
Florham Park, NJ 07932, USA
mohri@research.att.com

Abstract We describe the algorithmic and software design principles of a general grammar library designed for use in spoken-dialogue systems, speech synthesis, and other speech processing applications. The library is a set of general-purpose software tools for constructing and modifying weighted finite-state acceptors and transducers representing grammars.

The tools can be used in particular to compile weighted context-dependent rewrite rules into weighted finite-state transducers, read and compile, when possible, weighted context-free grammars into weighted automata, and dynamically modify the compiled grammar automata. The dynamic modifications allowed include: grammar switching, dynamic modification of rules, dynamic activation or non-activation of rules, and the use of dynamic lists. Access to these features is essential in spoken-dialogue applications.

7.1 Motivation

We describe the algorithmic and software design principles of a general grammar library (GRM library) designed for use in spoken-dialogue systems, speech synthesis, and other speech processing applications.

Most grammars used in speech processing applications are *weighted*. The use of grammar weights is crucial for building robust spoken-dialogue systems. In speech recognition, grammar weights are combined with acoustic weights to rank different hypotheses for a given speech utterance (Rabiner & Juang 1993, Jelinek 1998). Due to the high variability of the input speech signal, a robust grammar admits any hypothesis as a possible transcription with a certain degree of probability.

J.-C. Junqua and G. van Noord (eds.), Robustness in Languages and Speech Technology, 165–186.

Similarly, in speech synthesis (TTS) applications, weights are used to produce a single output even in presence of ambiguities. Grammar weights, weights of context-dependent morphological and phonemic rules, and other acoustic weights such as target and join costs in concatenative TTS systems (Beutnagel, Mohri & Riley 1999), are used to select the most plausible output speech signal. Thus, one crucial requirement for creating a grammar library for robust speech processing applications is that it allows one to create and use *weighted* rules and grammars.

The original goal of the GRM library was to provide algorithms and representations for constructing and using dynamic grammars in very large-vocabulary speech recognition. This led to the following requirements: generality, to support the representation and use of the various grammars in dynamic speech recognition, and efficiency, to support competitive large-vocabulary dynamic recognition using automata of more than 1M states and transitions.

Generality led us to define the most extended algebraic framework for each algorithm [1] as in a general-purpose finite-state machine library (FSM library) (Mohri, Pereira & Riley 2000). The weight set for the rules and grammars used in the GRM library can be an arbitrary semiring. [2] The implementation given for each algorithm is very general: it does not depend on the underlying semiring used. A general implementation also helps avoiding redundancy and reducing the amount of code.

The principle of efficiency led us to design and implement the best known algorithm for each problem, and to carefully reduce the memory requirement of each implementation. This was achieved in particular by devising dynamic or on-the-fly algorithms such as in our compilation of context-free grammars. Only the necessary part of the resulting weighted automaton is constructed or expanded for the specific input sequence. We present and illustrate in more detail the application of these principles in the design of this library.

7.2 Overview

The GRM library is a set of general-purpose software tools for constructing and modifying weighted finite-state acceptors and transducers representing grammars.

The tools can be used in particular to: compile weighted context-dependent rewrite rules into weighted finite-state transducers, read and compile, when possible, weighted context-free grammars into weighted automata, and dynamically modify the compiled grammar automata. The dynamic modifications allowed include: grammar switching, dynamic modification of rules,

1. See Mohri (1998) for the definition of these frameworks and corresponding generic algorithms for shortest-distance problems.
2. A *semiring* is a ring that may lack negation.

dynamic activation or non-activation of rules, and the use of dynamic lists. Access to these features is essential in spoken-dialogue applications.

The GRM library is a high-level library based on the FSM library (Mohri et al. 2000) of which it is using some of the functions. The format of the automata and transducers used in the GRM library is compatible with that of the FSM library. This allows one to use numerous utilities available with the FSM library to manipulate the machines created by the GRM library. In particular, thanks to this compatibility, one can use general algorithms, such as ϵ-removal, ϵ-normalization, determinization, and minimization of weighted transducers to optimize the result of the compilation of weighted rewrite rules, or that of weighted context-free grammars.

The GRM tools are accessible through three levels.

User-program level

The GRM user-program level consists of a set of programs that manipulate grammars by reading from and writing to files or pipelines, and the GRM textual file format specification. For example, `grmcdcompile` returns a weighted finite-state transducer that represents the context-dependent rewrite rule given by the input FSMs [3].

C library level

The GRM C library level consists of a library archive of C functions, which implement the user program level. Typically, a program from the user-level (e.g., `grmcdcompile`) is implemented by a call to a corresponding library function (e.g., `GRMCdCompile`), preceded by a GRM or FSM library function call to read input rules (`GRMRead`) or an input FSM (`FSMLoad`) into an internal object, and followed by the FSM library function call to write out the resultant FSM (`FSMDump`).

Definition level

The GRM definition level consists of the definition level of the FSM library as well as other low-level definitions of some FSMClass objects defined in the GRM library. Accessing this level is important if one wishes to know the finer points or extend the internals of the GRM library.

7.3 Compilation of weighted rewrite rules

The library includes a complete set of routines and utilities to compile *weighted context-dependent rules* into weighted finite-state transducers.

3. We will sometimes use the term FSM to refer to finite-state machines given in the format of the FSM library.

7.3.1 Weighted context-dependent rules

Context-dependent rewrite rules are used in many areas of text and speech processing, including syntax, morphology, and phonology. As shown by Johnson (1972), rewrite rules can be represented by finite-state transducers, under the condition that no rule be allowed to apply any more than a finite number of times to its own output. Kaplan & Kay (1994), or equivalently Karttunen (1995), presented an algorithm for compiling rewrite rules into finite-state transducers, under the equivalent condition that they do not rewrite their noncontextual part. We will also assume that condition in what follows.

In some applications such as those related to speech processing, in particular in speech synthesis, the input may allow many possible analyses. One can then use weights to assign some likelihood or uncertainty measure to each of these analyses. Such generalized rules are called *weighted context-dependent rules*. More formally, weighted context-dependent rewrite rules have the following general form:

$$\phi \rightarrow \psi / \lambda \underline{\quad} \rho$$

They can be interpreted in the following way: ϕ is to be replaced by ψ whenever it is preceded by λ and followed by ρ. λ and ρ represent the left and right contexts of application of the rules. The rewriting condition mentioned previously means that after a replacement of ϕ by ψ, ψ can still serve as a context for further applications of the rule, but it cannot be rewritten.

In general, ϕ, λ and ρ are all regular expressions over the alphabet of the rules, and ψ a weighted regular expression, or a rational power series (Berstel & Reutenauer 1988, Salomaa & Soittola 1978). The weight set for these power series can be any semiring $(\mathbb{K}, \oplus, \otimes, \overline{0}, \overline{1})$. In most cases, such as those considered in speech synthesis applications, the weights are interpreted as $-\log$ probabilities and a Viterbi approximation is assumed. The appropriate semiring $\mathbb{K}$ is then the tropical semiring $\mathcal{T} = (\mathbb{R}_+ \cup \{\infty\}, \min, +, \infty, 0)$.

Context-dependent rules can be used both in an optional and an obligatory mode depending on whether any occurrence of ϕ in the appropriate context is replaced by ψ (obligatory case) or not (optional case).

A rule can be applied from left to right, right to left, or in a simultaneous way. When applied from left to right, each replacement applies to the right of the output of the previous replacement. Similarly, in the case of a right-to-left application, it applies to the left of the output of the previous replacement. In the simultaneous case, the contexts of application of the rule are directly determined in the original string before any replacement.

As an example, the following weighted rule:

$$aa \rightarrow .25\ ba / a \underline{\quad} \epsilon$$

replaces each sequence of aa into ba with weight .25, when aa has left context a and right context ϵ. The effect of the rule on various strings is illustrated by

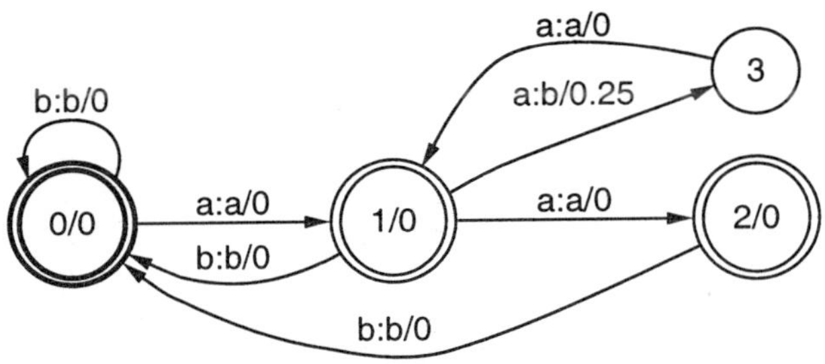

FIGURE 7.1: Weighted finite-state transducer realizing the weighted left-to-right obligatory rule $aa \rightarrow .25ba/a\_\_\epsilon$. The transducer is computed by the shell command: `grmcdcompile -i LAB AA BB A EPS`, where `AA`, `BB`, `A`, and `EPS` are automata representing aa, $.25\ ba$, a, and ϵ, and `LAB` the alphabet $\Sigma = \{a, b\}$. It has been further ϵ-removed and compacted using the utilities of the FSM library.

the following when it is interpreted as a left-to-right obligatory rule:

$$\begin{aligned} aaa &\rightarrow .25\,aba \\ aaaa &\rightarrow .25\,abaa \\ aaaaa &\rightarrow .50\,ababa \end{aligned}$$

Each application of the rule contributes weight .25 to the weight of the resulting string. By comparison, a right-to-left obligatory application of the rule gives:

$$\begin{aligned} aaa &\rightarrow .25\,aba \\ aaaa &\rightarrow .25\,aaba \\ aaaaa &\rightarrow .50\,ababa \end{aligned}$$

7.3.2 GRM utilities and illustration

Figure 7.1 shows the *weighted transducer* representing this rule when interpreted as left-to-right and obligatory. A weighted transducer is a labeled directed graph in which each transition is labeled with an input symbol, an output symbol and a weight in $\mathbb{K}$. A successful path is a path from an initial state (states in bold in the figure) to a final state (states represented by double circles in the figure). A successful path in a transducer pairs the concatenation of the input labels on its transitions with that of the corresponding output labels, assigning the pair the $\otimes$-product of the transition weights. The weight associated to a pair of input and output strings is obtained by taking the $\oplus$-sum of the weights of all successful paths labeled with that pair.

The transducer of figure 7.1 is obtained using the GRM utility `grmcdcompile` which compiles rules into weighted transducers. The syntax of the command is similar to the definition of the rules. `grmcdcompile` takes as

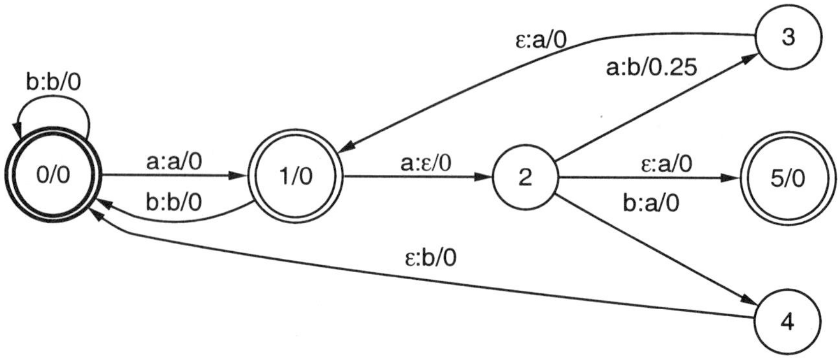

FIGURE 7.2: Subsequential transducer realizing the weighted left-to-right obligatory rule $aa \rightarrow .25\ ba/a\_\_\epsilon$ obtained from the transducer of figure 7.1 using the shell commands `fsmdeterminize`, `fsmpush` and `fsmcompact`. The transducer is subsequential and minimal with ϵ is considered as a regular symbol.

input four FSMs: f_1, f_2, f_3 and f_4, and returns a (weighted) transducer representing the context-dependent rewrite rule:

$$f_1 \rightarrow f_2/f_3\_\_f_4$$

The four input FSMs f_1, f_2, f_3 and f_4 must be acceptors. f_1, f_3 and f_4 must be costless. The rule expresses the fact that a string w recognized by f_1 is replaced by the (weighted) strings recognized by f_2 in the context where w is preceded by a string recognized by f_3 and followed by a string recognized by f_4.

The application of the rule proceeds from left to right by default or when the option `-l` is specified. It proceeds from right to left if `-r` is specified, and applies in parallel or simultaneously if the option `-s` is specified. If more than one option among these three is specified, all except the last one are ignored. The application of the rules is obligatory unless the option `-o` is specified, in which case the replacement is optional. The alphabet of the set of symbols of the rule is specified with the option `-i` symbols, which can be used to provide textual names for symbols. This feature can be used for instance to specify the initial non-terminal symbol with option `-s` using its textual name, S for instance, rather than the corresponding number associated to the non-terminal S.

Since the format of the finite-state machines produced by the GRM library is compatible with the FSM library, one can directly apply optimization algorithms such as weighted determinizaton and minimization to the output of `grmcdcompile`. The resulting minimal subsequential transducer is shown in figure 7.2. In some applications, it may be crucial to use the optimized transducer representing a weighted context-dependent rule both because of its efficiency of use and to save space.

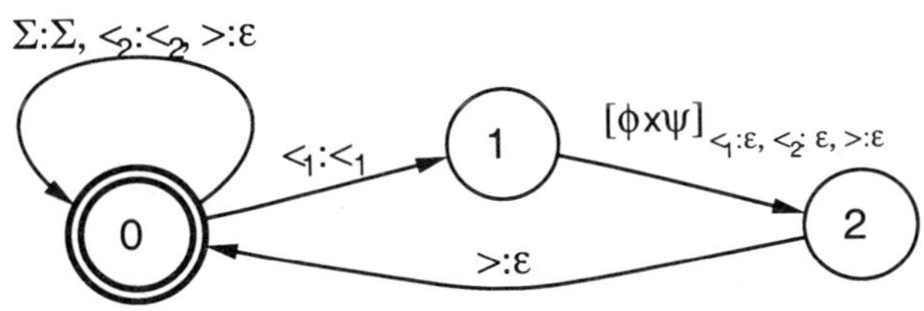

FIGURE 7.3: The construction of the transducer *replace*.

7.3.3 Algorithm

This section briefly presents and illustrates the compilation algorithm for weighted context-dependent rules implemented in the GRM library. We refer the reader to Mohri & Sproat (1996) for a more detailed description of the algorithm.

The algorithm is based on the use of *marking transducers*, that is transducers that introduce or delete auxiliary symbols marking either the beginning or the end of sequences described by a regular expression. These transducers are used to introduce auxiliary symbols only when necessary as opposed to the algorithm of Kaplan & Kay (1994) (KK for short) which introduces brackets everywhere only to restrict their occurrence subsequently. This makes intermediate results in the compilation of the rules easier to read and check. The number of intermediate transducers necessary in the construction of the rules is also smaller than in the KK algorithm, and each of the transducers can be constructed more directly and efficiently from the primitive expressions of the rule, ϕ, ψ, λ, ρ.

The construction of the marking transducers consists of the determinizations of three automata and other work linear in the size of those automata. The algorithm requires fewer compositions than the KK algorithm. On-the-fly composition of weighted transducers (Mohri, Pereira & Riley 1996) can be used to construct only the part of the compilation result needed for the specific input sequence considered. The algorithm was shown to be more efficient than the KK algorithm (Mohri et al. 1996).

For the sake of simplicity, we illustrate the algorithm in the case of an obligatory left-to-right rule:

$$\phi \rightarrow \psi / \lambda \_\_ \rho$$

Other combinations (optional, right-to-left, or simultaneous) can be easily derived from this case.

The following composition cascade gives a brief description of the compilation algorithm:

$$T = r \circ f \circ \mathit{replace} \circ l_1 \circ l_2$$

where r, f, *replace*, l_1, and l_2 are weighted transducers defined as follows. Transducer r is a marking transducer introducing an auxiliary left bracket symbol $>$ ($>$ does not belong to the alphabet Σ) before each occurrence of a se-

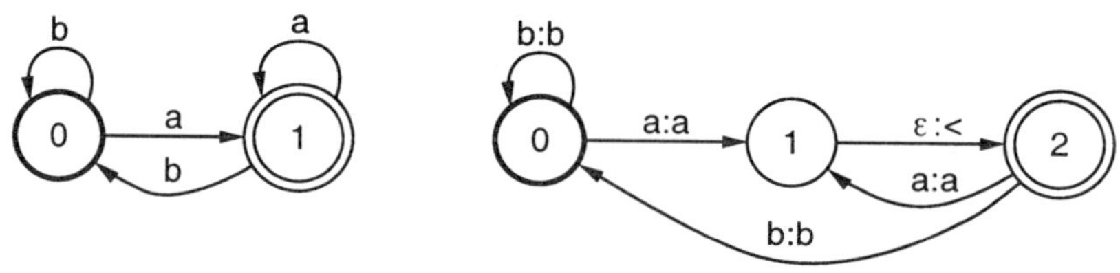

(A) (B)

FIGURE 7.4: (A) Minimal deterministic automaton representing $\alpha = (a+b)^*a$. (B) Transducer B post-marking by $<$ each occurrence of a string described by α.

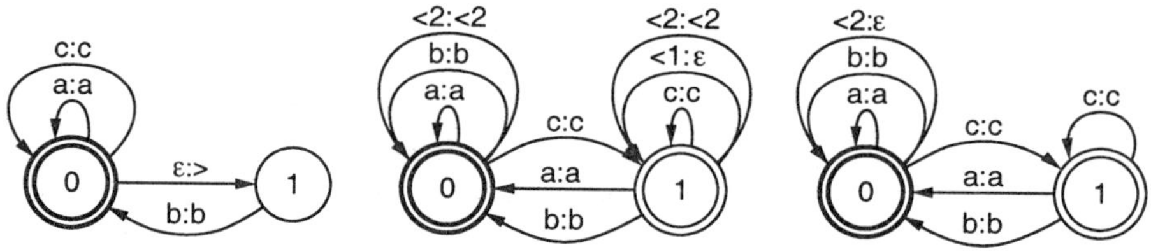

FIGURE 7.5: Marking transducers r, l_1, and l_2.

quence accepted by the regular expression ρ in the input. r can be described by the following left-to-right obligatory context-dependent rule:

$$r : \epsilon \rightarrow > /\Sigma^* \_\_\rho$$

The transducer f is a marking transducer introducing auxiliary right bracket symbols $<_1$ and $<_2$ before each occurrence of a sequence accepted by ϕ and followed by $>$ in a string over the alphabet $\Sigma \cup \{>\}$ with the auxiliary symbol $>$ freely appearing at any position ($>$ is ignored for matching purposes). The corresponding transduction is given by the left-to-right obligatory context-dependent rule over the alphabet $\Sigma \cup \{>\}$:

$$f : \epsilon \rightarrow (\{<_1\} \cup \{<_2\})/(\Sigma \cup \{>\})^* \_\_\phi_> >$$

where $\phi_>$ is the regular expression obtained from ϕ by freely allowing the symbol $>$ anywhere in ϕ. The use of $\phi_>$ instead of ϕ is necessary since $>$ should be ignored in the determination of the occurrences of ϕ.

Thus $r \circ f$ transforms an input string by introducing brackets (left bracket $<_1$ and $<_2$ and right bracket $>$) around sequences accepted by ϕ that are followed by a sequence accepted by ρ. The transducer *replace* replaces each of such ϕ sequences by all weighted sequences given by the rational power series ψ, additionally erasing the right bracket $>$. Figure 7.3 illustrates the construction of *replace*. The symbols $>$, $<_1$, and $<_2$ need to be ignored when determining an occurrence of ϕ, thus there are loops over the transitions

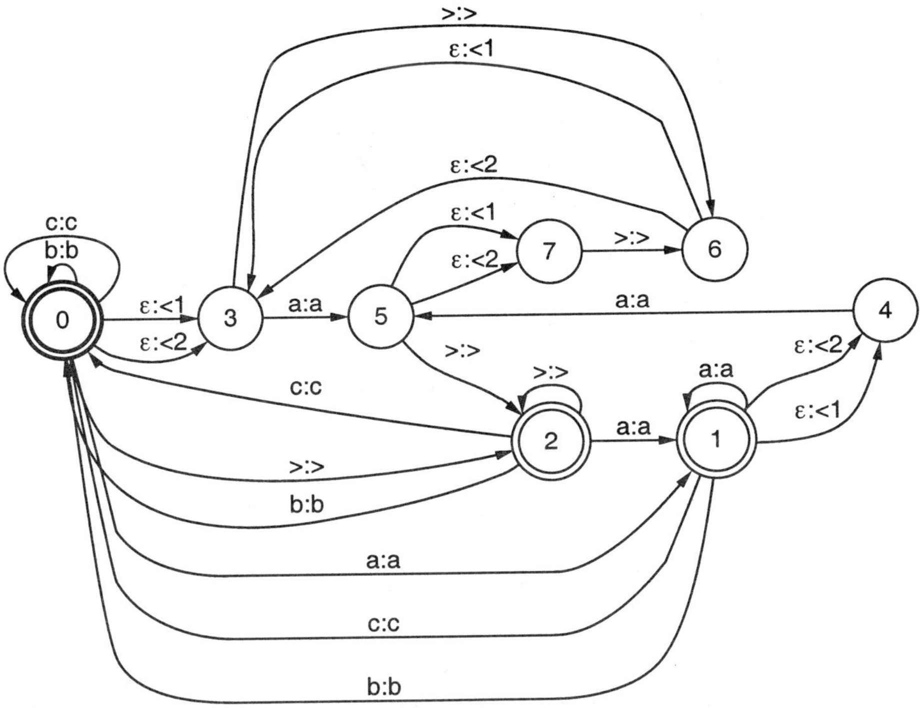

FIGURE 7.6: Transducer f.

$>: \epsilon, <_1: \epsilon, <_2: \epsilon$ at all states of the automaton accepting ϕ, or equivalently of the states of transducer realizing the cross product $\phi \times \psi$.

l_1 is a marking transducer that accepts only strings where a sequence accepted by λ is followed by $<_1$ and it erases that symbol in those strings. l_1 can be described by the context-dependent rule:

$$l_1 : <_1 \rightarrow \epsilon / \# \Sigma^* \lambda \_\_ \epsilon$$

where # denotes the beginning of the input sequence, with the symbol $<_2$ ignored everywhere in the input.

Similarly, the transducer l_2 is a marking transducer that only accepts strings where $<_2$ is preceded by a sequence not accepted by λ, namely a sequence accepted by the complement of $\Sigma^*\lambda$ denoted by $\overline{\Sigma^*\lambda}$, and erases that auxiliary symbol in those strings:

$$l_2 : <_2 \rightarrow \epsilon / \# \overline{\Sigma^* \lambda} \_\_ \epsilon$$

Thus $l_1 \circ l_2$ accepts only those strings where $<_1$ follows a sequence recognized by λ and $<_2$ a sequence not recognized by λ and erases these auxiliary symbols.

It is clear from the definition of these transducers that T realizes the weighted transduction defined by the weighted context-dependent rule. Note

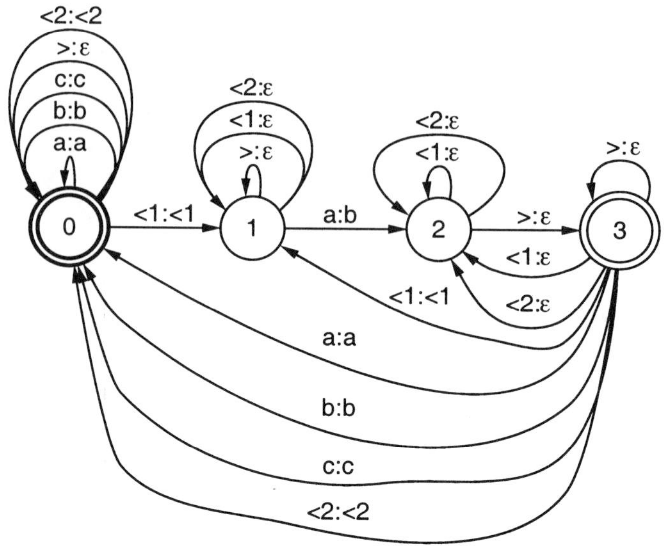

FIGURE 7.7: Transducer *replace*.

that the weighted transducer *replace* is independent of the left and right contexts λ and ρ, and thus can be computed independently. The construction of a marking transducer for a regular expression is based on a deterministic automaton representing that expression. We illustrate this by describing the construction of a marking transducer corresponding to the transduction:

$$\Sigma^*\alpha \to \Sigma^*\alpha <$$

where α is an arbitrary regular expression over the alphabet Σ. Let $A = (\Sigma, Q, I, F, E)$ be a deterministic automaton recognizing α, where Σ is the alphabet, Q the state set, $I \subseteq Q$ the set of initial states, $F \subseteq Q$ the set of final states and E the finite set of transitions of A.

A can be viewed as a transducer if we assign to each transition an output label equal to the input label. We can modify the corresponding transducer so that it inserts the symbol $<$ after each occurrence of a string accepted by α as follows:

1. all states in $Q - F$ are made final,
2. for each state state $q \in F$
   (a) a new state q' copy of q is created (q' has the same transitions as q, and q' is a final state)
   (b) q is made non-final,
   (c) outgoing transitions of q are removed, and replaced by a single transition from q to q' with input label ϵ and output $<$.

Figure 7.4 illustrates the construction of the resulting transducer B for the specific case of $\alpha = (a+b)^*a$. Figures 7.5, 7.6, and 7.7 give explicitly the transduc-

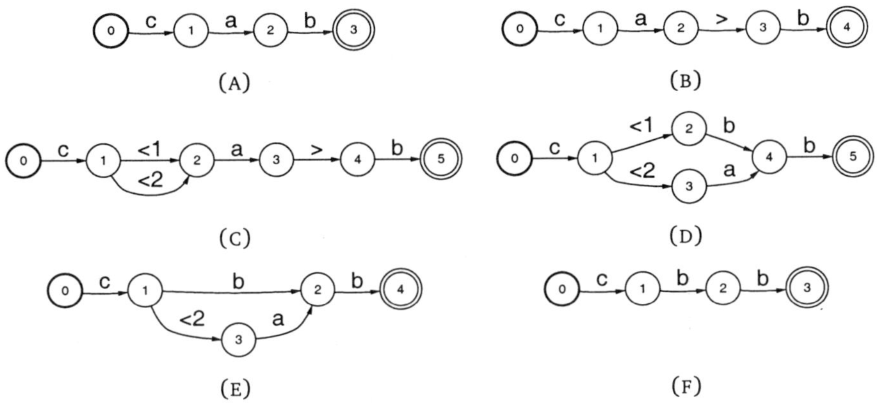

FIGURE 7.8: Intermediate results after application of each transducer of the rule $a \to b/c\_\_b$ to the input string *cab*. (A) Input automaton recognizing *cab*. (B) Result of the application of the transducer r. (C) Result after application of f. (D) Result after application of *replace*. (E) Result after application of l_1. (F) Final output resulting from the application of l_2.

ers used in the compilation of the rule $a \to b/c\_\_b$. The successive applications of these transducers to the input string *cab* are illustrated by figures 7.8 (a)–(f).

7.4 Compilation of weighted context-free grammars

The GRM library includes a general set of algorithms and utilities to compile weighted context-free grammars. The compilation algorithm can be executed on demand, as needed for the recognition of the actual input utterances, and the grammars can be dynamically modified.

7.4.1 Weighted context-free grammars

Context-free grammars (CFGs) are widely used in text and speech processing applications. In some applications such as speech recognition, it is necessary to use *weighted* CFGs by assigning some weights to grammar rules. These weights, often interpreted as probabilities, are then combined with acoustic weights to rank recognition transcriptions.[4]

The weight set of the weighted context-free grammar used can be any semiring $(\mathbb{K}, \oplus, \otimes, \overline{0}, \overline{1})$. In most cases, such as those considered in speech recognition applications, the weights are interpreted as $-\log$ probabilities and a Viterbi approximation is assumed. The appropriate semiring $\mathbb{K}$ is then the tropical semiring $\mathcal{T} = (\mathbb{R}_+ \cup \{\infty\}, \min, +, \infty, 0)$. For example, the following weighted CFG:

4. See Bahl, Jelinek & Mercer (1983) and Mohri & Riley (1999) for a full description of the way the grammar or language model can be combined with other components of a recognition system – pronunciation dictionary, phonemic context-dependency model, acoustic model.

$$\begin{array}{rcl} Z\ .1 & \rightarrow & XY \\ X\ .2 & \rightarrow & aY \\ Y\ .3 & \rightarrow & bX \\ Y\ .4 & \rightarrow & c \end{array}$$

FIGURE 7.9: Simple CFG grammar G.

assigns weight .1 to the rule $Z \rightarrow XY$, and weight .2 to $X \rightarrow aY$.

A weighted CFG $G = (V, P)$ over the alphabet Σ, with weights in a semiring $(\mathbb{K}, \oplus, \otimes, \overline{0}, \overline{1})$ consists of a finite alphabet V of variables or nonterminals disjoint from Σ, and a finite set $P \subseteq V \times \mathbb{K} \times (V \cup \Sigma)^*$ of productions or derivation rules (Autebert, Berstel & Boasson 1997). Given strings $u, v \in (V \cup \Sigma)^*$, and weights c and c', we write $(u, c) \stackrel{*}{\rightarrow} (v, c')$ when there is a derivation from u with weight c to v with weight c'. We denote by $L_G(X)$ the weighted language generated by a nonterminal X:

$$L_G(X) = \{(w, c) \in \Sigma^* \times \mathbb{K} : (X, 0) \stackrel{*}{\rightarrow} (w, c)\}$$

Context-free grammars used in spoken-dialogue applications often represent regular languages (Church 1983, Brown & Buntschuh 1994), either by construction or as a result of a finite-state approximation of a more general CFG (Pereira & Wright 1991, Nederhof 1998, Mohri & Nederhof 2000). This is because in most real-time applications, general CFGs are computationally too demanding.

The GRM library provides general tools for compiling a weighted CFG into a weighted finite automaton under certain general conditions described below. In addition, our compilation method allows the recognizer to dynamically *activate* or *deactivate* grammar rules and *substitute* a new regular language for some terminal symbols, depending on previously recognized inputs, all without recompilation. This is the terminology used in this field, a more appropriate expression would be dynamic activation or deactivation of *nonterminal symbols*. The following paragraphs define more precisely these dynamic modifications.

Dynamic activation or deactivation of rules

We define a set of *active nonterminals* which are the start symbols for derivations. Let $A \subseteq V$ be the set of active nonterminals, then the language generated by G is:

$$L_G = \bigcup_{X \in A} L_G(X)$$

Note that inactive nonterminals, and the rules involving them, can be used in derivations, they are just not available as start symbols. Dynamic rule activa-

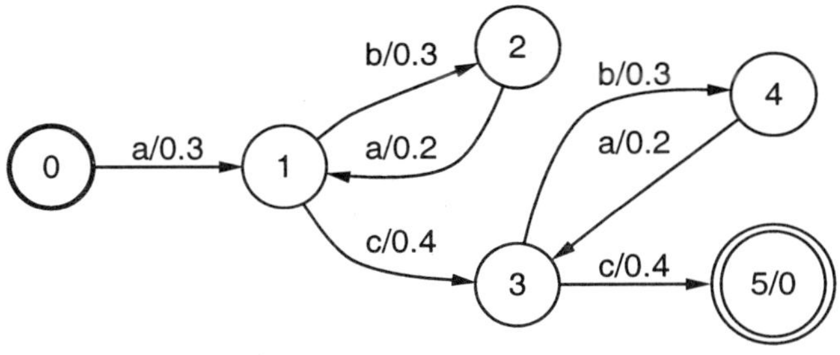

FIGURE 7.10: Weighted finite automaton representing the weighted CFG G (figure 7.9). The automaton is computed by the shell command:

```
grmread -i LAB -w R | grmcfcompile -i LAB -s Z
```

where LAB is the alphabet $\Sigma = \{a, b, c, X, Y, Z\}$, R the textual representation of the CFG. It has been further ϵ-removed and determinized and minimized using the utilities of the FSM library.

tion or deactivation consists exactly of the dynamic redefinition of the set A in successive uses of the grammar.

Dynamic substitution

Let σ be a weighted rational transduction of Σ^* to $\Delta^* \times \mathbb{K}$, $\Sigma \subseteq \Delta$, that is a regular weighted substitution (Berstel 1979). σ is a monoid morphism verifying:

$$\forall x \in \Sigma, \sigma(x) \subseteq Reg(\Delta^* \times \mathbb{K})$$

where $Reg(\Delta^* \times \mathbb{K})$ denotes the set of weighted regular languages over the alphabet Δ. Thus σ simply substitutes for each symbol $a \in \Sigma$ a weighted regular expression $\sigma(a)$. A dynamic substitution consists of the application of the substitution σ to Σ, during the process of recognition of a word sequence. Thus, after substitution, the language generated by the new grammar G' is:[5]

$$L_{G'} = \sigma(L_G)$$

The GRM compilation algorithm allows for both of those dynamic grammar changes without recompiling the grammar.

7.4.2 GRM utilities and illustration

The textual representation R of the grammar G of figure 7.9 in the GRM library is simply:

```
Z 1 X Y
X 2 a Y
Y 3 b X
```

5. σ can be extended as usual to map $\Sigma^* \times \mathbb{K}$ to $Reg(\Delta^* \times \mathbb{K})$.

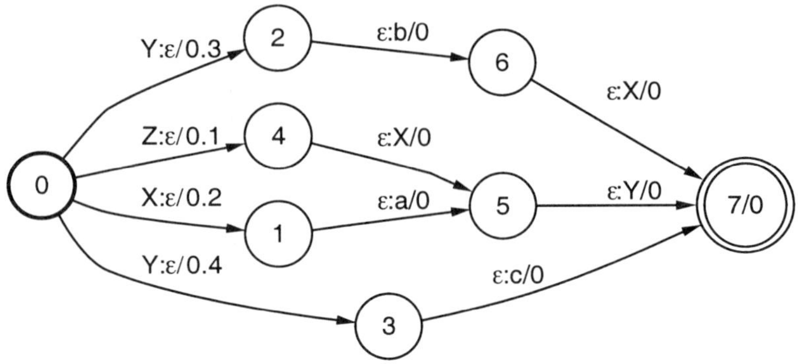

FIGURE 7.11: Weighted transducer output of `grmread`.

`Y 4 c`

Figure 7.10 shows the weighted automaton result of the compilation of the weighted CFG G with start symbol Z using the GRM utilities `grmread` and `grmcfcompile`.

`grmread` takes as input a textual representation of a weighted context-free grammar from file or standard input, and sends to standard output the binary encoding of a weighted transducer representing the grammar. The grammar can be represented using compound rules (by factorization of the left-hand side of the rules), if the option `-c` is used. For example, the following is a more compact representation of the grammar G:

```
Z 1 X Y
X 2 a Y
Y 3 b X | 4 c
```

Input symbols are represented in the input file by non-negative numbers, unless the option `-i` is used. This option allows non-terminal and terminal symbols to be given textual names, where the file symbols gives the translation from names to numbers. The input must be an unweighted context-free grammar, unless the option `-w` is given, in which case it must be a weighted context-free grammar.

The representation of the grammar as a weighted transducer is crucial because it can then be optimized (compacted, or determinized and minimized (Mohri 1997, Mohri 2000)). Such a preprocessing dramatically improves the time and space required by the main compilation step thanks to increased sharing among grammar rules that start or end the same way (Mohri & Pereira 1998). Figure 7.11 shows the output of the `grmread` for the grammar G after compaction.

`grmcfcompile` accepts as input a weighted transducer representing a context-free grammar and returns a delayed weighted acceptor recognizing that grammar. The input transducer must represent a context-free grammar for a weighted regular language. More precisely, the restriction on the rules of the context-free grammar is the following: the rules of a set of non-terminals that are mutually dependent (a set $\{X, Y\}$ for instance, with X depending on Y and Y depending on X) must be either right-linear or left-linear with respect to the elements of this set. For example, the following grammar:

$$\begin{array}{lcl} X & \rightarrow & aRY \\ Y & \rightarrow & bTVX \\ R & \rightarrow & cVTd \\ V & \rightarrow & TV \\ Y & \rightarrow & e \\ T & \rightarrow & f \\ V & \rightarrow & g \end{array}$$

is well-formed. The non-terminals X and Y depend on each other, and the rules defining X and Y are right-linear with respect to X and Y. The grammar below:

$$\begin{array}{lcl} X & \rightarrow & aRY \\ Y & \rightarrow & bTVX \\ R & \rightarrow & cVTd \\ V & \rightarrow & dX \\ Y & \rightarrow & e \\ T & \rightarrow & f \\ V & \rightarrow & g \end{array}$$

is not well-formed because X, Y, R and V depend on each other, but the rule $X \rightarrow aRY$ for instance, is neither left-linear nor right-linear with respect to these symbols.

The start symbol of the grammar can be specified using the option `-s`. The option `-i` is used to provide a textual name for the non-terminal symbol. This option is also used for error messages. The list of the non-terminals that are problematic is printed out when the input grammar is not well-formed.

When the grammar is made of only right-linear rules, or only left-linear rules, a more efficient compilation algorithm can be used. The option `-l` of `grmcfcompile` can be used for that purpose.

The object returned by `grmcfcompile` is a *delayed* acceptor which admits any FSM operation valid for acceptors. Only the necessary part of that acceptor (states and transitions) is expanded when using an FSM operation.

The delayed objects created by the GRM library are based on two FSM classes: the Edit class and the Replace class. [6] These are essential features that allow one to dynamically modify the input grammar.

---

6. In particular, the acceptor returned by `grmcfcompile` belongs to the Replace class.

Edit class

The Edit class is an FSM class defined by the GRM library. It allows users to dynamically modify the initial state of an FSM or the final cost of a state of an FSM. An Edit class object represents an FSM F defined by an input FSM S and a new initial state s, a state f, and a final cost c.

The FSM F is exactly the machine obtained from S by the following delayed operations: the state s becomes the new initial state, and the final cost c is assigned to f.

Replace class

The Replace class is an FSM class defined by the GRM library. It allows users to construct FSMs with labels referring to other FSMs. A Replace class object represents an FSM F defined by an FSM S that gives the general structure of F and an array of FSMs, $T[i]$, indexed with the set of all symbols i including the input labels of S.

The FSM F is exactly the result of the replacement by $T[i]$ of each label i of S for which $T[i]$ is defined. These replacements are done on-demand for a given input. In the case of Replace class acceptors representing context-free grammars, typically $T[i]$ is defined for each i corresponding to a non-terminal symbol, and S is an acceptor accepting exactly the set of initial non-terminals of the grammar.

Dynamic modifications

Dynamic modifications of the input grammar are essential in many spoken-dialogue systems. Indeed, those systems require grammars or language models to evolve as the dialogue proceeds, to exploit the information and context provided by previous interactions.

For instance, a previous request for an address would activate the address grammar and lexicon and inactivate the date grammar and lexicon in an automated reservations task. The use of such dynamic grammars can significantly improve the efficiency and accuracy of the system by considerably restricting the set of possible transcriptions.

The result of the compilation of CFG can be directly modified at the command level by the following dynamic modification utilities: `grmreplace`, `grmmutate`, and `grmsubstitute`. Note that these modifications do not require a recompilation of the grammar. Their results, just as their inputs, are delayed objects of the Replace class. They are possible thanks to the definition of the Replace class.

`grmreplace` takes an input finite-state machine S and returns an FSM F that belongs to the GRM Replace class.

`grmmutate` takes as input two FSMs f_1 and f_2 and returns the FSM result of the replacement in f_2 of the structure FSM by f_1. A typical application of `grmmutate` is the dynamic modification of the initial non-terminal symbols

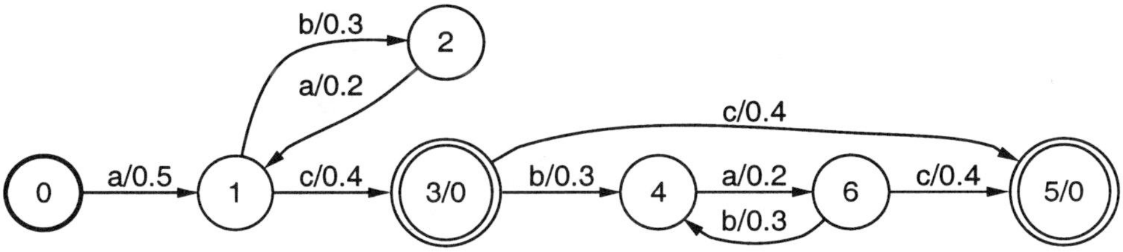

FIGURE 7.12: Dynamically modified grammar G with active non-terminals Z and X and weights .2 and .3. The automaton was generated by the following shell command:

```
grmmutate A F
```

where F is the delayed object output of `grmcfcompile` applied to the grammar G and A the activation acceptor.

of a context-free grammar (also called dynamic activation or inactivation of the rules of a context-free grammar) as defined above. Figure 7.12 illustrates the use of the `grmmutate` and the corresponding output automaton.

`grmsubstitute` takes as input a substitution file and a finite-state machine f and returns the FSM obtained by substituting FSMs for the symbols of f using the mapping defined in the substitution file. A typical application of `grmsubstitute` is the dynamic replacement of the terminal symbols of a context-free grammar (also called use of dynamic lists).

The GRM library has been used successfully to compile and dynamically modify context-free grammars of several hundred thousand rules.

7.4.3 Algorithm

This section briefly presents and illustrates the compilation algorithm for weighted CFGs as implemented in the GRM library. We refer the reader to Mohri & Pereira (1998) for a more detailed description of the algorithm.

The algorithm first gives a transducer representation of the input grammar which can be determinized and minimized to take advantage of more sharing among rules. The transducer representation is then used to:

1. build the dependency graph D_G of the input grammar G, that is the graph that has a node corresponding to each non-terminal and that has an edge from node X to node Y if non-terminal Y appears on the right hand-side of a rule with left hand-side X;
2. compute the strongly connected components of D_G, namely the sets of mutually dependent non-terminals; [7]
3. assuming that the grammar reduced to each SCC is either left-linear or right-linear, a weighted automaton $K(S)$ can be constructed for each

7. Recall that the strongly connected components, SCCs, of a directed graph are the equivalence classes of its nodes under the relation R defined by: $q\ R\ q'$ if q' can be reached from q and q from q' (Tarjan 1972).

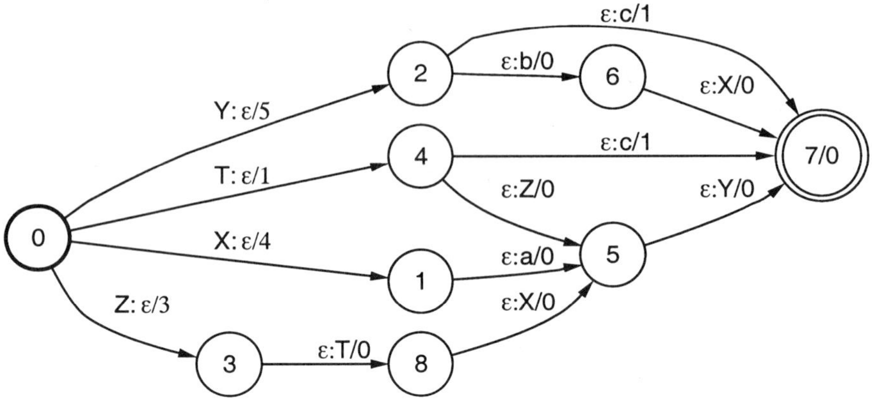

FIGURE 7.13: Compact transducer representation of the grammar G'.

SCC S, and for each non-terminal $X \in S$ a weighted automaton $M(X)$ can be built from $K(S)$; [8]

4. a simple automaton M_G accepting exactly the set of active nonterminals A is constructed;
5. the automaton is then expanded on-the-fly for each input string using lazy replacement and editing.

The consecutive steps of the compilation algorithm are illustrated for the following CFG:

$$
\begin{array}{lcl}
T\,1 & \rightarrow & ZY \\
T\,2 & \rightarrow & c \\
Z\,3 & \rightarrow & TXY \\
X\,4 & \rightarrow & aY \\
Y\,5 & \rightarrow & bX \\
Y\,6 & \rightarrow & c
\end{array}
$$

FIGURE 7.14: Grammar G'.

Figure 7.13 shows the representation of G' as a weighted transducer and the benefit of transducer determinization and minimization for sharing among rules starting or ending in the same way.

Figure 7.15 gives the dependency graph of the grammar G' which has two SCCs, S_1 and S_2.

8. The construction of each $M(X)$ from $K(S)$ is a delayed operation based on the Edit class, since $M(X)$ is derived from $K(S)$ just by picking out the state associated to X in $K(S)$.

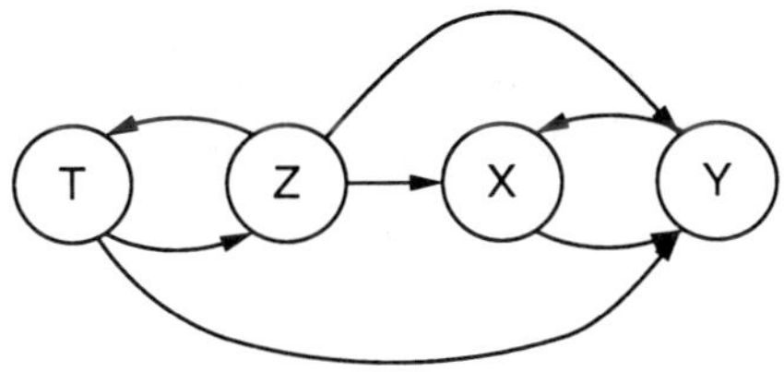

FIGURE 7.15: Dependency graph $D_{G'}$ of the grammar G'. The strongly connected components of $D_{G'}$ are $S_1 = \{T, Z\}$ and $S_2 = \{X, Y\}$.

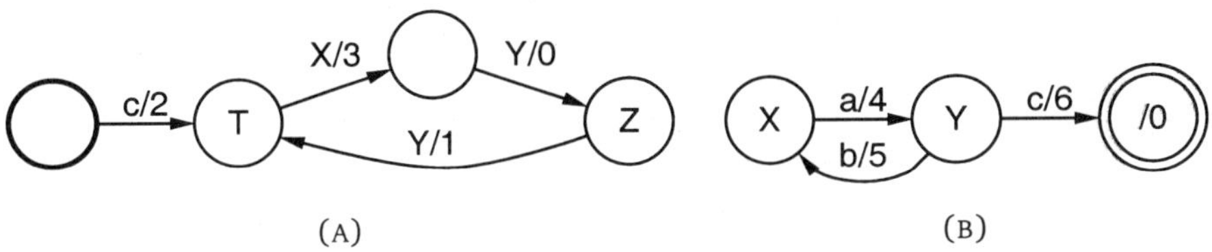

FIGURE 7.16: (A) Weighted automaton $K(S_1)$ for the SCC $S_1 = \{T, Z\}$. (B) Weighted automaton $K(S_2)$ for the SCC $S_2 = \{X, Y\}$.

The construction of automata representing left-linear or right-linear grammars is classical (Aho, Hopcroft & Ullman 1974). That construction can be easily extended to weighted grammars to build the weighted automata corresponding to the SCCs S_1 and S_2, see figure 7.16 (a) and (b).

Our compilation algorithm works only with the class of grammars such that the grammar restricted to each SCC is either left-linear or right-linear. This seems to cover a large set of useful grammars. Since there is no general algorithm for mapping an arbitrary CFG generating a regular language into a corresponding finite automaton (Ullian 1967), the use of a restriction is necessary.

The automaton corresponding to each non-terminal can be easily constructed from $K(S_1)$ and $K(S_2)$ by making final (left-linear case) or initial (right-linear case) the corresponding state of $K(S_1)$ or $K(S_2)$. As mentioned above, in our implementation, this operation is delayed. The automaton $M(T)$ for instance is only constructed and expanded if needed for the specific input string to parse.

The automaton $M_{G'}$ accepts exactly the set of active non-terminals. Figure 7.17 shows that automaton for the specific case of $A = \{T, X\}$. Note that this automaton can be dynamically modified without requiring a recompilation of the grammar (the steps 1-3 of the algorithm do not need to be repeated).

The automaton $M_{G'}$ can then be expanded for each input string using lazy replacement and editing: each non-terminal symbol B is replaced on-the-fly by its corresponding automaton $M(B)$ which is itself computed by a delayed editing operation of the automaton $K(S)$ with $B \in S$. Figure 7.18 shows the

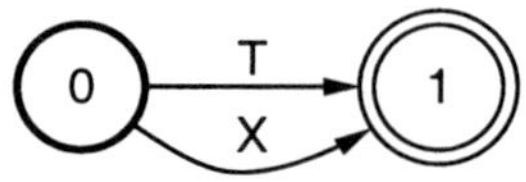

FIGURE 7.17: Automaton $M_{G'}$ accepting the set of active non-terminals $A = \{T, X\}$.

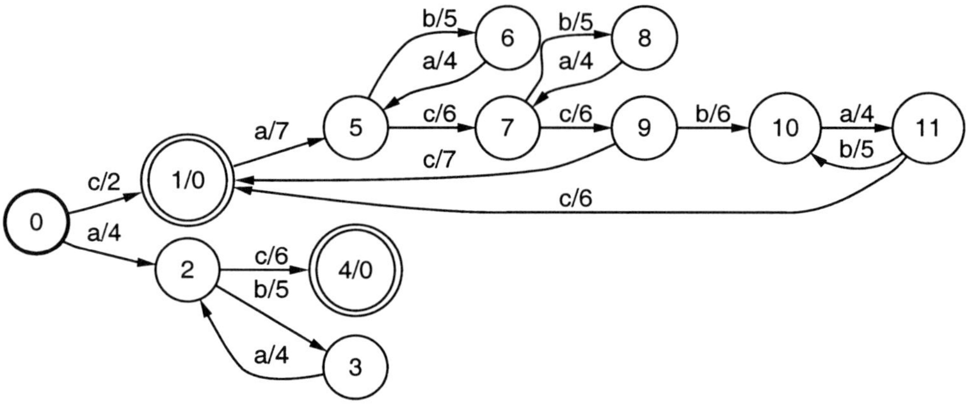

FIGURE 7.18: Automaton $M_{G'}$ after full expansion.

fully expanded automaton $M_{G'}$ after removal of ϵ's, determinization, and minimization.

Our experiments both with full bigram models of various vocabulary sizes, and with unweighted grammars derived by feature instantiation from hand-built feature-based grammars showed the efficiency of the algorithm (Mohri & Pereira 1998).

7.5 Conclusion

We presented a general weighted grammar library designed for use in spoken-dialogue systems, speech synthesis, and other speech processing applications. Since it is based on the notions of semiring and of rational power series, the same code can be used for a variety of different applications. The current version of the library is written in C, with the semiring operations defined in the FSM library.

The library is tested with randomly generated automata. Its efficiency was tested by compiling and dynamically modifying successfully very large grammars of more than several hundred thousand rules.

The GRM library is currently used in many projects and is included in the AT&T product WATSON. It is planned to be made available for academic use in the near future (Mohri 1996). It includes other utilities such as *local determinization* that can be useful in the context of speech recognition projects.

The compilation algorithm and programs for context-dependent rules have been shown to be efficient in several tasks. It is used by many in Lucent Bell Laboratories to develop TTS text analyzers for several languages including Spanish, Italian, French, Romanian, German, Russian, Mandarin and Japanese. It was also used in AT&T Labs – Research to construct efficiently several components of a speech understanding system based on finite-state transducers. The GRM utilities for compiling weighted context-dependent rules can also be used to compile decision trees into weighted transducers (Sproat & Riley 1996) and build general context-dependent model transducers for speech recognition systems.

The compilation algorithms and tools for weighted context-free grammars have been used in various speech processing projects including the construction of spoken dialog systems and speech translation systems at AT&T Labs – Research.

Future versions of the GRM library will include a compiler for rational power series, algorithms for constructing and compiling weighted approximations of weighted context-free grammars (Mohri & Nederhof 2000), and transducer learning algorithms and tools. Because of their generality, the Replace and Edit classes presented above will be directly incorporated in the FSM library in future versions of both libraries.

Acknowledgments

I thank Richard Sproat, Michael Riley, and Fernando Pereira for discussions about some of the utilities of this library, and Ted Roycraft and Bruce Buntschuh for their comments and help with defining the dynamic grammar features. Many of the GRM programs were kindly tested by Richard Sproat, Ted Roycraft, and Bruce Buntschuh in the context of various text and speech processing applications.

References

Aho, A. V., Hopcroft, J. E. & Ullman, J. D. (1974). *The Design and Analysis of Computer Algorithms*, Addison Wesley, Reading, MA.

Autebert, J.-M., Berstel, J. & Boasson, L. (1997). Context-free languages and pushdown automata, *in* G. Rozenberg & A. Salomaa (eds), *Handbook of Formal Languages*, Vol. 1, Springer, pp. 111–172.

Bahl, L. R., Jelinek, F. & Mercer, R. (1983). A maximum likelihood approach to continuous speech recognition, *IEEE Transactions on Pattern Analysis and Machine Intelligence (PAMI)* **5**(2): 179–190.

Berstel, J. (1979). *Transductions and Context-Free Languages*, Teubner Studienbucher, Stuttgart.

Berstel, J. & Reutenauer, C. (1988). *Rational Series and Their Languages*, Springer-Verlag, Berlin, New York.

Beutnagel, M., Mohri, M. & Riley, M. (1999). Rapid unit selection from a large speech corpus for concatenative speech synthesis, *Proceedings of the 6th European Conference on Speech Communication and Technology (Eurospeech '99)*, Budapest, Hungary, pp. 607–610.

Brown, M. K. & Buntschuh, B. M. (1994). A context-free grammar compiler for speech understanding systems, *Proceedings of the International Conference on Spoken Language Processing (ICSLP '94)*, Yokohama, Japan, pp. 21–24.

Church, K. W. (1983). A finite-state parser for use in speech recognition, *21st Meeting of the Association for Computational Linguistics (ACL '83), Proceedings of the Conference*, ACL, pp. 91–97.

Jelinek, F. (1998). *Statistical Methods for Speech Recognition*, The MIT Press, Cambridge, MA.

Johnson, C. D. (1972). *Formal Aspects of Phonological Description*, Mouton, Mouton, The Hague.

Kaplan, R. M. & Kay, M. (1994). Regular models of phonological rule systems, *Computational Linguistics* **20**(3): 331–378.

Karttunen, L. (1995). The replace operator, *33rd Meeting of the Association for Computational Linguistics (ACL 95), Proceedings of the Conference, MIT*, ACL, Cambridge, MA, pp. 16–23.

Mohri, M. (1996). *General-Purpose Weighted Grammar Tools: The GRM Library*, AT&T Labs — Research. http://www.research.att.com/sw/tools/grm.

Mohri, M. (1997). Finite-state transducers in language and speech processing, *Computational Linguistics* **23**(2): 269–311.

Mohri, M. (1998). General algebraic frameworks and algorithms for shortest-distance problems, *Technical Memorandum 981210-10TM*, AT&T Labs - Research, 62 pages.

Mohri, M. (2000). Minimization algorithms for sequential transducers, *Theoretical Computer Science* **234**.

Mohri, M. & Nederhof, M.-J. (2000). Regular approximation of context-free grammars through transformation, *in* J.-C. Junqua & G. van Noord (eds), *Robustness in Language and Speech Technology*, Kluwer Academic Publishers, The Netherlands. This volume.

Mohri, M. & Pereira, F. C. N. (1998). Dynamic compilation of weighted context-free grammars, *36th Meeting of the Association for Computational Linguistics (ACL '98), Proceedings of the Conference*, ACL, Montréal, Québec, Canada, pp. 891–897.

Mohri, M., Pereira, F. C. N. & Riley, M. (1996). Weighted automata in text and speech processing, *Proceedings of the 12th biennial European Conference on Artificial Intelligence (ECAI-96), Workshop on Extended finite state models of language*, ECAI, Budapest, Hungary, pp. 46–50.

Mohri, M., Pereira, F. C. N. & Riley, M. (2000). The design principles of a weighted finite-state transducer library, *Theoretical Computer Science* **231**: 17–32.

Mohri, M. & Riley, M. (1999). Integrated context-dependent networks in very large vocabulary speech recognition, *Proceedings of the 6th European Conference on Speech Communication and Technology (Eurospeech '99)*, Budapest, Hungary.

Mohri, M. & Sproat, R. (1996). An efficient compiler for weighted rewrite rules, *34th Meeting of the Association for Computational Linguistics (ACL '96), Proceedings of the Conference*, ACL, Santa Cruz, CA, pp. 231–238.

Nederhof, M.-J. (1998). Context-free parsing through regular approximation, *Proceedings of the International Workshop on Finite State Methods in Natural Language Processing*, Ankara, Turkey, pp. 13–24.

Pereira, F. C. N. & Wright, R. N. (1991). Finite-state approximation of phrase structure grammars, *29th Annual Meeting of the Association for Computational Linguistics (ACL '91), Proceedings of the Conference*, ACL, Berkeley, CA, pp. 246–255.

Rabiner, L. & Juang, B.-H. (1993). *Fundamentals of Speech Recognition*, Prentice-Hall, Englewood Cliffs, NJ.

Salomaa, A. & Soittola, M. (1978). *Automata-Theoretic Aspects of Formal Power Series*, Springer-Verlag, New York.

Sproat, R. & Riley, M. (1996). Compilation of weighted finite-state transducers from decision trees, *34th Meeting of the Association for Computational Linguistics (ACL '96), Proceedings of the Conference, Santa Cruz, California*, ACL, pp. 215–222.

Tarjan, R. E. (1972). Depth first search and linear graph algorithm, *SIAM Journal of Computing* **1**(2): 146–160.

Ullian, J. S. (1967). Partial algorithm problems for context free languages, *Information and Control* **11**: 90–101.

Chapter 8

Robust Parsing and Beyond

Jean-Pierre Chanod
Xerox Research Centre Europe
6, chemin de Maupertuis
38240 Meylan, France
Jean-Pierre.Chanod@xrce.xerox.com
http://www.xrce.xerox.com/

Abstract Robust analysis covers not only processing partially ungrammatical input (Carbonell & Hayes 1983, Weng 1993), but more generally unrestricted text as actually produced by end-users in various situations.

While early work in robust analysis dates back to the seventies, the topic has recently gained broader attention, due to the large quantities of textual data now available in electronic form (e.g. technical documentation, news, web pages, email, voice system).

This article first introduces the notion of robust parsing in natural language processing. It then reviews a series of linguistic phenomena that illustrate the need for specific approaches to robust parsing. Finally, it mentions some of pioneer and recent work in robust parsing.

8.1 Introduction

The growing need to process large quantities of text and speech now widely available in electronic format has boosted the effort to produce robust language processing systems.

In that context, corpus linguistics has gained deeper attention in the language processing community (Garside, Leech & Sampson 1987, Wilks 1993). The most spectacular effect of this shift has been the push of statistical-based methods, with speech recognition, part-of-speech taggers, information retrieval systems reaching higher and higher levels of accuracy (Church 1988, Brown, Della-Pietra, Jelinek, Mercer & Roossin 1988, Briscoe & N. Waegner 1992, Cutting, Kupiec, Pedersen & Sibun 1992, Charniak 1993).

In contrast, such characteristics underline the weaknesses of more traditional linguistic approaches, especially their inability to process large quantities

J.-C. Junqua and G. van Noord (eds.), Robustness in Languages and Speech Technology, 187–204.

of data and, as a natural consequence, their lack of consideration for evaluation measures.

Still robustness is not about statistical vs. rule-based methods. It is not about virtual core languages, be they defined by principles or by frequency counts. It is not even about the quantity of unrestricted text that can be parsed by a given system. Robustness is about exploring all constructions humans actually produce, be they grammatical, conformant to formal models, frequent or not. Linguistic phenomena, regardless of their oddity or frequency, account for the meaning of whatever segment of text they appear in. Rare or deviant phenomena may become predominant in certain domains or genres (e.g. telegraphic style in maintenance reports or news dispatches, dates and addresses in agenda management applications, imperative elliptic statements in user interfaces, lists of complex noun phrases in medical presciptions, etc.).

In this view, robustness is a matter of broadth and depth of linguistic analysis. Altogether.

8.2 Linguistic Descriptions and Robustness

This section addresses the issue of linguistic descriptions confronted with robustness. It investigates how the variety and complexity of linguistic constructions found in unrestricted texts may reduce the domain of validity of NLP descriptions.

In order to illustrate this purpose, a few characteristic examples are selected and discussed, where linguistically motivated propositions are confronted with a series of surface phenomena. Linguistic phenomena *in situ* (i.e. in corpora) undergo multiple variations and distortions that core descriptions do not capture easily. This does not necessarily undermine the value of principles from a pure linguistic perspective (Chomsky 1981, Berwick 1991), but it shows that linguistically sound principles do not necessarily translate into effective computational models (Jensen 1988, Nagao 1992). Such primary facts, on the other hand, are essential in the conception of robust systems.

8.2.1 Robustness and Unification of Lexical Features

Unification of lexical features has been a productive research paradigm in syntax and has led to various formalisms such as GPSG, LFG or TAG (Joshi, Levy & Takahashi 1975, Kaplan & Bresnan 1982, Gazdar, Klein, Pullum & Sag 1985). Unification presents a computationally elegant way of handling such phenomena as agreement or subcategorisation. Nevertheless, linguistic facts collected across languages present a wide variety of situations where the underlying principles behind unification do not properly account for agreement or verb subcategorisation. This does not reduce the beauty of the underlying principles, but creates an obstacle to the implementation of robust parsers. Indeed, one basic requirement for a robust parser is to handle standard phenomena (i.e. linguistic phenomena that fall under core linguistic principles) and non-standard phenomena (i.e. the complementary set of standard phenomena!) with the same care and depth.

The case of agreement constraints illustrates perfectly why such requirements must be taken into consideration. If one considers standard agreement principles such as *noun modifiers must agree in number, gender and case with the head noun they modify* or *the subject agrees in number with the verb*, one may identify various causes for agreement failure (not even considering grammar errors made by the speaker). Such causes include:

- phonological constraints;
- lexical deviations;
- ellipsis, deletion;
- interaction of syntactic phenomena imposing conflicting constraints;
- semantic constraints prevailing over syntactic constraints.

Although, more than one factor may contribute to the failure of standard agreement rules, the following examples are each attached to a single dominant factor for illustration purposes.

Phonological Constraints

In French, masculine possessive adjectives may be substituted to expected feminine, if the following word, (generally a noun or an adjective) begins with a vowel, as in:

> *Mon adorable chatte*
> My [masc] adorable[] cat[fem]

The possessive *Mon* does not agree with the feminine noun *chatte*, and indeed the phrase:

> **Mon chatte*
> My [masc] cat[fem]

is ungrammatical.

Only the presence of the adjective *adorable* makes the phrase:

> *Mon adorable chatte*

acceptable. Although this kind of phenomena can certainly be described by computational means, it does not fit with a general principle that imposes unification of gender and number.

One easy way to handle this construction (including in a unification-based approach), is to attach both masculine and feminine gender to the possessive adjective mon. However, this will lead to the acceptance of ungrammatical expressions such as:

> *Mon chatte*

Ungrammaticality and Constraints Relaxation

This raises the issue of the traditional role assigned to parsers, i.e. to discriminate between grammatical and ungrammatical utterances. Assigning such a role to a parser makes sense if the parser is assumed to simulate the human capacity to utter grammaticality judgements, for instance an issue relevant to the study of human language learning. But there is no reason why this should be a requirement for a parser aiming at determining the structure of a given input actually produced by a human. It can even be argued that humans have the capacity to reconstruct the meaning of ungrammatical input, e.g. as produced by children or non-native speakers. Although this may be irrelevant to the study of human language learning, this is still a fascinating and useful research area. Another argument against parsers accepting ungrammatical input is around reversibility. If the same grammar were to be used for parsing and for generating utterances, grammaticality control would alledgedly become a necessity, to avoid the production of ungrammatical input during the generation phase. Nevertheless, this notion of reversibility and grammaticality control, fairly academic at this stage of research as far as robust reversible parsers are concerned, does not necessarily holds. Again, it can be argued that humans have the capacity of parsing ungrammatical input while producing only grammatical utterances. Parsers, seen as computational devices, can be freed from the requirement to simulate psycholinguistic behaviour, just as electronic calculators do not simulate human counting processes. But it might be also that some traditional views on the need for grammaticality control in parsing are not even psycho-linguistically founded.

Robust parsers accept ungrammatical utterances as a way to increase coverage and hence robustness. This is indeed connected to a central notion in robust parsing: *relaxation of constraints*. To get back to the example above, accepting that the masculine possessive adjective *mon* modifies all French nouns regardless of their gender has two simultaneous consequences. The grammar will accept the non-standard grammatical phrase *Mon adorable chatte* as well as the ungrammatical one **Mon chatte*. One might constrain further the rule, so that the non-standard construction would apply only if the following word starts with a vowel. At first sight, this would satisfy grammaticality control, at a cost of greater complexity in rule writing and lesser computational efficiency. But worse than that, as far as robustness is concerned, this additional constraint would eliminate other non-standard constructions, where determiners and head nouns do not agree in gender, as in:

> *Mon*[masc] *chèvre* [fem]
> (lit.) My goat (= my goat cheese)

This illustrates the power of constraint relaxation, i.e. the fact of introducing minimal rather than full constraints with respect to a given phenomena. The relaxation allows for more constructions to be recognised, than one has initially in mind when describing a particular one. The primary role of constraints is thus not to prune out ungrammatical input, but rather to eliminate spurious, unwanted analyses when a preferred analysis holds. In that respect, core agreement rules are useful, but not as a core principle. The role of core principles is to select preferred analyses whenever they are relevant. The role of constraints relaxation is to allow for a wider range of non-standard constructions when core principles do not hold.

Ellipsis

It could be argued that the example above (*Mon chèvre*) could be nicely resolved at lexical level (i.e. *chèvre* having two entries, a feminine one as an animal, a masculine one as a kind of cheese).

However, language (and not only French) provides systematic mechanisms that cause surface strings to depart from core syntactic rules. A dominant one is ellipsis. Ellipsis is certainly one of the most systematic ways of disrupting the application of core parsing rules. The deletion of any word displaces the linguistic interpretation from the explicit to the implicit field. As parsers basically rely on explicit background (the input surface string, to start with, and ambiguous lexical information attached to individual tokens), they poorly recognise constructions whose explicit properties do not match implicit interpretations.

Here are some examples where, due to ellipsis, the surface strings no longer match expectations derived from agreement control:

> *un cinq tonnes* (French)
> lit. a [masc, sing] five-ton [fem,plur] (a five-ton truck)

This example is fairly productive in French and can be generalised according to the following pattern: The underlying expression *det noun1 prep num noun2* can be rewritten as *det num noun2* after deletion of the head noun and of the preposition (*noun1 prep*). As a result, at surface level, the determiner of the elliptic string (*det num noun2*) seems to be governed by the PP object *noun2*, while indeed it agrees with the deleted head noun *noun1* (*un camion de cinq tonnes*, a truck of five tons).

This can also be compared with the English expressions *Three pints is not enough* or *A good twenty minutes*.

Again, one way to parse such non-standard constructions, is to relax the agreement constraints, at the possible cost of accepting and parsing ungrammatical structures such as:

> *Le cinq livres que Jean lit*
> The[sing] five books that Jean reads.

Similar elliptical phenomena occur frequently with denomination, i.e. when the internal syntactic structure of a denominating string is superimposed (Shaumyan & Segond 1992) by an overall different syntactic structure. The superimposition may imply a change of syntactic category, esp. with titles and quotations:

I never saw Gone with the wind

or in French:

des tee shirts avec buvez Coca-Cola écrit dessus
Tee shirts with drink Coca-Cola written on them

This quite frequently involves a disruption of agreement rules as predicted by lexical features as in:

J'ai perdu mon Madame Bovary (re. the book)
I lost my [masc] Madame [fem] Bovary

Le France s'appelle aujourd'hui le Norway (re. the cruising ship *France*)
The [masc] France [fem] is now called the Norway

Les Misérables est un roman de Victor Hugo
Les Misérables [plur] is [sing] a novel by Victor Hugo.

Fish and chips is getting expensive.

8.2.2 Syntactic Diversity: Local Grammars

To complicate further the task of robust parsers, one may observe in corpora a massive range of marginal phenomena such as postal addresses, money, dates, bibliographical citations, lists, etc. that induce additional complexity. To some extent, these phenomena may be described individually by local grammars, but most of the computational complexity results from the integration of all the specific descriptions into a single parser. In particular, one often needs some syntactic evidence about a given construction before triggering a specific local grammar, but the evidence may itself result form the prior application of the local grammar. If one adds that non-essential complements, esp. adverbials, can be inserted in almost any position, this gives an overall idea of the domain of complexity primarily addressed by robust parsers. Even in the case of agreement, one may find situations that require very specific syntactic treatment, to be embedded in the overal parser.

An interesting case of apparent subject-verb disagreement in French is commonly found in constructions like:

Reste trois jours
(lit.) Remains three days

where the plural subject *trois jours* disagrees with the singular verb *reste* (besides appearing in a non-standard verb-subject order). This construction is to be connected to the following construction:

Il reste trois jours
(lit.) It remains three days

a common construction in French (esp. with movement verbs, e.g. *il arrive un train*) where the pronominal syntactic subject *il* does agree with the verb, and introduces a nominal sequence as the actual deep subject. However, unlike in other romance languages, the subject pronoun cannot normally be dropped in French so that the phrase *reste trois jours* cannot easily be captured with standard verb-subject agreement rules.

Another example of complex interaction between syntactic constraints is found in Russian with numerical adjectives attached to subject nouns. *Dva* (two) governs the singular genitive case, as in: *dva slona* (two elephants [sing, gen]). However, adjectives attached to a masculine noun in this context are genitive plural:

dva rozovykh slona sidiat
two pink[gen,plur] elephants [gen, sing] are seated.

To complicate further the matter, adjectives attached to feminine nouns in a similar context are nominative plural:

dve belie sobaki sidiat
two white[nom,plur] dogs[fem,gen,sing] are seated

while in both cases the resulting subject NP is to be treated as nominative plural.

This illustrates that morpho-syntactic properties of phrases cannot be derived by mere unification of the lexical features of their internal sub-constituents.

This can be further illustrated in Hungarian (Jászó 1994), where, in case of a co-ordinated subject, the number of the subject constituent closer to the verb determines the verb-subject agreement:

Bejött a kutya és a macskák.
(lit.) Has entered the dog and the cats

Bejöttek a macskák és a kutya.
(lit.) Have entered the cats and the dog

**Bejöttek a kutya és a macskák.*
(lit.) Have entered the dog and the cats

Interestingly, the proximity principle for verb-subject agreement in Hungarian extends to apposition. The verb agrees not with the subject but with its appositive postmodifier, if the postmodifier is closer to the verb than the subject itself, as in:

A közönség — arra tévedt járókelõk — biztatták a verekedõket.
(lit.) The public [sing] — pedestrians [plur] who happened to be there — encouraged [plur] the fighters.

Co-ordination is another cause for the apparent lack of agreement between constituents that are normally expected to agree. This is in particular the case when the co-ordination allows for the factorisation of identical nouns, i.e. when all but one occurrence of the noun are deleted, as in:

Les premier et dernier chapitres
The first [sing] and last [sing] chapters [plur]

ie. an expression derived from:

Les premier (chapitre) et dernier chapitre
The [plur] first [sing] chapter and last [sing] chapter [plur]

after deletion of one occurrence of *chapitre* and pluralisation of the factorised noun *chapitres*. A construction to be compared with English:

Good and bad taste are inculcated by example

This results in a surface expression where co-ordinated singular adjectives are attached to a plural noun. Actually, the number of the adjectives is determined by semantic and pragmatic constraints and does not even require parallelism, as shown by:

Les représentants occidentaux et russe
The [plur] Western [plur] and Russian [sing] representatives [plur]

8.2.3 Semantic and Pragmatic Constraints

In addition to the consideration stated above, syntactic rules can be disrupted by semantic constraints. This is for instance the case with collective nouns that represent a group of persons and objects. They form plural NPs whose head is singular, as in (esp. British) English (Swan 1980).

The team are full of enthusiasm.

or in French:

La salle était pleine d'étudiants. Une bonne moitié sont partis avant la fin.
The room was filled with students. A good[sing] half [sing] left [plur] before the end.

This is actually a fairly common phenomenon across languages, as shown by the following Berber example:

Axxam-negh add-assen azekka
(lit.) House [sing] our will [plur] come tomorrow.

Another case is shown with the French impersonal pronoun *on* (one) which is third person singular as far as the subject-verb agreement is concerned. However, its semantic value is more flexible, as it may refer to one or more persons (including 1st or 2nd in certain contexts). While the verb agrees with the syntactic gender and number, past participle or predicative adjectives do agree with the semantic gender and number:

On est contentes
One [= we] is [sing] happy [fem,plur]

To conclude this section, here follows some considerations that prelude to the construction of a robust parser:

- Even principles that tend to be generally valid such as number, gender and case agreement undergo various exceptions that need to be taken into consideration by a robust grammar.
- Many other syntactic phenomena could exemplify the fact that core linguistic descriptions do not properly account for structures actually found in unrestricted texts. [Sells 85] provides a good illustration of the previous point, about sub-categorisation. The sentence:

  Reagan thinks banana

  is considered ungrammatical because of the sub-categorisation of the verb think; but this sentence becomes acceptable (due to ellipsis) as soon as, for instance, the following context is given:

  What is Kissinger's favorite fruit?
  Reagan thinks banana.

  Syntactic information such as subcategorisation gives only partial evidence of what is the underlying syntactic structure and cannot be reliably used to determine deep structures. This is all the more true when input sentences are extracted from real texts, where phenomena such as ellipsis, insertion of non-essential complements and coordination open the door to multiple variations of core structures.
- When general principles are embedded in the core of parsing formalism, the description of exceptions becomes fairly complicated if possible at all; indeed, using a computational grammar as a system to discriminate between grammatical and ungrammatical input leads to additional complexity when a large set of exceptions has to be described.

8.3 Some Properties of Robust Parsers

This section describes some of the major properties shared by most robust parsers. These properties basically result from the two following facts:

- The automatic description of syntactic structures is constrained by the limited background knowledge available at any given stage of the process, no matter how deep the linguistic insight humans may have about the underlying structure. Typically, syntactic structures that require a full semantic interpretation or more generally a level of disambiguation that cannot be derived from the mere knowledge of the surface string to be parsed are only partially analysed by robust parsers.
- Robust parsers describe constructions whose features go beyond the limited scope of selected core principles. They must take into account the wide variety of linguistic phenomena that appear in unrestricted texts *simultaneously*, while avoiding the multiplication of irrelevant parses. In order to handle a variety of non-core phenomena and the many exceptions to general principles, robust parsers must allow for a tractable relaxation mechanism.

Ejerhed (1993) observed similar properties for some the early work on broad-coverage parsing, such as FIDDITCH (Hindle 1983), the French PLNLP grammar (Heidorn 1975, Jensen, Heidorn, Richardson & Haas 1986); or probabilistic parsers (Church 1988, Ejerhed 1988). Similar properties are still observed in later work (Hobbs, Appelt, Bear, Israel, Kameyama & Tyson 1993, Grefenstette 1996, Federici, Montemagni & Pirrelli 1996).

- The analysis is incomplete and incremental: robust parsers do not aim at providing a full syntactic analysis (in the traditional sense in generative grammar) in one go. They rather aim at resolving syntactic structures with various degrees of depth, depending on the complexity and the availibility the background information required for the resolution. This does not imply that a deeper analysis is not desirable nor achievable, but it should be attempted only once a preliminary syntactic organisation is computed. In that sense, robust parsers tend to be syntactic preprocessors rather than parsers, the degree of pre-processing depending on various aspects such as background information in use, implementation framework, coverage, precision, language.

  One may see robust parsing as heterogeneous in terms of depth rather than just incomplete. The heterogeneity of the analysis depends not only on the nature of the syntactic constructions considered, but also on the specific input to be processed. For instance, the depth of analysis for a given construction may well depend on lexical ambiguity or on the presence of adverbials that prevent the parser to make a reliable decision. Nevertheless, the parser might resolve a similar construction in the absence of ambiguous attachments or ambiguous lexical forms.
- The analysis provides a partial constituent structure; robust parsers tend to identify parts-of-speech (if this is not done separately by a part-of-speech tagger) and a low level syntactic category to some sequences of words, which may or may not represent syntactic phrases in the traditional sense. Such groups are commonly named chunks (Abney 1991)

to establish a distinction with standard phrases. This distinction is in itself a specific property of robust parsers. A typical example deals with PP attachment, which cannot be resolved on purely syntactic ground. Even when the PP matches lexical constraints defined by a subcategorisation frame, the constituent structure cannot be reliably decided on that ground (due to the possible conflict between arguments and adverbials that share the same superficial structure). Similar constituent underspecification applies to a wide range of constructions, such as adjectival modifiers or adverbials, coordinated structures or elliptic constructions.
Indeed, the set of core constructions that are computationally tractable is not defined by syntactic theory alone, but by language specific parameters that more or less overlap with the semantic limitations of computational systems. This can be illustrated by cross-linguistic considerations. For instance, while functional analysis (e.g. subject and object assigment) may be seen as a minimal requirement for a parser, it is clear from various examples (e.g. Italian or Spanish, with free word order, subject pronoun deletion and no case marking) that subject assignment is only possible with semantic and pragmatic analysis, unless when syntactic constraints do provide evidence for a reliable subject assignment.

Combinatorial explosion: A robust parser must not only describe a wide range of constructions. It must also address the issue of construction control, i.e. restrict the average number of parses per sentence, in order to keep the output usable for further processing. As, at the same time, robust parsers describe a far wider range of constructions than toy grammars; this makes combinatorial explosion a central issue for robust parsing. It is not enough to build a large collection of construction-specific algorithms to achieve broad coverage; one must also prevent construction-specific algorithms from applying in a parasite mode, i.e. to surface strings that coincidentally match an extraneous construction. Ambiguity compression through syntactic underspecification and controlled relaxation are key elements to achieve this goal. Incidentally, the length of the input string (usually a sentence) has an important impact on robustness. It affects not only the parsing time, but also the parsing complexity. Chances that multiple constructions occur simultaneously grow with the size of the input, as well as possibility to find a non-core construction in it.

8.4 Some Implementation Frameworks

Although robust parsers share some major common properties and goals, fairly different architectures have been proposed. The differences depend mainly on the overall strategy to provide for robustness, and the underlying descriptive formalism.

8.5 The Constructivist Approach: PLNLP

The constructivist approach is based on the description of large collections of syntactic patterns. This is in continuity with the early work in generative grammar, and has inspired most attempts at parsing, be they driven by formal or broad-coverage approaches.

As far as robust parsing is concerned, one of the most significant constructivist systems was built with PLNLP (Heidorn 1975).

PLNLP is a bottom-up parser where strings and substrings are represented by records, i.e. collections of feature-value pairs, where values can be strings, numbers, records, lists, etc. Records are derived from production rules, initial records being derived from lexical entries associated with minimal tokens (cf. terminals). Rules are controlled by logical conditions on the value of features, thanks to a collection of basic operators, including disjunction and negation. As such, the parser is fairly general and not committed to any particular linguistic theory.

Nevertheless, PLNLP became strongly connected with broad-coverage robust parsing with PLNLP English Grammar (PEG) (Jensen et al. 1986, Jensen, Heidorn & Richardson 1992). The linguistic strategy used in PEG, and later for many other languages can be summarised as follows:

- Broad coverage is achieved through the relaxed approach. Constraints described in rules are not meant to restrict language, but rather at identifying the most likely interpretation of any given input; in particular only very limited use of sub-categorisation is made at this early stage of parsing;
- Reduction of ambiguity is achieved by collapsing multiple analyses into a single conventional compact representation: e.g. PPs are conventionally attached to the nearest left candidate, while other potential attachements are only marked by pointers. The task of assigning the actual attachment is delayed to a later stage, during which the parse is re-evaluated with additional background information.
- Rules are mostly binary rules, thus allowing for a greater combination of primary records into higher level units. This improves the coverage, esp. as combinations of records are not restricted by a priori patterns. For instance, an input string made of a NP followed by an adverbial as in French *le diable probablement* (the devil likely) can be parsed by a specific binary rule, although standard grammar does not allow for the combination of NPs and adverbials. Such hybrid phrases (NP ADV) can be combined into more complex hybrid phrases at later stages of analysis, e.g. to handle gapping in coordinated structures (Steedman 1990, Brun 1998).
- Conditions in rules do not constraint syntactic structures in terms of grammaticality. They mostly disallow constructions if the underlying string can be represented by a more stable construction. For instance, if a

string includes a finite verb, a full parse including a subject is preferred, but, if no subject can be found, a full parse can still be constructed. The presence of a subject is preferred, but not required.

- In case of parse failure, a second pass is performed with some specific conditions being relaxed in the rules. The resulting parse keeps track of which rules and which conditions have been relaxed, so that a diagnosis for relaxation is available.
- In case of parse failure after the second pass, a specific procedure (fitted parse) selects partial parses to reconstruct an analysis as a sequence of partical optimal analyses.
- In case of multiple parses, a ranking procedure is applied (P-metric) (Heidorn 1982). It is based on assigning weights to constituents, based on the value of selected features.

The result of this parsing phase is a preliminary structure, called syntactic sketch. It can be further enriched through additional transformations. This includes computing predicate-argument structure or disambiguating PP-attachment using dictionary information. Broad coverage PLNLP grammars have been written for about ten different languages.

8.5.1 A Reductionist Approach: ENGCG

Another important paradigm in robust parsing was introduced by the work of Karlsson (1990) as constraint grammars. This approach is based on the consideration that explicitly describing all syntactic constructions is an almost impossible task. To overcome this difficulty, the reductionist approach relies on the idea that a grammarian is more apt at identifying conditions when a given syntactic construction cannot occur than at listing and describing all possible constructions. Consequently, the constraint-based, or reductionist approach, consists in building an initial set of all possible syntactic combinations attached to a given input string, and at pruning out those constructions which are not possible. The constraints may be expressed by a set of elimination rules applied in a sequence (Voutilainen & Tapanainen 1993) or by a set of restrictions applied in parallel (Koskenniemi, Tapanainen & Voutilainen 1992). The pruning is not based on normative grammar, but rather on fairly general incompatibilities, especially between sequences of categories and functions. For instance, in English, one cannot accept a determiner or a preposition immediately followed by a finite verb. This would eliminate the sequence determiner + verb for the string *the book*. Similarly, at the functional level, all subject verb subject sequences are discarded from the initial set of all possible syntactic combinations, and so on. Once all constraints have been applied, the remaining combinations represent the set of accepted parses. A constraint relaxation mechanism can be activated when no parse remains after all constraints have applied.

ENGCG (for English Constraint Grammar) has been implemented under this framework, using a specific set of operations to discard unwanted sequences from the initial set and led to very good performances in terms of speed and coverage.

A finite-state implementation of a French Constraint Grammar is described in Chanod & Tapanainen (1996). In this implementation, the result of the lexical analysis is a finite-state network, where each path represents an alternative analysis for each token. The analyses are concatenated to form a single finite-state network that contains all possible readings for the sentence. The initial sentence network is actually built by appending the various lexical strings associated with each token and by inserting word boundaries between every pair of tokens. Word boundaries include simple boundaries between two words of a same clause, or various types of clause boundaries (e.g. endings, clausal subjects).

The finite-state syntax rules (constraints) are expressed as regular expressions and compiled to finite-state automata. These constraints are intersected with the initial sentence network. The resulting network represent all the acceptable parses, i.e. the set of paths that abide by all the constraints specified in the grammar.

The rules express local constraints (bearing on tokens in an immediate neighbourhood) as well as long distance constraints (bearing on a segment of unrestricted length at sentence level). Constraints are expressed as regular expressions defining logical conditions about any combination of symbols within the sentence: the symbols can represent morpho-syntactic features (e.g. gender, number, person, tense), or parts-of-speech, base forms, functional tags, subcategorisation features, punctuation, or word boundaries.

8.5.2 A Hybrid Approach: IFSP

A third approach to robust parsing consists in merging constructivist and reductionist approaches depending on the syntactic phenomena to be described. It is implemented in IFSP (Incremental Finite-State Parsing, Aït Mokhtar & Chanod 1997a, Aït Mokhtar & Chanod 1997b). In IFSP, each input token is assigned a single tag, generally representing the part-of-speech and some limited morphological information. The sentence is initially represented as a sequence of wordform-plus-tag pairs.

The incremental parser consists of a sequence of transducers, compiled from regular expressions. Each of these transducers adds syntactic information represented by reserved symbols (annotations), such as brackets and names for segments and syntactic functions. Actually, the added syntactic information may represent temporary hypotheses and be later deleted if specific constraints so require. This double process of syntactic annotation and controlled deletion makes the approach hybrid with respect to constructivism/reductionism and actually combines the advantages of both. Syntactic annotation can be performed on the basis of explicit constructivist statements when the context so permits (i.e. when there is enough contextual evidence to identify a given construction). The annotation can also be performed in a reductionist way, i.e. by introducing hypothetical constructions and applying them sets of constraints.

Each transducer performs a specific linguistic task. For instance, some networks identify segments for NPs, PPs, APs (adjective phrases) and verbs, while others are dedicated to subject or object pick-up. The ordering of the transducers is syntactically significant. Annotations related to a given phenomenon are determined by the syntactic information gathered across the prior levels of analysis. The same task (e.g. subject assignment or verb segmentation) may be performed by more than one transducer. The additional information provided at each stage of the sequence is instrumental in the definition of the later stages of the sequence. Networks are ordered in such a way that the most standard constructions are addressed first. For instance, in French, if a left subject is already attached to a verb at a given level (SVO sequences are standard in French although VS sequences are possible in various contexts), then a the Verb Subject rule cannot apply. In other words, the Verb Subject rule does not require an exhaustive description of the context in which it is allowed. The rule control is provided by the ordering.

More generally, the contextual constraints taken in consideration at any given level of analysis are incrementally relaxed. This incrementality is the key property that allows for relaxation and robustness without mutiplying the number of unwanted analysis. This may be in a sense related to ordering in optimality theory, although in a very different environment.

8.6 Conclusion

The absence of a clear epistemological status for robust parsers has occasionaly been noted: are their purely engineering devices, ad hoc combinations of heuristics, unprincipled experimental devices? Such interrogations and doubts can certainly be put in relation with the central role of formal syntax in computational linguistics. Robust parsers represent a paradigm that does not reflect the major focus of formal syntax, such as language acquisition, grammaticality, core language description. On this epistemological ground, robust parsers could be compared with part-of-speech taggers, rather than with parsers inspired by formal syntax. Part-of-speech disambiguation does not correspond to a traditional level of linguistic description (part-of-speech is a traditional notion, but not the disambiguation process). Nevertheless, part-of-speech disambiguation has proven tractable, robust and useful from a computational viewpoint. In that respect, robust parsers are fairly orthogonal to principled-based parsers. Or to quote J. Hobbs: "We were struck by the strong performance in MUC-3 that the group at the University of Massachusetts got out of a fairly simple system. It was clear they were not doing anything like the depth of preprocessing, syntactic analysis, or pragmatics that was being done by the systems at SRI, General Electric, or New York University. They were not doing a lot of processing. But they were doing the right processing for the task."

As long as the attention is directed to formal properties of languages, i.e. to abstract generalisations and ways to manipulate such abstractions, the quantity of data processed is not necessarily relevant. Properties demonstrated at a certain level of abstraction apply to all surface forms that reduce to the same abstraction. This is a strong argument for principle-based approaches to natural language processing, as they provide the adequate level of abstraction for repeatability, regardless of the particularities of surface forms across languages. Narrow coverage may impede the development of real life applications, but it does not ipso facto invalidate the underlying linguistic principles of a syntactic theory.

However, the absence of systematic evaluation for such systems is a scientific weakness. Evaluation is one essential way to provide for falsification, in the Popperian sense, i.e. an experiment to demonstrate the falsehood of theoretical principles. The lack of evaluation is a fault in the attempt to develop a science of language. Building this science cannot be restricted to proposals for a mathematical representation of language. Equally important for the theory to evolve is the proper implementation of a falsification process, as a means to explore the limits of theoretical principles. Besides their practical applications, robust parsers may be seen as experimental devices for exploring the truth domain in which computational linguistic descriptions hold. Robust parsing is the locus of linguistic description confronted with the fundamental limitations of computational description.

References

Abney, S. (1991). Parsing by chunks, *in* R. Berwick, S. Abney & C. Tenny (eds), *Principled-Based Parsing*, Kluwer Academic Publishers, Dordrecht, pp. 257–278.

Aït Mokhtar, S. & Chanod, J.-P. (1997a). Incremental finite-state parsing, *Proceedings of Applied Natural Language Processing*, Washington, DC, pp. 72–79.

Aït Mokhtar, S. & Chanod, J.-P. (1997b). Subject and object dependency extraction using finite-state transducers, *ACL Workshop on Automatic Information Extraction and Building of Lexical Semantic Resources for NLP Applications*, Madrid.

Berwick, R. (1991). Principles of principled-based parsing, *in* R. Berwick, S. Abney & C. Tenny (eds), *Principled-Based Parsing*, Kluwer Academic Publishers, pp. 1–37.

Briscoe, T. & N. Waegner, N. (1992). Robust stochastic parsing using the inside-outside algorithm, *AAAI Workshop on Probabilistically-Based Natural Language Processing Techniques*, pp. 39–53.

Brown, P., Della-Pietra, J., Jelinek, F., Mercer, R. & Roossin, P. (1988). A statistical approach to language translation, *Twelfth Conference on Computational Linguistics*, Budapest, pp. 71–76.

Brun, C. (1998). *Étude et implantation de la coordination en vue de l'analyse automatique du français écrit dans le cadre de la Grammaire Lexicale Fonctionnelle*, Doctorat en informatique et communication.

Carbonell, J. G. & Hayes, P. J. (1983). Recovery strategies for parsing extragrammatical language, *American Journal of Computational Linguistics* **9**(3–4): 123–146.

Chanod, J.-P. & Tapanainen, P. (1996). Finite-state based reductionist parsing for French, *in* A. Kornai (ed.), *Extended Finite State Models of Language*, Cambridge University Press, pp. 72–85.

Charniak, E. (1993). *Statistical Language Learning*, The MIT Press, Cambridge, MA.

Chomsky, N. (1981). Markedness and core grammar, *in* A. Belletti, L. Brandi & L. Rizzi (eds), *Theory of Markedness in Generative Grammar, Proceedings of the 1979 GLOW Conference*, Scuola Normale Superiore di Pisa, Pisa.

Church, K. (1988). A stochastic parts program and noun phrase parser for unrestricted text, *Proceedings of the Second Conference on Applied NLP*, ACL, Austin, pp. 136–143.

Cutting, D., Kupiec, J., Pedersen, J. & Sibun, P. (1992). A practical part-of-speech tagger, *ANLP-92*, Trento, pp. 133–140.

Ejerhed, E. (1988). Finding clauses in unrestricted text by finitary and stochastic methods, *2nd Conference on Applied Natural Language Processing*, ACL, Texas, pp. 133–140.

Ejerhed, E. (1993). Nouveaux courants en analyse syntaxique, *Traitement automatique des langues* **34**(1): 61–82.

Federici, S., Montemagni, S. & Pirrelli, V. (1996). Shallow parsing and text chunking: A view on underspecification in syntax, *ESSLLI'96 Workshop on Robust Parsing*, Prague, pp. 35–44.

Garside, R., Leech, G. & Sampson, G. (eds) (1987). *The Computational Analysis of English: A Corpus-Based Approach*, Longman, London.

Gazdar, Klein, Pullum & Sag (1985). *Generalized Phrase Structure Grammar*, Harvard University Press.

Grefenstette, G. (1996). Light parsing as finite-state filtering, *Proceedings ECAI'96 Workshop on Extended Finite-State Models of Language*, Budapest, pp. 20–25.

Heidorn, G. (1975). Augmented phrase-structure grammars, *in* R. Shank & B. Nash-Webber (eds), *Theoretical Issues in Natural Language Processing*, Association for Computational Linguistics.

Heidorn, G. (1982). Experience with an easily computed metric for ranking alternative parses, *20th Annual Meeting of the ACL*, Toronto, pp. 82–84.

Hindle, D. (1983). User manual for Fidditch, *Technical memorandum 7590-142*, Naval Research Laboratory.

Hobbs, J., Appelt, D., Bear, J., Israel, D., Kameyama, M. & Tyson, M. (1993). FASTUS. a system for extracting information from text, *Proceedings, Human Language Technology*, Princeton, New Jersey, pp. 133–137.

Jászó, A. (1994). *A Magyar Nyelv Könyve*, Trezor Kiadó, Budapest.

Jensen, K. (1988). Why computational grammarians can be skeptical about existing linguistic theories, *COLING*, Budapest, pp. 448–449.

Jensen, K., Heidorn, G. & Richardson, S. (eds) (1992). *Natural Language Processing: The PLNLP Approach*, Kluwer Academic Publishers.

Jensen, K., Heidorn, G., Richardson, S. & Haas, N. (1986). PLNLP, PEG and CRITITQUE: Three contributions to computing in the humanities, *Proceedings of the 8th International Conference on Computers and the Humanities*, Toronto.

Joshi, A., Levy, L. & Takahashi, M. (1975). Tree adjoining grammar, *Journal of the Computer and System Science* **10**(1).

Kaplan, R. & Bresnan, J. (1982). Lexical functional grammar: A formal system for grammatical representation, *in* J. Bresnan (ed.), *The Mental Representation of Grammatical Relations*, The MIT Press, Cambridge, MA, pp. 173–282.

Karlsson, F. (1990). Constraint grammar as a framework for parsing running text, *13th Conference on Computational Linguistics*, Helsinki, pp. 168–173.

Koskenniemi, K., Tapanainen, P. & Voutilainen, A. (1992). Compiling and using finite-state syntactic rules, *Proceedings of the Fourteenth International Conference on Computational Linguistics COLING*, Nantes, pp. 156–162.

Nagao, M. (1992). Are the grammars so far developed appropriate to recognize the real structure of a sentence?, *Fourth International Conference on Theoretical and Methodological Issues in Machine Translation*, Montreal, pp. 127–137.

Shaumyan, S. & Segond, F. (1992). Discontinuous constructions and applicative universal grammar, *MOL3 (Mathematics of Language) conference*, Austin, Texas.

Steedman, M. (1990). Gaping as constituent coordination, *Linguistics and Philosophy* **13**: 207–263.

Swan, M. (1980). *Practical English Usage*, Oxford University Press.

Voutilainen, A. & Tapanainen, P. (1993). Ambiguity resolution in a reductionistic parser, *Proceedings of the Sixth Conference of the European Chapter of the Association for Computational Linguistics*, Utrecht, pp. 394–403.

Weng, F. L. (1993). Handling syntactic extra-grammaticality, *Proceedings of the 3rd International Workshop on Parsing Technologies*, Tilburg, pp. 319–332.

Wilks, Y. (1993). Corpora and machine translation, *Proceedings of Machine Translation Summit IV*, Kobe.

Chapter 9

Robust Parsing of Word Graphs

Gertjan van Noord
Rijksuniversiteit Groningen
vannoord@let.rug.nl

Abstract In spoken language understanding systems, the interface between automatic speech recognition on the one hand and natural language understanding on the other hand often consists of word graphs: a compact representation of the sequences of words that the automatic speech recognition component hypothesises for a given utterance.

In this chapter, it is shown how 'standard' parsing algorithms (which normally assume a string as input) can be generalized straightforwardly to accept such a word graph as their input.

Furthermore, a general model for robust parsing is presented in which a word graph is annotated with (perhaps partial) results of the parser. A search algorithm is defined which selects the best sequence of analyses from this annotated word graph, provided an appropriate multi-dimensional score function is defined.

Finally, a number of techniques is discussed which approximate the fully general search algorithm and experimental results are provided indicating that efficient approximations are possible with little or no reduction in concept accuracy.

9.1 Introduction

In spoken language understanding systems, the interface between automatic speech recognition on the one hand and natural language understanding on the other hand often consists of word graphs: a compact representation of the sequences of words that the automatic speech recognition component hypothesises for a given utterance. Each sequence is a possible hypothesis. The reason that the automatic speech recognition component produces multiple hypotheses is that for many applications the quality of current automatic speech recognition is not good enough. It is hoped that later phases of processing are able to pick out the appropriate hypothesis from a number of competing ones, taking

J.-C. Junqua and G. van Noord (eds.), Robustness in Languages and Speech Technology, 205–238.

into account knowledge sources that are not available to the automatic speech recognition component.

A natural language understanding component needs to assign a representation to the user utterance indicating the *meaning* of that utterance. Such a representation can take the form of a formula in a logical language specifically designed for the application, or perhaps a syntax tree from which such a meaning representation can easily be derived. In most systems, the natural language understanding component consists of at least two parts. The grammar *specifies* in a declarative way the utterances accepted by the system and the corresponding meaning representations. The parser is a component which *computes* for a given user utterance on the basis of the grammar the applicable meaning representation(s).

In the case of a word graph interface the task of the parser becomes somewhat more complicated. One possible approach would be to generate all paths through a given word graph and then apply the parser to each of these paths. It is obvious that such a method is impracticable since the number of paths can be astronomic. Luckily, as we explain below, it is straightforward to extend parsing algorithms for sequences of words to word graphs.

A further complication arises in that typically the parser will not assign a single analysis to a given word graph. For instance, there might be several paths through the word graph that each give rise to a complete analysis. Thus, apart from the lexical and structural ambiguity introduced by the dictionary and grammar the word graph introduces a new kind of ambiguity.

The parser faces the opposite type of problem if no path in the word graph gives rise to a complete analysis. Although the same problem occurs for text-based systems, the problem is much more prominent in a spoken dialogue system because of the quality of the speech recognizer (sometimes the actual utterance is not a path in the word graph), and because of spoken language characteristics (hesitations, repetitions, false starts and other disfluencies).

Below we discuss a very general approach to treat both of these problems. We will assume that the parser is generalized somewhat in that it finds all analyses *anywhere in the input word graph*. The word graph is then annotated with each of these analyses. The final result is then given by the best path through the annotated word graph. The notion 'best' is defined with respect to a scoring function which takes into account acoustic scores, Ngram scores and possibly other preference scores.

The efficiency and adequacy of the proposed model and a number of variants is evaluated with respect to its performance on the OVIS2 test set. This test set is developed in the context of the NWO Priority Programme *Language and Speech Technology*. Below we describe this test set and some other characteristics related to the evaluation procedure.

9.2 Word graphs

The input to the parser consists of word graphs produced by the speech recogniser (Oerder & Ney 1993). A word graph is a compact representation for all sequences of words that the speech recogniser hypothesises for a spoken utterance. The states of the graph represent points in time, and a transition between two states represents a word that may have been uttered between the corresponding points in time. Each transition is associated with an *acoustic score* representing a measure of confidence that the word perceived there was actually uttered. These scores are typically derived from negative logarithms of probabilities and therefore require addition as opposed to multiplication in order to combine two scores. An example of a typical word graph is given as the first graph in figure 9.1.

A word graph is a directed acyclic graph $G = \langle \Sigma, V, v_s, T, F \rangle$ where Σ is a set of words (the labels), V is a set of states (vertices), $v_s \in V$ is the start state, T is a set of transitions and F is a set of final states. Transitions are tuples $\mathit{trans}(v_i, v_j, w, a)$ for a transition from v_i to v_j $(v_i, v_j \in V)$ with label $w \in \Sigma$ and acoustic score a. Finals are tuples $\mathit{final}(v_i, a)$ where $v_i \in V$ indicating that v_i is a final state with associated acoustic score a.

At an early stage, the word graph is normalised to eliminate the *pause transitions*. Such transitions represent periods of time for which the speech recogniser hypothesises that no words are uttered. Depending on the details of the grammar, filled pause transitions and noise transitions can be treated as pauses as well. The motivation for pause transition elimination is that the grammar will not anticipate such transitions. The elimination might also result in graphs with a smaller number of states (but typically with a larger number of transitions). Finally, pause transition eliminiation turns out to be more efficient than extending lexical lookup in a way which would extend lexical entries to include any neighbouring pause transitions.

For a given input word graph $G = \langle \Sigma, V, v_s, T, F \rangle$ and a set of pause labels Σ_{pause}, pause transition elimination is the word graph $G' = \langle \Sigma - \Sigma_{pause}, V, v_s, T', F' \rangle$ such that:

1. $T' = \{\mathit{trans}(v_i, v_k, w, a + \text{pause}(v_i, v_j)) | \mathit{trans}(v_j, v_k, w, a) \in T\}$
2. $F' = \{\mathit{final}(v_i, a + \text{pause}(v_i, v_f)) | \mathit{final}(v_f, a) \in F\}$

For each two states $v_i, v_j \in V$ the function $\text{pause}(v_i, v_j)$ is the score of the best path from v_i to v_j using pause transitions only. The computation of this function reduces to a well-known problem for graphs, namely the all-pairs shortest path problem for directed acyclic graphs (the graph made up of all pause transitions).

Note that the resulting word graph could contain non-start states without any incoming transitions. These states can be removed (as well as any transitions leaving these states). As an example of pause transition elimination, refer to figure 9.1. From now on, we will assume word graphs are pause eliminated.

After pause transition elimination, the word graph contains exactly one start state and one or more final states, associated with a score, representing a measure of confidence that the utterance ends at that point. In figure 9.1, the second word graph is the pause free variant of the first word graph.

9.3 Parsing as Intersection

Ordinarily, the input for parsing is a sequence of atomic symbols, words. In the case of word graphs the input is a compact representation of a finite number of such sequences of words.[1] Thus a naive solution would be to parse each of these sequences in turn. Such an approach, however, is impracticable because the number of paths in a given word graph can be enormous. It turns out that it is possible to generalize parsing algorithms in such a way that the compact word graph representation can be used directly as input. In fact, parsing algorithms for context-free grammars (CFG) can be generalized to have *finite state automata* (FSA) as their input. In such a setup, parsing is the computation of the *intersection* of the input (FSA) and the grammar.

In this section, we briefly review why it might be interesting to generalize the parsing problem to FSA. This generalization not only naturally subsumes wordgraph parsing, but also enables a number of techniques to deal with ill-formed input, and techniques to treat *haplology* phenomena. Some examples are provided in subsection 9.3.1. In subsection 9.3.2 we then show how the intersection of a context-free grammar (CFG) and a FSA can be computed (both in theory and in practice). In subsection 9.3.3 we consider the case of *Definite Clause Grammars* (DCG) (Pereira & Warren 1980), rather than CFG. The DCG formalism is a simple example of a family of constraint-based grammar formalisms that are widely used in (computational) linguistics, because such formalisms are much better equiped to deal with the subtleties and idiosyncracies of the syntax of natural languages than CFG. In fact, CFG have been shown to be insufficient in general for the description of natural languages (Huybregts 1984, Shieber 1985). It turns out, however, that even if we consider the *off-line parsable* subset of DCGs the intersection with FSA is undecidable. This follows from the fact that the emptyness problem for off-line parsable DCGs is undecidable. The proof is given in subsection 9.3.3. As we discuss in subsection 9.3.4, we therefore assume input FSA are acyclic; i.e. we consider parsing as the intersection of acyclic FSA and off-line parsable DCG (section 9.3.4).

9.3.1 Finite state automata as input for parsing

Following Bernard Lang we consider parsing as the computation of the intersection of FSA and CFGs (Lang 1988, Lang 1989). Word graphs are FSA with

1. For the moment we ignore the complication introduced by the acoustic scores.

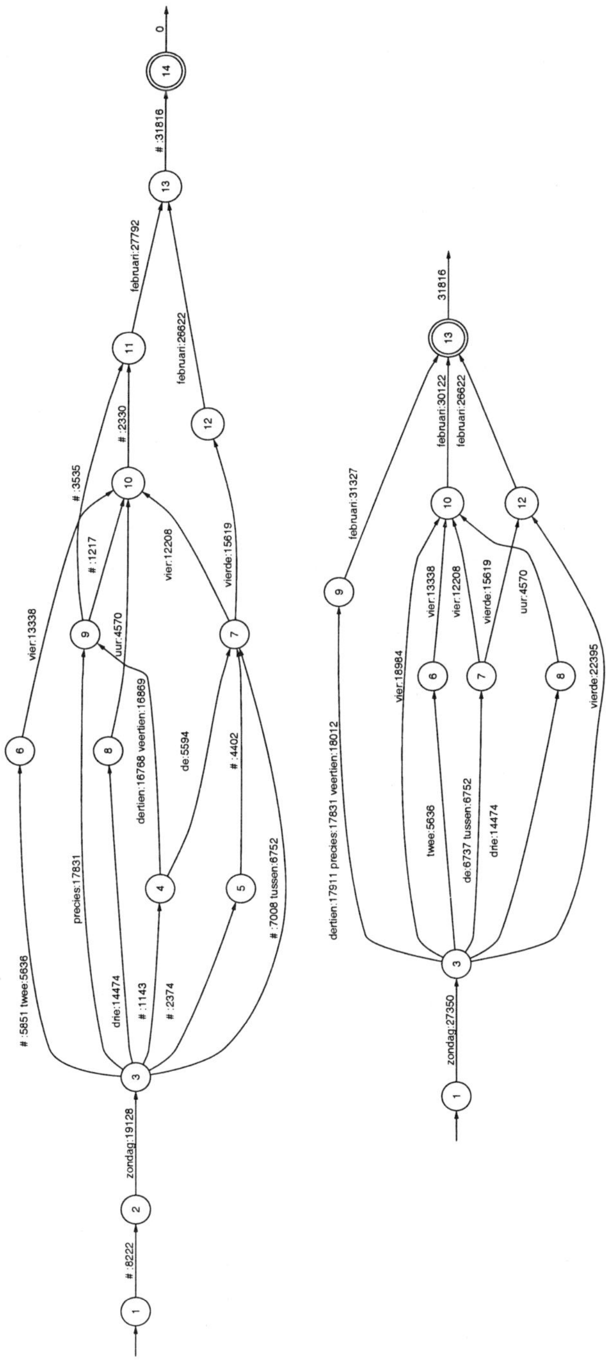

FIGURE 9.1: Word graphs for the utterance *Zondag vier februari* (Sunday Februari fourth). The special label # in the first graph indicates a pause transition. These transitions are eliminated in the second graph.

the additional constraint that no cycles occur. It has been argued that FSA are a good input format in order to model cases of uncertain input, both in case of ill-formed textual input and in case of speech input. *Haplology* phenomena are a further area in which FSA input seems useful.

For example, certain techniques to deal with ill-formed input can be characterized as finite state transducers (Lang 1989); the composition of an input string with such a finite state transducer results in a finite state automaton that can then be input for syntactic parsing. Such an approach allows for the treatment of missing, extraneous, interchanged or misused words (Teitelbaum 1973, Saito & Tomita 1988, Nederhof & Bertsch 1996), and is a generalization of the *word skipping* techniques described in Rosé & Lavie (2000). Similar techniques can be used to treat false starts and repairs (Heeman & Allen 1994, Carter 1994, Heeman & Allen 2000). As an example, consider the utterance:

When eeh what time leaves the train

One of the rules treating hesitations could be a transducer which maps every word to itself, and which optionally maps a sequence of words followed by a hesitation marker to the empty string (ϵ) (in such a transducer the special symbol ? is used to indicate any symbol):

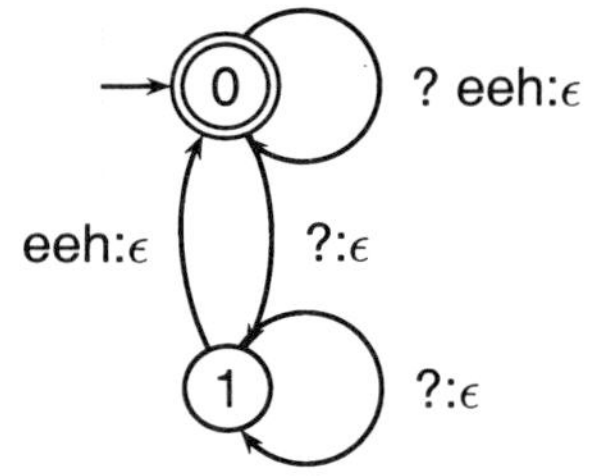

Composing the input with this transducer yields the following automaton:

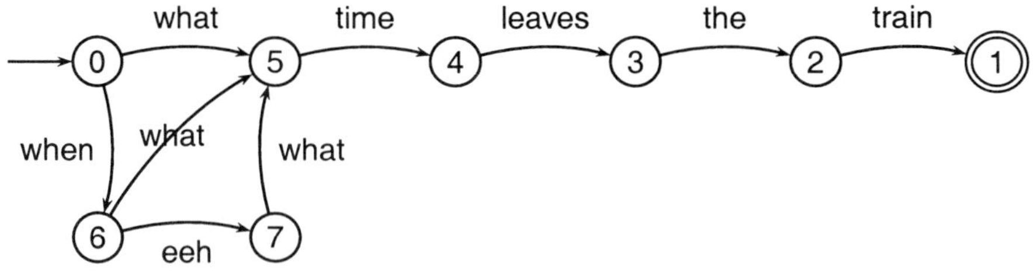

Note that we allow the input to be a full finite state automaton (possibly including cycles) since some of the above-mentioned techniques indeed result in cycles. Cycles might emerge to treat unknown sequences of words, i.e. sentences with unknown parts of unknown lengths (Lang 1988). Another phenomenon which might give rise to cycles in the input is *haplology*, if modeled by a finite state transducer. An example is given by "'s" in English sentences like

We were at a restaurant called "Joe's". The food was great, but the chef at Joe's wife kept complaining about the noise we made.

The second occurrence of the "'s" morpheme supposedly functions both as part of the name "Joe's" and as a genitive marker on the phrase "the chef at Joe's". It is possible to write a transducer which maps every ordinary word to itself, and which maps a word such as "'s" to a sequence of "'s":

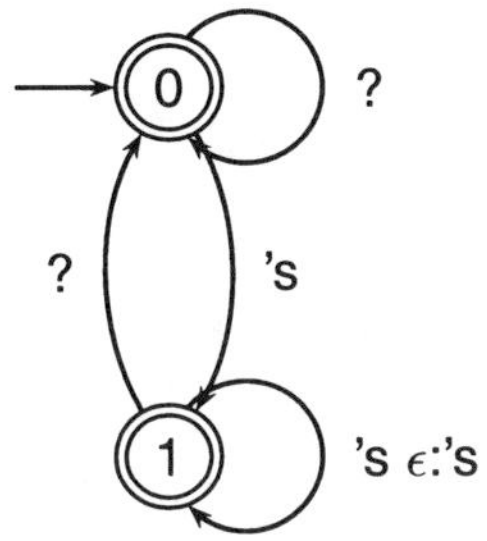

This would also lead to a finite state automaton with cycles, if applied to an input containing "'s":

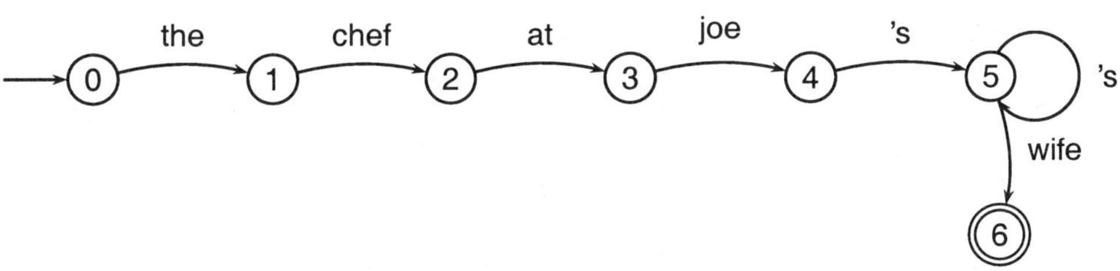

Another example of haplology is the Dutch particle "er". The particle occurs both in indefinite passive contexts (b. versus a.), and in pronominal numerical contexts (d. versus c.):[2]

a. dat de tien soldaten werden gezien
that the ten soldiers were seen
that the ten soldiers were seen

b. dat er tien soldaten werden gezien
that **er** ten soldiers were seen
that ten soldiers were seen

c. dat ik tien soldaten heb gezien
that I ten soldiers have seen
that I have seen ten soldiers

d. dat ik er tien heb gezien
that I **er** ten have seen
that I have seen ten of them

2. This example is simplified for expository purposes, for an overview refer to Bennis (1986) and Model (1991).

If the two contexts are combined in root sentences, then two occurrences of the particle are possible (a.); however if the contexts are combined in such a way that the particles end up in the same place, only a single occurrence remains (b. and c.).

a. er werden er tien gezien
er were **er** ten seen
ten of them were seen

b. *dat er er tien werden gezien

c. dat er tien werden gezien
that **er** ten were seen
that ten of them were seen

Such a syntactic puzzle is solved by assuming a transducer which maps every occurrence of the particle "er" in the input to a sequence of "er". The grammar of Dutch could be kept simple as if no haplology of "er" occurred. A phonological rule performing such a transduction is for example proposed in Bennis (1986). Again, this would yield a cyclic FSA as input for the parser.

9.3.2 The intersection of a CFG and a FSA

If we view parsing as the computation of the intersection of a FSA and a CFG, then the result of parsing will itself be a CFG. This grammar is such that it generates exactly the strings in the intersection, and moreover it assigns the appropriate parse trees to each of these. The resulting intersection grammar, therefore, is a compact representation of all parse trees, i.e. a *parse forest* (Lang 1974, Billot & Lang 1989).

It can be shown that the computation of the intersection of a FSA and a CFG requires only a minimal generalization of existing parsing algorithms. We simply replace the usual string positions with the names of the states in the finite state automaton (this is explained in more detail below). It is also straightforward to show that the complexity of this process is cubic in the number of states of the finite state automaton (in the case of ordinary parsing the number of states equals $n + 1$ where n is the number of words of the input sentence) (Lang 1974, Billot & Lang 1989).

Let a finite-state machine M be specified by a tuple $(Q, \Sigma, \delta, Q^S, Q^F)$ where Q is a finite set of states, Σ is a finite alphabet, δ is a function $Q \times \Sigma \rightarrow 2^Q$. Furthermore, $Q^S \subseteq Q$ is the set of start states, and $Q^F \subseteq Q$ is the set of final states. [3]

Following Bar-Hillel, Perles & Shamir (1961) the CFG defining this intersection is constructed by keeping track of the state names in the non-terminal category symbols. For each rule $X_0 \rightarrow X_1 \ldots X_n$ there are rules $\langle X_0 q_0 q \rangle \rightarrow \langle X_1 q_0 q_1 \rangle \langle X_2 q_1 q_2 \rangle \ldots \langle X_n q_{n-1} q \rangle$, for all $q \in Q$. Furthermore for

3. For simplicitly we do not allow ϵ-moves in the finite state automaton, although the construction could easily be extended to treat those.

FSA: $ab^*(aab^*)^*$

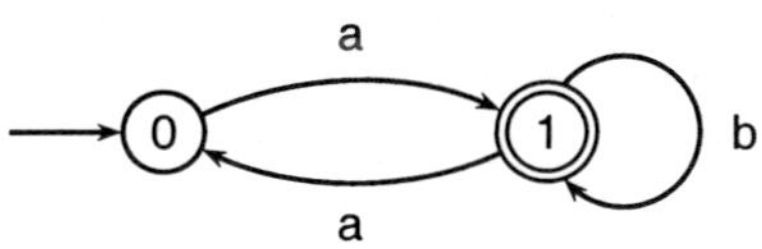

CFG: $a^n b^n$

$$
\begin{array}{lcl}
S & \rightarrow & \epsilon \\
 & | & aSb
\end{array}
$$

FIGURE 9.2: Example FSA and CFG to illustrate the construction of the intersection of a FSA and a CFG. For this example the intersection is the language consisting of an odd number of a's followed by the same number of b's.

$$
\begin{array}{rcl}
S & \rightarrow & \langle S\ 0\ 1\rangle \\
\langle S\ 0\ 0\rangle & \rightarrow & \epsilon \\
\langle S\ 1\ 1\rangle & \rightarrow & \epsilon \\
\langle S\ 0\ 0\rangle & \rightarrow & \langle a\ 0\ 0\rangle\langle S\ 0\ 0\rangle\langle b\ 0\ 0\rangle \\
 & | & \langle a\ 0\ 1\rangle\langle S\ 1\ 0\rangle\langle b\ 0\ 0\rangle \\
 & | & \langle a\ 0\ 1\rangle\langle S\ 1\ 1\rangle\langle b\ 1\ 0\rangle \\
 & | & \langle a\ 0\ 0\rangle\langle S\ 0\ 1\rangle\langle b\ 1\ 0\rangle \\
\langle S\ 0\ 1\rangle & \rightarrow & \langle a\ 0\ 0\rangle\langle S\ 0\ 0\rangle\langle b\ 0\ 1\rangle \\
 & | & \langle a\ 0\ 1\rangle\langle S\ 1\ 0\rangle\langle b\ 0\ 1\rangle \\
 & | & \langle a\ 0\ 1\rangle\langle S\ 1\ 1\rangle\langle b\ 1\ 1\rangle \\
 & | & \langle a\ 0\ 0\rangle\langle S\ 0\ 1\rangle\langle b\ 1\ 1\rangle \\
\langle S\ 1\ 0\rangle & \rightarrow & \langle a\ 1\ 0\rangle\langle S\ 0\ 0\rangle\langle b\ 0\ 0\rangle \\
 & | & \langle a\ 1\ 1\rangle\langle S\ 1\ 0\rangle\langle b\ 0\ 0\rangle \\
 & | & \langle a\ 1\ 1\rangle\langle S\ 1\ 1\rangle\langle b\ 1\ 0\rangle \\
 & | & \langle a\ 1\ 0\rangle\langle S\ 0\ 1\rangle\langle b\ 1\ 0\rangle \\
\langle S\ 1\ 1\rangle & \rightarrow & \langle a\ 1\ 0\rangle\langle S\ 0\ 0\rangle\langle b\ 0\ 1\rangle \\
 & | & \langle a\ 1\ 1\rangle\langle S\ 1\ 0\rangle\langle b\ 0\ 1\rangle \\
 & | & \langle a\ 1\ 1\rangle\langle S\ 1\ 1\rangle\langle b\ 1\ 1\rangle \\
 & | & \langle a\ 1\ 0\rangle\langle S\ 0\ 1\rangle\langle b\ 1\ 1\rangle \\
\langle a\ 0\ 1\rangle & \rightarrow & a \\
\langle a\ 1\ 0\rangle & \rightarrow & a \\
\langle b\ 1\ 1\rangle & \rightarrow & b
\end{array}
$$

FIGURE 9.3: Applying the construction of Bar-Hillel, Perles and Shamir (1961) to the example in figure 9.2.

each $q_k \in \delta(q_i, \sigma)$ we have a rule $\langle \sigma q_i q_k \rangle \rightarrow \sigma$. Finally, there are rules rewriting the start symbol S into tuples $\langle S\ q_s\ q_f \rangle$ for each $q_s \in Q^S, q_f \in Q^F$. The intersection of a CFG and a FSA is thus given as a CFG for the strings in the intersection.

Although this construction shows that the intersection of a FSA and a CFG is itself a CFG, it is not of practical interest. The reason is that this construction typically yields an enormous amount of rules with *useless* symbols (Hopcroft & Ullman 1979), i.e. symbols which are not derivable from the start state, and/or from which no sequence of terminal symbols is derivable. In fact the (possibly enormously large) parse forest grammar might define an empty language (if the intersection was empty). Luckily 'ordinary' parsers for CFG can be generalized to construct this intersection yielding (in typical cases) a much smaller grammar. Pure bottom-up parsers ensure that the symbols in the parse forest grammar do indeed derive a number of terminal symbols; pure top-down parsers ensure that the symbols in the parse forest grammar are derivable from the start symbol. For some parsers (e.g. a generalization of the parser described in Earley (1970)) it can be shown that the parse forest grammar will never contain useless symbols. Checking whether the intersection is empty or not then reduces to the question whether the parse forest grammar contains rules or not.

As an example, consider in figure 9.4 the simple bottom-up inactive chart parser presented abstractly in the style of Shieber, Schabes & Pereira (1994) and Sikkel (1993). The parser maintains items of the form $[p_i, X, p_j]$ indicating that a category X has been shown to exist between position p_i and p_j. The inference rules indicate how new items can be constructed from existing items. Side-effects are attached to these inference rules which assert the existence of rules in the parse forest grammar. The intersection is non-empty if and only if the parse forest grammar contains a rule rewriting the start symbol.

In figure 9.5 this parser is generalized to take a FSA as its input. Instead of string positions, the items now keep track of the state names of the FSA. Applying this parser to the example of figure 9.2 yields the parse forest grammar given in figure 9.6. Other parsing algorithms can be generalized in the same manner. In van Noord (1997a) we propose in detail a left/head-corner parser with selective memorization and goal-weakening which performs very well on word graphs.

Admittedly, some differences between parsing of strings and parsing of FSA are ignored because of the abstract presentation chosen here. For instance, the parser in figure 9.4 can be implemented under a left-to-right processing regime in such a way that for each item i that is constructed it is only necessary to consider the completion inference rule using i as the right-most antecedent item (such a simplication is possible if the grammar does not contain ϵ productions). In constrast, for the parser in figure 9.5 an item i could trigger the completion inference rule as any one of the antencedent items. Note however that in the special case of word graphs a left-to-right processing regime is equally possible by a topological sort of the state names.

9.3.3 The intersection of a DCG and an FSA

The DCG formalism is a simple example of a family of constraint-based grammar formalisms that are widely used in computational linguistics. DCG is more powerful than context-free grammars because instead of atomic symbols the grammar formalism employs first-order terms of arbitrary complexity. Such first-order terms allow for a succint treatment of various linguistic phenomena including *agreement*, *subcategorisation* and *non-local dependencies* as well as for the construction of semantic representations.

First note that the problem of calculating the intersection of a DCG and a FSA can be solved trivially by a generalization of the construction by Bar-Hillel et al. (1961). However, if we use that method we will end up, as before, with an enormously large number of rules in the parse forest grammar.[4] Therefore, we are interested in methods which will only generate a subset of these rules by the requirement that each symbol must be *useful*.

4. In fact, the standard compilation of DCG into Prolog clauses does something similar using variables instead of actual state names. This also illustrates that this method is not very useful yet; all the work still has to be done.

Scan $\frac{}{[p_i, w, p_{i+1}]}$

condition: w is the $i+1$th word of the string
side-effect: $\langle w\ p_i\ p_{i+1}\rangle \rightarrow w$

Complete $\frac{[p_k, X_1, p_{k'}][p_{k'}, X_2, p_{k''}]\dots[p_{m'}, X_l, p_m]}{[p_k, X_0, p_m]}$

condition: $X_0 \rightarrow X_1, X_2 \dots X_l$
side-effect: $\langle X_0\ p_k\ p_m\rangle \rightarrow \langle X_1\ p_k\ p_{k'}\rangle\langle X_2\ p_{k'}\ p_{k''}\rangle \dots \langle X_l\ p_{m'}\ p_m\rangle$

Finish $\underline{[p_0, S, p_n]}$

condition: n is the length of the string
side-effect: $S \rightarrow \langle S\ 0\ n\rangle$

FIGURE 9.4: Specification of a simple bottom-up inactive chart parser. The parse forest grammar is constructed as a side-effect. It contains derivations if and only if a rule rewriting the start symbol has been constructed.

Scan $\frac{}{[q_i, w, q_j]}$

condition: $q_j \in \delta(q_i, w)$
side-effect: $\langle w\ q_i\ q_j\rangle \rightarrow w$

Complete $\frac{[q_k, X_1, q_{k'}][q_{k'}, X_2, q_{k''}]\dots[q_{m'}, X_l, q_m]}{[q_k, X_0, q_m]}$

condition: $X_0 \rightarrow X_1, X_2 \dots X_l$
side-effect: $\langle X_0\ q_k\ q_m\rangle \rightarrow \langle X_1\ q_k\ q_{k'}\rangle\langle X_2\ q_{k'}\ q_{k''}\rangle \dots \langle X_l\ q_{m'}\ q_m\rangle$

Finish $\underline{[q_s, S, q_f]}$

condition: $q_f \in Q^F, q_s \in Q^S$
side-effect: $S \rightarrow \langle S\ q_s\ q_f\rangle$

FIGURE 9.5: Specification of a simple bottom-up inactive chart parser, generalized to compute the intersection of a FSA and a CFG. As before, the parse forest grammar is constructed as a side-effect and it contains derivations if and only if a rule rewriting the start symbol has been constructed.

$$
\begin{array}{lcl}
S & \rightarrow & \langle S\ 0\ 1\rangle \\
\langle S\ 0\ 0\rangle & \rightarrow & \epsilon \\
\langle S\ 1\ 1\rangle & \rightarrow & \epsilon \\
\langle S\ 0\ 1\rangle & \rightarrow & \langle a\ 0\ 1\rangle\langle S\ 1\ 1\rangle\langle b\ 1\ 1\rangle \\
\langle S\ 1\ 1\rangle & \rightarrow & \langle a\ 1\ 0\rangle\langle S\ 0\ 1\rangle\langle b\ 1\ 1\rangle
\end{array}
\qquad
\begin{array}{lcl}
\langle S\ 1\ 0\rangle & \rightarrow & \langle a\ 1\ 0\rangle\langle S\ 0\ 1\rangle\langle b\ 1\ 1\rangle \\
\langle a\ 0\ 1\rangle & \rightarrow & a \\
\langle a\ 1\ 0\rangle & \rightarrow & a \\
\langle b\ 1\ 1\rangle & \rightarrow & b
\end{array}
$$

FIGURE 9.6: Resulting parse forest grammar after applying the simple bottom-up chart parser to the example FSA and CFG.

A straightforward approach is to generalize existing recognition algorithms. The same techniques that are used for calculating the intersection of a FSA and a CFG can be applied in the case of DCGs.

But we are also confronted with an undecidability result: the recognition problem for DCGs is undecidable (Pereira & Warren 1983). A yes-no problem is *undecidable* (Hopcroft & Ullman 1979, pp.178-179) if there is no algorithm that takes as its input an instance of the problem and determines whether the answer to that instance is 'yes' or 'no'. An instance of a problem consists of a particular choice of the parameters of that problem. An algorithm is required to terminate for all possible inputs. The undecidability of the recognition problem for DCG thus implies that it is impossible to define a parser for DCGs which terminates for all inputs. A fortiori the problem of deciding whether the intersection of a FSA and a DCG is empty is undecidable (because that problem subsumes the recognition problem).

This undecidability result is usually circumvented by considering subsets of DCGs which can be parsed effectively. For example, we can restrict the attention to DCGs of which the context-free skeleton does not contain cycles. Recognition for such 'off-line parsable' grammars is decidable (Pereira & Warren 1983). Most existing constraint-based parsing algorithms will terminate for grammars that exhibit the property that for each string there is only a finite number of possible derivations (from this perspective, both rules with an empty right hand side and rules with a right hand side containing a single daughter are suspicious). Note that off-line parsability is one possible way of ensuring that this is the case.

This observation is not helpful in establishing insights concerning interesting subclasses of DCGs for which termination can be guaranteed in the case of FSA input. The reason is that the emptyness problem for off-line parse DCGs is undecidable as well: there is no algorithm to compute whether a given DCG generates any strings at all. We provide a proof of this fact below. The emptyness problem can be seen as a specific instance of the intersection problem, namely the intersection with Σ^*: the universal language consisting of all strings over the alphabet Σ. Therefore, undecidability of the intersection problem of FSA and off-line parsable DCG follows from the undecidability of the emptyness problem for off-line parsable DCG. This indicates that we cannot hope to reach our initial goal (a procedure which returns a DCG without useless sym-

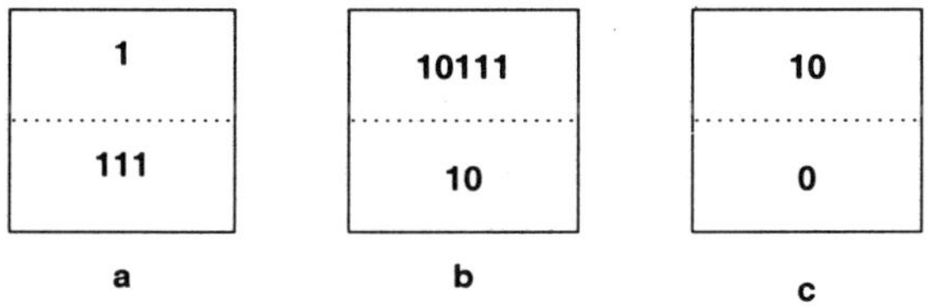

FIGURE 9.7: Instance of a Post's Correspondence Problem: is there a way to place a number of a, b and c blocks in such a way that the sequence of 0 and 1 read from left to right on the top halves of the blocks is the same as the sequence read on the bottom halves?

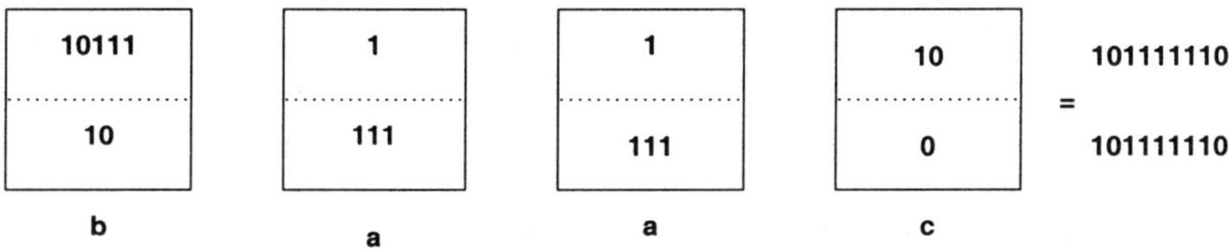

FIGURE 9.8: A solution to the example PCP: the sequence of blocks $baac$ indeed yields the same sequence on the top halves and on the bottom halves.

bols deriving the intersection of a given FSA and an off-line parsable DCG).

I use Post's Correspondence Problem (PCP) as a well-known and simple undecidable problem. I show that if the emptyness problem for off-line parsable DCG were decidable, then we could solve the PCP too. Since the latter is known to be undecidable this shows that the former cannot be decidable. The following definition and example of a PCP are taken from Hopcroft & Ullman (1979)[chapter 8.5].

An instance of PCP consists of two lists, $A = v_1 \dots v_k$ and $B = w_1 \dots w_k$ of strings over some alphabet Σ. This instance has *a solution* if there is any sequence of integers $i_1 \dots i_m$, with $m \geq 1$, such that

$$v_{i_1}, v_{i_2}, \dots, v_{i_m} = w_{i_1}, w_{i_2}, \dots, w_{i_m}.$$

The sequence $i_1, \dots, i_m$ is a solution to this instance of PCP. As an example, assume that $\Sigma = \{0, 1\}$. Furthermore, let $A = \langle 1, 10111, 10 \rangle$ and $B = \langle 111, 10, 0 \rangle$. A solution to this instance of PCP is the sequence 2,1,1,3 (obtaining the sequence 101111110). For an illustration, cf. the figures 9.7 and 9.8.

Some PCP's do not have a solution. Assume again that $\Sigma = \{0, 1\}$. Furthermore let $A = \langle 1 \rangle$ and $B = \langle 0 \rangle$. Clearly this PCP does not have a solution. In general, however, the problem whether some PCP has a solution or not is not decidable. This result is proved by Hopcroft & Ullman (1979) by showing that the halting problem for Turing Machines can be encoded as an instance of Post's Correspondence Problem.

I now give a simple algorithm to encode any instance of a PCP as an off-line parsable DCG, in such a way that the question whether there is a solution

$s \rightarrow \text{series}(X, [\,], X, [\,])$.

$\text{series}(A0, A, B0, B) \rightarrow \text{block}(A0, A, B0, B)$.
$\text{series}(A0, A, B0, B) \rightarrow \text{block}(A0, A1, B0, B1), \text{series}(A1, A, B1, B)$.

$\text{block}([1|A], A, [1, 1, 1|B], B) \rightarrow [x]$.
$\text{block}([1, 0, 1, 1, 1|A], A, [1, 0|B], B) \rightarrow [x]$.
$\text{block}([1, 0|A], A, [0|B], B) \rightarrow [x]$.

FIGURE 9.9: Equivalent off-line parsable DCG for the PCP given in figure 9.7.

to this PCP is equivalent to the question whether the DCG generates any strings at all. This suffices to show that the question whether an off-line parsable DCG generates the empty set is undecidable.

Encoding of PCP

1. For each $1 \leq i \leq k$ (k the length of lists A and B) define a DCG rule (the i-th member of A is the string $a_1 \ldots a_m$, and the i-th member of B is $b_1 \ldots b_n$):
   $\text{block}([a_1 \ldots a_m|A], A, [b_1 \ldots b_n|B], B) \rightarrow [x]$.
2. In addition, the DCG contains the following rules:
   $\text{series}(A_0, A, B_0, B) \rightarrow \text{block}(A_0, A_1, B_0, B_1), \text{series}(A_1, A, B_1, B)$.
   $\text{series}(A_0, A, B_0, B) \rightarrow \text{block}(A_0, A, B_0, B)$.
   $s \rightarrow \text{series}(X, [\,], X, [\,])$.
3. s is the start category of the DCG.

Observe that the DCG is off-line parsable.

For each pair of strings from the lists A and B there will be one lexical entry (deriving the terminal x) where these strings are represented by a difference-list encoding. Furthermore there is a general combination rule that simply concatenates A strings and concatenates B strings. Finally the rule for the start symbol s states that in order to construct a succesful top category the A and B lists must match. The DCG for the example in figure 9.7 is given in figure 9.9.

9.3.4 Intersection of acyclic FSA and off-line parsable DCG

Summarizing, it is impossible to decide whether the intersection of a FSA and an off-line parsable DCG is empty or not; and it is impossible to produce a parse forest grammar for this intersection without useless rules. In van Noord (1995) we discussed various alternative approaches to deal with this undecidability result. For purposes of this paper it suffices to ignore the possibility that the input FSA contains cycles because word graphs are acyclic.

Thus, rather than assuming the input for parsing is a FSA in its full generality, we might assume that the input is acyclic. This implies that the techniques for robust processing that give rise to such cycles (as discussed in

$$\textbf{Scan} \quad \frac{}{[q_i, w, q_j]}$$

condition: $q_j \in \delta(q_i, w)$

$$\textbf{Complete} \quad \frac{[q_k, X_1, q_{k'}][q_{k'}, X_2, q_{k''}] \dots [q_{m'}, X_l, q_m]}{[q_k, \sigma(X_0), q_m]}$$

condition: $X_0 \rightarrow X'_1, X'_2 \dots X'_l$
and $\sigma = \text{mgu}((X'_1, X'_2 \dots X'_l), (X_1, X_2 \dots X_l))$

Finish $\underline{[q_s, S', q_f]}$

condition: $q_f \in Q^F, q_s \in Q^S, \sigma = mgu(S', S)$
side-effect: $\sigma(S : \text{sem})$

FIGURE 9.10: Specification of a simple bottom-up inactive chart parser, generalized to compute the intersection of an acyclic FSA and an off-line parsable DCG.

subsection 9.3.1) cannot be used.

Note that it is easy to verify that the question whether the intersection of an acyclic FSA and an off-line parsable DCG is empty or not is decidable since it reduces to checking whether the DCG derives any one of a finite number of strings. A practical implementation will not use this method, but instead use a generalized DCG parser along the lines of the generalizations discussed in subsection 9.3.2. An example of such a parser is given in figure 9.10.

For simplicity we ignore the construction of the parse forest grammar and exploit the possibility of the DCG formalism to encode information in the categories. In particular, we assume that we are interested in obtaining a semantic representation which is available as part of the start category S; we refer to this part of a category using the notation S : sem. As a result, this parser produces a set of semantic representations for all strings in the intersection. Here, mgu(X, X') is the most general unifier of X and X', i.e. the least specific substitution of the variables in X and X' making X and X' identical.

9.4 Robustness and Word Graph Parsing

9.4.1 Introduction

In the ideal case, the parser will find a path in the word graph that can be assigned an analysis according to the grammar, such that the path covers the complete time span of the utterance, i.e. the path leads from the start state to a final state. The analysis gives rise to a semantic representation which is then passed on to the next processing component.

However, often no such paths can be found in the word graph, due to:

- errors made by the speech recognizer,
- linguistic constructions not covered in the grammar, and
- irregularities in the spoken utterance.

Even if no full analysis of the word graph is possible, it is often the case that useful information can be extracted from the word graph. Consider for example the utterance:

Ik	reizen	van	Assen	naar	Amsterdam
I	to travel	from	Assen	to	Amsterdam

I want to travel from Assen to Amsterdam

The grammar might not assign an analysis to this utterance due to the missing finite verb. However, it would be useful if the parser discovered the prepositional phrases *van Assen* and *naar Amsterdam* since in that case the important information contained in the utterance can still be recovered. Thus, in cases where no full analysis is possible we would like to fall back on an approach reminiscent of concept spotting. The following proposal implements this idea.

Firstly, the grammar is defined in such a way that each *maximal projection* such as S, NP, PP, etc., can be analysed as a top category. This is well-motivated because utterances very often consist of a single NP or PP anyway.

Secondly, the parser is required to discover all instances of the top category *anywhere in the word graph*, i.e. for all partial paths in the word graph. This has the desired effect for the example on page 220: both PPs will be found by the parser (as well as the noun phrases "Ik", "Assen" and "Amsterdam"). The consequences for parsing *any* sequence in the word-graph, rather than a sequence from the beginning to the end, are discussed in section 9.4.2.

Thirdly, an appropriate search algorithm will find a non-overlapping sequence of such top categories connected by uncovered words in the word-graph. For the example on page 220 this results for instance in the sequences:

⟨NP reizen van NP naar NP⟩
⟨NP reizen van NP PP⟩
⟨NP reizen PP naar NP⟩
⟨NP reizen PP PP⟩

Using appropriate heuristics, including a heuristic to minimize the number of uncovered words, we are then able to pick out the last path as the best path in this particular case.

Thus, we are interested in paths from the start state to the final state consisting of a number of categories and transitions in the word graph (the latter are called *skips*). The problem is to find the optimal path, according to a number of criteria. This problem is formalized by defining the annotated word graph as a directed acyclic graph (section 9.4.3). The vertices of this graph are the states of the word graph; the edges are the transitions of the word graph and the categories found by the parser.

The criteria which are used to favor some paths over other paths are expressed as a weight function on the edges of the graph. The criteria we might take into account are discussed in section 9.4.4. For instance, a typical criterion will favor paths consisting of a small number of categories, and a small number of skips. The case in which the parser found a full analysis from the start state of the word graph to a final state then reduces to a special case: the analysis solely consisting of that category will be favored over sequences of partial analyses.

Obviously, it is not a good idea to generate all possible sequences of categories and skips, and then to select the best path from this set: in typical word graphs there are simply too many different paths. If a certain uniformity requirement on weights is met, however, then efficient graph search algorithms are applicable. The particular algorithm discussed in section 9.4.5 is a variant of the DAG-SHORTEST-PATH algorithm (Cormen, Leiserson & Rivest 1990).

The result of the search algorithm consists of a sequence of semantic representations. Depending on the nature of the task, these representations can be sent directly to the next processing phase, or a post-processor combines this sequence into a single representation.

The criteria used to determine the best path may also include Ngram statistics. It turns out that in those cases some complications arise in the definition of the annotated word graph. This is explained in section 9.4.6.

A full parse of the word graph is impracticable for large word graphs. Therefore we consider (in subsection 9.4.7) techniques to filter the word graph in a pre-processing phase.

In section 9.4.8 we discuss an approximation of the search algorithm (augmented to treat Ngrams) which is not guaranteed to find the best result, but which is much more efficient than the sound algorithm.

9.4.2 Parsing everywhere

We require that the parser finds all instances of the start category anywhere in the word graph. If a purely bottom-up parser is used, then this modification is straightforward. In figure 9.11 we have adapated the bottom-up chart parser given earlier to the present purpose. The parser maintains items of the form $[v_i, X, v_j, a, w]$ indicating that a category X has been shown to exist between states v_i and v_j, with accumulated acoustic score a and sequence of words w. As a side-effect the parser creates a number of tuples $parsed(v_i, v_j, x, a, q)$ indicating an instance of the top category from word graph state v_i to word graph state v_j, with associated sequence of words x, and accumulated acoustic score a; q is the associated semantic representation.

As an example, the result of applying a parser for Dutch to the normalized word graph of figure 9.1 is given in figure 9.12.

9.4.3 Annotated word graph

An annotated word graph is a word graph annotated with the results of the parser. Such an annotated word graph is defined with respect to an input word

Scan $\dfrac{}{[v_i, w, v_j, a, w]}$

condition: trans(v_i, v_j, w, a)

Complete $\dfrac{[v_k, X_1, v_{k'}, a_1, w_1][v_{k'}, X_2, v_{k''}, a_2, w_2] \ldots [v_{m'}, X_l, v_m, a_l, w_l]}{[v_k, \sigma(X_0), v_m, a_1 + a_2 \ldots a_l, w_1 \cdot w_2 \ldots w_l]}$

condition: $X_0 \rightarrow X'_1, X'_2 \ldots X'_l$
and $\sigma = \text{mgu}((X'_1, X'_2 \ldots X'_l), (X_1, X_2 \ldots X_l))$

Finish $\dfrac{[v_i, S', v_j, a, w]}{}$

condition: $\sigma = mgu(S', S)$
side-effect: parsed$(v_i, v_j, w, a, \sigma(S : \text{sem}))$

FIGURE 9.11: Specification of a simple bottom-up inactive chart parser, generalized to find all instances of the start symbol anywhere in the word graph (keeping track of the associated acoustic score as well as the sequence of words).

1.	parsed(1,	3,	⟨zondag⟩,	27350,	sunday	)
2.	parsed(3,	9,	⟨precies⟩,	17831,	exactly	)
3.	parsed(3,	10,	⟨drie,uur⟩,	19044,	3 o'clock	)
4.	parsed(3,	12,	⟨de,vierde⟩,	22356,	the fourth	)
5.	parsed(3,	13,	⟨de,vierde,februari⟩,	48978,	february 4th	)
6.	parsed(3,	13,	⟨dertien,februari⟩,	49238,	february 13	)
7.	parsed(3,	13,	⟨veertien,februari⟩,	49239,	february 14	)
8.	parsed(3,	13,	⟨vier,februari⟩,	49106,	february 4	)
9.	parsed(6,	13,	⟨vier,februari⟩,	43460,	february 4	)
10.	parsed(7,	13,	⟨vier,februari⟩,	42330,	february 4	)

FIGURE 9.12: Output of the parser given in figure 9.11 with input the normalized word graph of figure 9.1. Each item lists respectively the source state, the target state, the sequence of words and the accumulated acoustic score. The fifth argument normally is the associated semantic representation; we provide the English translation of the sequence of words instead.

graph (given by the functions *trans* and *final*) and with respect to the results of parsing (given by the function *parsed*).

The *annotated word graph* is a directed acyclic graph (V, E, v_s, v_f) where

- V is the set of vertices consisting of the states of the word graph $v_0 \ldots v_n$, and a new vertex v_{n+1}. v_s is the start state of the word graph and the start state of the annotated word graph. $v_f = v_{n+1}$ is the single final state of the annotated word graph.
- E is the set of edges consisting of:
  1. *skip* edges. For all *trans*(v_i, v_j, w, a) there are edges $(v_i, v_j, w, a, \epsilon)$.
  2. *category* edges. For all *parsed*(v_i, v_j, x, a, q) there are edges (v_i, v_j, x, a, q).
  3. *stopping* edges. For all final(v_i, a) there are edges $(v_i, v_{n+1}, \epsilon, a, \epsilon)$.

9.4.4 Weights

The weights that are associated with the edges of the graph can be sensitive to the following factors.

- Acoustic score. Obviously, the acoustic score present in the word graph is an important factor. The acoustic scores are derived from probabilities by taking the negative logarithm. For this reason we aim to minimize this score. If edges are combined, then we have to sum the corresponding acoustic scores.
- Number of 'skips'. We want to minimize the number of skips, in order to obtain a preference for the maximal projections found by the parser. Each time we select a skip edge, the number of skips is increased by 1.
- Number of maximal projections. We want to minimize the number of such maximal projections, in order to obtain a preference for more extended linguistic analyses over a series of smaller ones. Each time we select a category edge, this number is increased by 1.
- Ngram statistics. We have experimented with bigrams and trigrams. Ngram scores are expressed as negative logarithms of probabilities. This implies that combining Ngram scores requires addition, and that lower scores are to be preferred.

The only requirement we make to ensure that efficient graph searching algorithms are applicable is that weights are *uniform*. This means that a weight for an edge leaving a vertex v_i is independent of how state v_i was reached.

In order to be able to compute with such multi-dimensional weights, we express weights as tuples $\langle c_1, \ldots, c_k \rangle$. For each cost component c_i we specify an initial weight, and we need to specify for each edge the weight of each cost component. To specify how weights are updated if a path is extended, we use the function *uw* that maps a pair of a multi-dimensional weight and an edge to a multi-dimensional weight. [5] Moreover, we need to define an ordering on such tuples. In order to experiment with different implementations of this idea we refer to such a collection of specifications as a *method*. Summarizing, a weight method is a triple $W = \langle ini, uw, \prec \rangle$ where

1. *ini* is the initial weight;
2. *uw* is the update weight function;
3. $\prec$ is an ordering on weights

Example: speech method As a trivial example of such a method, consider the problem of finding the best path through the word graph ignoring all aspects but the acoustic scores present in the word graph. In order to implement a method W^{speech} to solve this problem, we define weights using a unary tuple $\langle c \rangle$. The initial weight is $ini = \langle 0 \rangle$ and uw is defined as follows:

$$uw(\langle c \rangle, (v_i, v_j, w, a, q)) = \begin{cases} \langle c + a \rangle & \text{for skip edges} \\ \langle \infty \rangle & \text{for category edges} \\ \langle c + a \rangle & \text{for stopping edges} \end{cases}$$

Note that we specify an infinite weight for category edges because we want to ignore such edges for this simple method (i.e. we are simply ignoring the results of the parser). We define an ordering $\prec$ on such tuples, simply by stating that $\langle w \rangle \prec \langle w' \rangle$ iff $w < w'$.

Example: nlp_speech method A more interesting example is provided by the following method which not only takes into account acoustic scores, but also the number of skip edges and category edges. Weights are expressed as $\langle c_1, c_2, c_3 \rangle$, where c_1 is the number of skips, c_2 is the number of categories, and c_3 is the acoustic score.

We define $ini = \langle 0, 0, 0 \rangle$ and uw is defined as follows.

$$uw(\langle c_1, c_2, c_3 \rangle, (v_i, v_j, w, a, q)) : \begin{cases} \langle c_1 + 1, c_2, c_3 + a \rangle & \text{for skip edges} \\ \langle c_1, c_2 + 1, c_3 + a \rangle & \text{for category edges} \\ \langle c_1, c_2, c_3 + a \rangle & \text{for stopping edges} \end{cases}$$

Finally, we define the ordering on such tuples:

$$\langle c_1, c_2, c_3 \rangle \prec \langle c_1', c_2', c_3' \rangle \text{ iff} : \begin{cases} c_1 < c_1' \text{ or} \\ c_1 = c_1' \text{ and } c_2 < c_2' \text{ or} \\ c_1 = c_1' \text{ and } c_2 = c_2' \text{ and } c_3 < c_3' \end{cases}$$

5. We do not define a weight function on edges, but we specify how weights are updated if a path is extended, for generality. This approach allows e.g. for the possibility that different cost components employ different operations for combining weights. For example, some cost components may use addition (e.g. for weights which are expressed as negative logarithms derived from probabilities), whereas other cost components may require multiplication (e.g. for probabilities).

path	#skips	#cats	acoustic
$\langle c_1, c_5, s\rangle$	0	2	108144
$\langle c_1, c_8, s\rangle$	0	2	108272
$\langle c_1, c_6, s\rangle$	0	2	108404
$\langle c_1, c_7, s\rangle$	0	2	108505
$\langle(1, 3, zondag, 27350, \epsilon), c_5, s\rangle$	1	1	108144

FIGURE 9.13: Example of the application of the *nlp_speech* method to the normalized wordgraph of figure 9.1 annotated with the results of the parser given in figure 9.12. A term c_i refers to a category edge based on parsed item numbered i in that figure. The terms s is the single stopping edge $(13, 14, \epsilon, 31816, \epsilon)$. The table lists the best five results.

As an example, consider the application of this method to the normalized wordgraph of figure 9.1 annotated with the results of the parser given in figure 9.12. The best five paths according to this method are given in figure 9.13. The user utterance corresponds to the second-best path.

9.4.5 Search algorithm

The robustness component can be characterised as a search in the annotated word graph. The goal of the search is the best path from v_0 to v_{n+1}. This search reduces to a well-known graph search problem, namely the problem of finding the shortest path in a directed acyclic graph with *uniform* weights.

We use a variant of the DAG-SHORTEST-PATH algorithm (Cormen et al. 1990). This algorithm finds shortest paths for uniformly weighted directed acyclic graphs. The first step of the algorithm is a topological sort of the vertices of the graph. In this ordering, a vertex v_i precedes v_j if there is no path from v_j to v_i. The topological sort is required because in the second step each edge with source state v_i is treated in turn using the fact that all edges ending in v_i have already been treated. The second step of the algorithm maintains an array A which records for each state v_k the weight associated with the best path known from v_0 to v_k. Another array, P, is used to represent for each state the history of this best path, as a sequence of edges. The first three lines of the algorithm initialises these arrays. After this initialisation phase the algorithm treats each edge of the graph in topologically sorted order of the source vertex, and applies the function relax. The function *relax* is defined on edges and updates the arrays if a better path to a vertex has been found. The algorithm is presented in figure 9.14.

When the algorithm finishes, $P[v_{n+1}]$ constitutes the sequence of edges associated with the best path found. The weight of this path is given by $A[v_{n+1}]$. This algorithm is efficient. Its running time is $O(V+E)$, where V is the number of vertices and E is the number of edges. Therefore, it can be expected that this part of processing should not decrease parsing efficiency too much, since the number of edges is $O(V^2)$. For a more detailed account of the correctness and

```
function dag-shortest-path (V, E, v_s, v_f)
    foreach v_i ∈ V
        A[v_i] = ∞; P[v_i] = NIL
    A[v_s] = ini; P[v_s] = ε
    foreach state v_i ∈ top-sort(V)
        foreach edge (v_i, v_j, w, a, q) ∈ E
            relax (v_i, v_j, w, a, q)
    return A[v_f], P[v_f]

function relax (v_i, v_j, w, a, q)
    if uw(A[v_i], (v_i, v_j, w, a, q)) ≺ A[v_j] then
        A[v_j] =: uw(A[v_i], (v_i, v_j, w, a, q))
        P[v_j] =: P[v_i].(v_i, v_j, w, a, q)
```

FIGURE 9.14: The DAG-SHORTEST-PATH algorithm.

complexity of this algorithm, see Cormen et al. (1990).[6]

A simple generalisation of the algorithm can be implemented in order to obtain the P best paths. In this case we maintain in the algorithm for each vertex the P best paths found so far. Such a generalisation increases the worst-case complexity by only a constant factor. This could be used to pass on the P best paths to the next processing stage. It is very useful for development work as well, and it allows for the implementation of the 'filtering' methods discussed in subsection 9.4.7.

9.4.6 Extension for Ngrams

In this section we want to extend the *nlp_speech* method to take into account Ngram probabilities. Obviously, we can extend the weight tuple with a further argument which expresses the accumulated weight according to the Ngram probabilities. However, a potential problem arises. If we extend a given path by using a transition labeled w, then we take into account the probability of reading this word w given the previous $N-1$ words. However note that in the definition of the annotated word graph given above these words are not readily available. Even if we make sure that each path is associated with the last words seen so far, we must be careful that weights remain uniform.

The solution to this problem is to alter the definition of the graph, in such a way that the relevant $N-1$ words are part of the vertex. If we take into account Ngram probabilities (for some fixed $N = 2, 3, \ldots$), then the vertices of the graph will be tuples $(v, w_1 \ldots w_{N-1})$ where v is a state of the word graph as before, and $w_1 \ldots w_{N-1}$ are the previous $N-1$ words. For example, in the case of

6. Note that the algorithm is different from the Viterbi algorithm. The latter algorithm finds the best path through a possibly cyclic weighted graph, *for a given sequence of observed outputs.* In the current application we require an algorithm to find the best path in an *acyclic* weighted graph (without an additional observed output sequence).

bigrams ($N = 2$), vertices are pairs consisting of a word graph state and a word (the previous word). A number of special symbols $y_{N-1} \dots y_1$ is introduced as beginning-of-sentence markers. The start vertex is now $(v_0, y_{N-1} \dots y_1)$. The notation $x : k$ is used to refer to the last k elements of x.

The *annotated word graph for Ngrams (with fixed N)* is a directed acyclic graph (V, E, v_s, v_f) such that:

- V is a set of pairs $(v, w_1 \dots w_{N-1})$ where v is a word graph state and w_i are labels in the word graph. The start vertex is $v_s = (v_0, y_{N-1} \dots y_1)$; the final vertex $v_f = (v_{n+1}, \epsilon)$.
- E is the set of edges consisting of:
  1. *skip* edges. For all *trans*(v_i, v_j, w, a), and all vertices $V_i = (v_i, x)$ and $V_j = (v_j, xw : N-1)$, there are edges $(V_i, V_j, w, a, \epsilon)$.
  2. *category* edges. For all *parsed*(v_i, v_j, x_2, a, q), and for all vertices $V_i = (v_i, x_1)$ and $V_j = (v_j, x_1x_2 : N-1)$, there are edges (V_i, V_j, x_2, a, q).
  3. *stopping* edges. For all final(v_i, a) and for all vertices $V_i = (v_i, x)$ there are edges $(V_i, (v_{n+1}, \epsilon), \epsilon, a, \epsilon)$.

Example: nlp_speech_trigram method For this method we fix $N = 3$, and therefore the start state of the graph search now is the vertex (v_0, y_2y_1). Weights are expressed as 4-tuples by extending the triples of the *nlp_speech* method with a fourth component expressing trigram weights. These trigram weights are expressed using negative logarithms of (estimates of) probabilities. Let tri be the function which returns for a given sequence of three words this number. Moreover, the definition of this function is extended for longer sequences of words in the obvious way by defining $tri(w_0w_1w_2x) = tri(w_0w_1w_2) + tri(w_1w_2x)$.

The initial weight is defined as *ini* $= \langle 0, 0, 0, 0 \rangle$. Weights are updated as follows:

$$uw(\langle c_1, c_2, c_3, c_4 \rangle, ((v_i, w_0w_1), (v_j, y), x, a, q)) = \begin{cases} \langle c_1 + 1, c_2, c_3 + a, c_4 + tri(w_0w_1x) \rangle & \text{for skip edges} \\ \langle c_1, c_2 + 1, c_3 + a, c_4 + tri(w_0w_1x) \rangle & \text{for category edges} \\ \langle c_1, c_2, c_3 + a, c_4 \rangle & \text{for stopping edges} \end{cases}$$

Finally, we define an ordering on such tuples. The function *total* is defined on tuples as follows. Here k_{nlp} and k_{wg} are constants.

$$total(\langle c_1, c_2, c_3, c_4 \rangle) = c_4 + k_{\text{nlp}} * (c_1 + c_2) + k_{\text{wg}} * c_3.$$

We then define the ordering on tuples C, C' simply as:

$$C \prec C' \text{ iff } total(C) < total(C').$$

9.4.7 Filtering Word Graphs

The method *nlp_speech_trigram*, involving a full parse of the word graph, is impracticable for large word graphs: both CPU-time and memory requirements become too demanding.

For this reason we have experimented with methods in which the word graph is filtered before it is passed on to the parser. Thus we make two passes in the word graph. In the first pass the best P paths are computed using Ngram statistics and acoustic scores. Transitions not taking part in any of these paths are removed from the word graph. In the second pass the *nlp_speech_trigram* method is applied to the filtered word graph. We refer to these methods as B-P and T-P, for various values of P:

B-P In the first pass this method selects the P best paths using bigram statistics. The second pass applies the *nlp_speech_trigram* method.

T-P In the first pass this method selects the P best paths using trigram statistics. The second pass applies the *nlp_speech_trigram* method.

9.4.8 A More Efficient Approximation

It turns out that the filtering methods discussed in the previous section still require unattractive processing times for larger word graphs. Therefore, this section introduces an efficient approximation technique. Rather than complicating the vertices structure of the search graph in order to be able to work with Ngram statistics (as we proposed in subsection 9.4.6), we now describe a variant of the search algorithm in which the search graph remains as simple as before. This variant is not sound, since it does not always yield the best path possible. In practice, however, it turns out (as we will show in a later section) that the resulting implementation almost always produces the same results, using only a fraction of the processing time required by the sound method.

The approximation is very similar to beam search techniques in that a constant b (the "beam") is assumed which indicates the maximum number of paths associated with a given vertex. Just like in the generalization for all P best paths defined in section 9.4.5, we maintain for each vertex the best b paths to that vertex. But in case Ngrams are added to the weight function, we cannot guarantee that we find the best path, because weights are not uniform anymore. It may be possible that a path ending in some vertex v_i is discarded by the algorithm, whereas the last sequence of words of that path would ensure that extending that path yields a very good score and in fact even better than any of the scores that result from extending any of the paths ending in v_i which were maintained by the algorithm.

Larger values for b entail more chance of obtaining the best path (therefore it makes sense to talk about approximation here); and for every word graph there will be some b such that the best path is produced.

Note that the complexity of the search algorithm is linearly dependent on b. Finally, b is always at least as great as P, i.e. the number of solutions we want to obtain.

We write M, b for a method M with beam b. For example, the method B-4, 10 is the method which filters the word-graph in the first phase using the approximation search with a beam $b = 10$ to select the best 4 paths using bigram statistics (and acoustic scores). The second phase, as always, consists of the (sound) application of the *nlp_speech_trigram* method.

9.4.9 Related Work

Partial parsing techniques have been proposed many times before; two overviews of partial parsing and related techniques are Carbonell & Hayes (1983) and Abney (1996); some related proposals in the context of spoken dialogue systems are Stallard & Bobrow (1992), Dowding, Gawron, Appelt, Bear, Cherny, Moore & Moran (1993), Ehrlich & Hanrieder (1996), and Lavie (1996).

Note that partial parsing does not neccessarily require the use of a purely bottom-up parser. In van Noord (1996) and van Noord (1997a) we showed that in a logic programming setting, the use of underspecification of the state names associated with the top-most goal obtains the desired effect, even for parsing algorithms which employ top-down filtering. Thus, rather than asking the parser to find an instance of the start category from the start state of the word graph to a final state of the word graph, we now ask the parser to find an instance of the start category between any two word graph states.

In Nederhof, Bouma, Koeling & van Noord (1997) a similar robust parsing method for word graphs is described, but instead of the DAG-SHORTEST-PATH algorithm, Dijkstra's algorithm is employed (Dijkstra 1959, Nilsson 1971). We prefer the current method because it is simpler and allows edges with negative scores.

In Kasper, Kiefer, Krieger, Rupp & Worm (1999) a very similar robust parsing method for wordgraphs is presented, which in addition permits anytime processing.

The search procedure yields a sequence of semantic representations. A post-processor can be defined which attempts to construct a single analysis from this sequence. This possibility is discussed for instance in Weischedel, Ayuso, Bobrow, Boisen, Ingria & Palmucci (1991) and Worm & Rupp (1998).

9.5 Experimental Results

9.5.1 Test material

The efficiency and adequacy of the proposed model and a number of variants is evaluated with respect to its performance on the OVIS2 test set. This test set is developed within the NWO Priority Programme *Language and Speech Technology*: a 5-year research programme aiming at the development of spoken language information systems for Dutch. Its immediate goal is the development of

TABLE 9.1: **Characterisation of test set (1).** This table lists the number of transitions, the number of states, the number of words of the actual utterances, the average number of transitions per word, the maximum number of transitions, and the maximum number of states. The first row provides those statistics for the input word graph; the second row for the pause eliminated word graph. The number of transitions per word is an indication of the extra ambiguity for the parser introduced by the word graphs in comparison with parsing of an ordinary string. These numbers indicate that the difficulty of this test-set lies mostly in the ambiguity in the word graphs; the utterances themselves are fairly simple (average length only 3.2 words per utterance).

	graphs	trans	states	words	t/w	max(t)	max(s)
input	1000	48215	16181	3229	14.9	793	151
normalized	1000	73502	11056	3229	22.8	2943	128

a demonstrator of a public transport information system, which operates over ordinary telephone lines. This demonstrator is called OVIS, Openbaar Vervoer Informatie Systeem (*Public Transport Information System*).

In this Programme, two alternative NLP modules are developed in parallel: a grammar-based (conventional, rule-based) module and a data-oriented (memory-based, stochastic, DOP) module. In order to compare the NLP modules, a formal evaluation has been carried out three years after the start of the Programme. The results of the evaluation are described in van Zanten, Bouma, Sima'an, van Noord & Bonnema (1999). The test set used for that evaluation is used here to illustrate the difference in performance of various variations of the algorithms.

The OVIS2 test set consists of 1000 word graphs (collected with a preliminary version of the system and typical for the application) annotated with the actual user utterance and the corresponding semantic representation. Some indication of the difficulty of the set of 1000 word graphs is presented in table 9.1 and figure 9.15.

9.5.2 Evaluation Criteria

For the experiments below we express the quality in terms of string accuracy (comparison of the best sentences with the test sentences), semantic accuracy (comparison of the best semantic representations with the test semantic representations) and computational resources.

String accuracy measures the distance of the test sentence and the best sentence, and is expressed in terms of word accuracy (WA). The string comparison on which word accuracy is based is defined by the minimal number of substitutions, deletions and insertions of words that is required to turn the best sentence into the test sentence (Levenshtein distance). Word accuracy is defined as

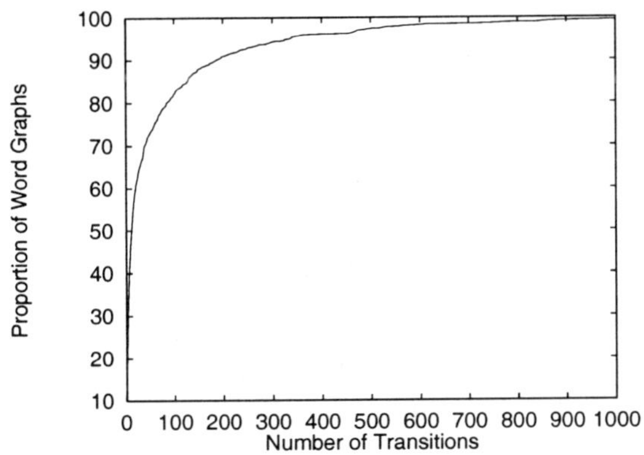

FIGURE 9.15: **Characterisation of test set (2).** The graph displays the distribution of pause eliminated word graph sizes (in number of transitions).

$$\text{WA} = 1 - \frac{d}{n}$$

where n is the length of the actual utterance and d is the Levenshtein distance. For example, if the analysis gives 'a b a c d' for the utterance 'a a c e', then the Levenshtein distance is 2, hence the WA is 1-2/4 is 50%.

Following a somewhat similar proposal in Boros, Eckert, Gallwitz, Görz, Hanrieder & Niemann (1996), we translate each semantic representation into a set of "semantic units". The syntax of the semantic unit language and the translation to semantic units is defined in van Noord (1997b).

Semantic accuracy (also called *concept accuracy*) is defined as in Boros et al. (1996):

$$\text{CA} = 1 - \frac{SU_S + SU_I + SU_D}{SU}$$

where SU is the total number of semantic units in the corpus annotation, and SU_S, SU_I, and SU_D are the number of substitutions, insertions, and deletions that are necessary to make the set of semantic units of the analysis equivalent to the set of semantic units of the corpus annotation.

In order to measure computational efficiency, we list the total amount of CPU-time, the maximum amount of CPU-time per input.

A further indication of the difficulty of this set of word graphs is obtained if we look at the word accuracy obtained by a number of simple methods (figure 9.2). The method *speech* only takes into account the acoustic scores found in the word graph. No language model is taken into account. The method *possible* assumes that there is an oracle which chooses a path such that it turns out to be the best possible path. This method can be seen as a natural upper bound of what can be achieved.

TABLE 9.2: **Characterization of test set (2).** Word accuracy based on acoustic score only (speech); using the best possible path through the word graph, based on acoustic scores only (possible).

method	WA %
speech	69.8
possible	90.5

For the construction of N-gram statistical models about 100K transcribed user utterances were available (collected with a preliminary version of the system, and typical for the intended application).

9.5.3 Results for Filtering Methods

The first set of experiments compares the accuracy of the *nlp_speech_trigram* method and the more efficient filtering methods defined in subsection 9.4.7. In order to find out how much quality we lose in the filtering process, consider the results for the subset of 822 word graphs with at most 100 transitions, presented in table 9.3. In this table the method *nlp* stands for the *nlp_speech_trigram* method applied to the full word graph. As is clear from the table, it turns out that the full nlp method performs better than any of the word graph filtering methods, but for larger values of P the filtering methods approximate the nlp method. Furthermore, the full nlp method is considerably slower. As expected, the filtering methods using trigrams work somewhat better, but slower, than the corresponding bigram filtering methods.

The table also lists the result for a version of the system in which no robustness component is assumed. In that case a result is obtained only if a parse exists from the beginning of the word graph to the end of word graph (using the acoustic score to differentiate between competing analyses). This result illustrates the importance of a robustness component (at least for this application).

The results for the filtering methods on the full test set are given in table 9.4. As is clear from the results, the CPU-time requirements increase rapidly for trigram filtering methods as well. The results, however, of the trigram filtering methods are better than for the corresponding bigram methods. This suggests that improvements could be obtained if it were possible to apply the trigram filtering methods more efficiently. The next section describes one such proposal.

9.5.4 Results for Approximation Methods

The second set of experiments compares the sound implementation with the approximation defined in subsection 9.4.8. In table 9.5 we compare the results for the approximating filtering methods for different values of b. The tables also list the results for the sound implementation. As can be observed, a

TABLE 9.3: Results for the subset of 822 word graphs which have at most 100 transitions.

Method	CPU		WA	CA
	mean	max		
	msec	sec	%	%
B-1	16	0.1	90.9	88.2
B-2	20	0.2	91.8	90.4
B-4	26	0.3	92.8	91.2
B-8	34	0.4	93.2	91.9
B-16	45	0.6	93.2	92.0
T-1	19	0.1	92.8	91.0
T-2	24	0.2	93.0	91.9
T-4	30	0.3	92.8	91.6
T-8	40	0.5	93.2	92.0
T-16	53	0.8	93.2	92.0
nlp	76	2.7	93.2	92.0
no robust	54	1.6	57.1	64.7

TABLE 9.4: Results for full test set: filtering methods.

Method	CPU		WA	CA
	mean	max		
	msec	sec	%	%
B-1	38	1.1	81.3	78.9
B-2	56	2.2	82.4	80.9
B-4	87	4.4	83.4	81.8
B-8	148	9.4	83.9	83.0
B-16	300	24.0	84.4	83.6
T-1	168	22.1	84.5	82.4
T-2	320	47.8	84.9	83.4
T-3	449	67.3	85.0	83.4
T-4	638	98.8	85.0	83.4
T-10	1801	339.6	85.0	83.5

TABLE 9.5: Results for the full test set. The top rows of the table compare for T-1 the approximating filtering method (using various beams) with the sound implementation. Using a beam of $b = 4$ already obtains results similar to the sound implementation; the CPU-time requirements are far more modest though. The lower rows of the table compare for B-8 the approximating filtering method (using various beams) with the sound implementation. The results indicate that using larger beams has less effect for larger values of P.

Method	CPU		WA	CA
	mean	max		
	msec	sec	%	%
T-1,1	26	0.4	79.6	78.1
T-1,2	30	0.6	82.2	80.6
T-1,3	33	0.8	83.7	82.1
T-1,4	37	1.0	84.0	82.3
T-1,5	40	1.2	84.0	82.3
T-1,10	58	2.7	84.3	82.4
T-1,20	102	6.5	84.4	82.4
T-1	168	22.1	84.5	82.4
B-8,8	73	2.2	83.7	82.8
B-8,12	89	3.0	83.8	83.0
B-8,16	105	4.3	83.8	83.0
B-8,20	125	5.8	83.9	83.0
B-8	148	9.4	83.9	83.0

value of $b = 4$ yielded for T-1 a concept accuracy which was already close to the sound implementation (82.3% versus 82.4%). The difference in speed was considerable. The difference in maximum amount of CPU-time was even more dramatic. The second part of the table shows that for methods in which more than a single result is obtained it hardly makes a difference to increase b.

In table 9.6 the results for various values of M, b are presented. These experiments clearly indicate that it is indeed possible to use an approximation to obtain useful results: similar results for word-accuracy can be obtained in only a fraction of the CPU-time. In fact, if we concentrate on the results for concept accuracy we can even perform somewhat better! The reason is that the faster implementation allows us to filter the best paths (e.g. the best 10 paths) from the word graph according to acoustic and Ngram scores, and then use the parser to select the best path from that.

This suggests that the use of the linguistic information, as given by the categories found by the parser, in fact enables a somewhat better concept accuracy. However, in the sound implementation we are not able to take full advantage of this effect because word graphs are often too large to be parsed completely, and at the same time we cannot select more than the best one or two paths from that word graph using acoustic and Ngram scores.

The differences can be illustrated if we display the results using an ex-

TABLE 9.6: Results for full test set: approximated filtering methods.

Method	CPU mean msec	CPU max sec	WA %	CA %
T-3	449	67.3	85.0	83.4
T-3,3	45	1.1	84.4	83.1
T-3,6	56	1.6	84.9	83.7
T-3,10	71	2.8	84.9	83.7
T-3,20	118	7.1	85.0	83.6
T-5,5	58	1.3	84.5	83.4
T-5,10	76	2.8	84.8	83.8
T-5,15	97	4.7	84.9	83.8
T-10,10	89	2.9	84.9	83.8
T-10,20	135	7.1	84.9	83.6
T-10,30	194	12.8	85.0	83.7
T-20,20	164	8.0	84.9	83.6
T-20,40	285	19.1	85.0	83.7
T-20,60	452	37.3	85.0	83.7

perimental setup in which a time-out t is assumed, with various values for t. We can then calculate concept accuracy for a method and a given t by assuming that for all word graphs that took longer than t msec CPU-time to process, an empty set of semantic units is assumed. Figure 9.16 compares the best scoring sound methods and the best scoring approximations for various time-outs.

9.6 Final Remarks

In this chapter it was shown how standard parsing algorithms can be generalized for word graph input. In particular, we concentrated on parsing algorithms for off-line parsable Definite Clause Grammars. Although we identified reasons why it might be interesting to generalize the input even further to finite-state automata, we showed that such a generalization can be considered only for grammatical formalisms for which the emptyness problem is decidable; for off-line parsable DCG the emptyness problem was shown to be undecidable.

In many cases, the parser will not assign a single analysis to a word graph. In some cases, the parser will find too many analyses (in case of ambiguity), whereas in other cases too few analyses result (in case of speech recognition errors, disfluencies in the input, gaps in the grammar). Therefore, the parser was extended to find all meaningful phrases anywhere in the word graph. A shortest path algorithm has been presented which finds the best sequence of partial parses, incorporating disambiguation techniques in a multidimensional weight function. The resulting algorithm is accurate but slow if this weight function takes Ngram statistics into account. A very efficient approxima-

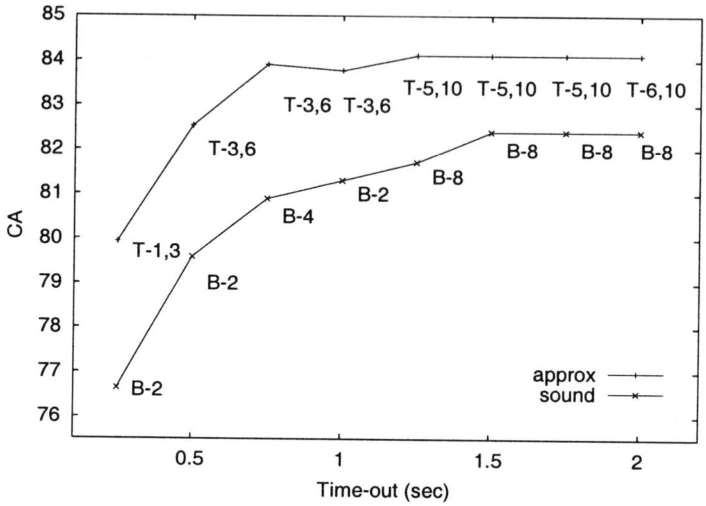

FIGURE 9.16: Comparison of the best scoring sound methods versus the best scoring approximation methods (concept accuracy), using a time-out. The graph shows that for various realistic time-outs the approximated methods perform about 3 percent better (which is an error reduction of 15%).

tion is possible by filtering the word graph using an approximation search; the resulting smaller word graph can then be analysed accurately and efficiently. The accuracy of the approximation turns out to be very good in practice.

Acknowledgments

This research was carried out within the framework of the Priority Programme Language and Speech Technology (TST). The TST-Programme is sponsored by NWO (Dutch Organization for Scientific Research).

References

Abney, S. (1996). Tagging and partial parsing, *in* K. Church, S. Young & G. Bloothooft (eds), *Corpus-Based Methods in Language and Speech*, An ELSNET volume. Kluwer Academic Publishers, Dordrecht, pp. 118–136.

Bar-Hillel, Y., Perles, M. & Shamir, E. (1961). On formal properties of simple phrase structure grammars, *Zeitschrift für Phonetik, SprachWissenschaft und Kommunicationsforschung* **14**: 143–172. Reprinted in Bar-Hillel's Language and Information — Selected Essays on their Theory and Application, Addison Wesley series in Logic, 1964, pp. 116–150.

Bennis, H. (1986). *Gaps and Dummies*, PhD thesis, Katholieke Hogeschool Tilburg.

Billot, S. & Lang, B. (1989). The structure of shared parse forests in ambiguous parsing, *27th Annual Meeting of the Association for Computational Linguistics*, Vancouver, pp. 143–151.

Boros, M., Eckert, W., Gallwitz, F., Görz, G., Hanrieder, G. & Niemann, H. (1996). Towards understanding spontaneous speech: Word accuracy vs. concept accuracy, *Proceedings of the Fourth International Conference on Spoken Language Processing (ICSLP 96)*, Philadelphia.

Carbonell, J. G. & Hayes, P. J. (1983). Recovery strategies for parsing extragrammatical language, *American Journal of Computational Linguistics* **9**(3–4): 123–146.

Carter, D. (1994). Chapter 4: Linguistic analysis, *in* M.-S. Agnäs, H. Alshawi, I. Bretan, D. Carter, K. Ceder, M. Collins, R. Crouch, V. Digalakis, B. Ekholm, B. Gambäck, J. Kaja, J. Karlgren, B. Lyberg, P. Price, S. Pulman, M. Rayner, C. Samuelsson & T. Svensson (eds), *Spoken Language Translator: First Year Report*, SICS Sweden / SRI Cambridge. SICS research report R94:03, ISSN 0283-3638.

Cormen, Leiserson & Rivest (1990). *Introduction to Algorithms*, MIT Press, Cambridge, MA.

Dijkstra, E. W. (1959). A note on two problems in connexion with graphs, *Numerische Mathematik* **1**: 269–271.

Dowding, J., Gawron, J. M., Appelt, D., Bear, J., Cherny, L., Moore, R. & Moran, D. (1993). Gemini: A natural language system for spoken-language understanding, *31th Annual Meeting of the Association for Computational Linguistics*, Columbus Ohio, pp. 54–61.

Earley, J. (1970). An efficient context-free parsing algorithm, *Communications of the ACM* **14**: 94–102.

Ehrlich, U. & Hanrieder, G. (1996). Robust speech parsing, *in* J. Carroll (ed.), *Workshop on Robust Parsing*, Prague, pp. 26–34. These proceedings are also available as Cognitive Science Research Paper #435; School of Cognitive and Computing Sciences, University of Sussex.

Heeman, P. & Allen, J. (1994). Detecting and correcting speech repairs, *32th Annual Meeting of the Association for Computational Linguistics*, New Mexico, pp. 295–302.

Heeman, P. & Allen, J. (2000). Improving robustness by modeling spontaneous speech events, *in* J.-C. Junqua & G. van Noord (eds), *Robustness in Language and Speech Technology*, Kluwer.

Hopcroft, J. E. & Ullman, J. D. (1979). *Introduction to Automata Theory, Languages and Computation*, Addison Wesley.

Huybregts, R. (1984). The weak inadequacy of context-free phrase structure grammars, *in* G. de Haan, M. Trommelen & W. Zonneveld (eds), *Van Periferie naar Kern*, Foris, pp. 81–99.

Kasper, W., Kiefer, B., Krieger, H.-U., Rupp, C. J. & Worm, K. L. (1999). Charting the depths of robust speech parsing, *37th Annual Meeting of the Association for Computational Linguistics*.

Lang, B. (1974). Deterministic techniques for efficient non-deterministic parsers, *in* J. Loeckx (ed.), *Proceedings of the Second Colloquium on Automata, Languages and Programming*, pp. 255–269. Also: Rapport de Recherche 72, IRIA-Laboria, Rocquencourt (France).

Lang, B. (1988). Parsing incomplete sentences, *Proceedings of the 12th International Conference on Computational Linguistics (COLING)*, Budapest, pp. 365–371.

Lang, B. (1989). A generative view of ill-formed input processing, *ATR Symposium on Basic Research for Telephone Interpretation (ASTI)*, Kyoto Japan.

Lavie, A. (1996). GLR*: A robust parser for spontaneously spoken language, *in* J. Carroll (ed.), *Workshop on Robust Parsing*, Prague, pp. 45–53. These proceedings are also available as Cognitive Science Research Paper #435; School of Cognitive and Computing Sciences, University of Sussex.

Model, J. (1991). *Grammatische Analyse. Syntactische verschijnselen van het Nederlands en Engels*, ICG Publications, Dordrecht, Netherlands.

Nederhof, M.-J. & Bertsch, E. (1996). Linear-time suffix recognition for deterministic languages, *Journal of the ACM* **43**(3): 524–554.

Nederhof, M.-J., Bouma, G., Koeling, R. & van Noord, G. (1997). Grammatical analysis in the OVIS spoken-dialogue system, *Proceedings of the ACL/EACL Workshop on Spoken Dialog Systems*, Madrid, Spain, pp. 66–73. cmp-lg/9705003.

Nilsson, N. (1971). *Problem Solving Methods in Artificial Intelligence*, McGraw-Hill.

Oerder, M. & Ney, H. (1993). Word graphs: An efficient interface between continuous-speech recognition and language understanding, *International Conference on Acoustic, Speech and Signal Processing (ICASSP)*, Vol. 2, pp. 119–122.

Pereira, F. C. N. & Warren, D. (1980). Definite clause grammars for language analysis — A survey of the formalism and a comparison with augmented transition networks, *Artificial Intelligence* **13**: 231–278.

Pereira, F. C. N. & Warren, D. (1983). Parsing as deduction, *21st Annual Meeting of the Association for Computational Linguistics*, Cambridge, MA, pp. 137–144.

Rosé, C. P. & Lavie, A. (2000). Balancing robustness and efficiency in unification-augmented context-free parsers for large practical applications, *in* J.-C. Junqua & G. van Noord (eds), *Robustness in Language and Speech Technology*, Kluwer.

Saito, H. & Tomita, M. (1988). Parsing noisy sentences, *Proceedings of the 12th International Conference on Computational Linguistics (COLING)*, Budapest, pp. 561–566.

Shieber, S. M. (1985). Evidence against the non-context-freeness of natural language, *Linguistics and Philosophy* **8**: 333–343.

Shieber, S. M., Schabes, Y. & Pereira, F. C. N. (1994). Principles and implementation of deductive parsing, *Technical Report TR-11-94*, Center for Research in Computing Technology, Division of Applied Sciences, Harvard University. CMP-LG 9404008.

Sikkel, K. (1993). *Parsing Schemata*, PhD thesis, Twente University, Enschede. Published in 1997 by Springer Verlag, Texts in Theoretical Computer Science, An EATCS Series.

Stallard, D. & Bobrow, R. (1992). Fragment processing in the DELPHI system, *Proceedings of the Speech and Natural Language Workshop*, Harriman, NY, pp. 305–310. DARPA workshop.

Teitelbaum, R. (1973). Context-free error analysis by evaluation of algebraic power series, *Proceedings of the Fifth Annual ACM Symposium on Theory of Computing*, Austin, Texas, pp. 196–199.

van Noord, G. (1995). The intersection of finite state automata and definite clause grammars, *33th Annual Meeting of the Association for Computational Linguistics*, MIT Cambridge Mass., pp. 159–165. cmp-lg/9504026. http://www.let.rug.nl/~vannoord/papers/.

van Noord, G. (1996). Robust parsing with the head-corner parser, *in* J. Carroll (ed.), *Workshop on Robust Parsing*, Prague, pp. 83–92. These proceedings are also available as Cognitive Science Research Paper #435; School of Cognitive and Computing Sciences, University of Sussex.

van Noord, G. (1997a). An efficient implementation of the head corner parser, *Computational Linguistics* **23**(3): 425–456. cmp-lg/9701004.

van Noord, G. (1997b). Evaluation of OVIS2 NLP components, *Technical Report 46*, NWO Priority Programme Language and Speech Technology.

van Zanten, G. V., Bouma, G., Sima'an, K., van Noord, G. & Bonnema, R. (1999). Evaluation of the nlp components of the OVIS2 spoken dialogue system, *in* F. van Eynde, I. Schuurman & N. Schelkens (eds), *Computational Linguistics in the Netherlands 1998*, Rodopi Amsterdam, pp. 213–229.

Weischedel, R., Ayuso, D., Bobrow, R., Boisen, S., Ingria, R. & Palmucci, J. (1991). Partial parsing: A report on work in progress, *Proceedings Speech and Natural Language Workshop*, Pacific Grove, CA, pp. 204–209. DARPA workshop.

Worm, K. L. & Rupp, C. J. (1998). Towards robust understanding of speech by combination of partial analyses, *Proceedings of the 13th EACI*, Brighton, UK, pp. 190–194.

Chapter 10

BALANCING ROBUSTNESS AND EFFICIENCY IN UNIFICATION-AUGMENTED CONTEXT-FREE PARSERS FOR LARGE PRACTICAL APPLICATIONS

Carolyn Penstein Rosé
University of Pittsburgh
rosecp@pitt.edu

Alon Lavie
Carnegie Mellon University
alavie@cs.cmu.edu

Abstract Large practical NLP applications require robust analysis components that can effectively handle input that is disfluent or extra-grammatical. The effectiveness and efficiency of any robust parser are a direct function of three main factors: (1) Flexibility: what types of disfluencies and deviations from the grammar can the parser handle?; (2) Search: How does the parser search the space of possible interpretations, and what techniques are applied to prune the search space?; and (3) Parse Selection and Disambiguation: What methods and resources are used to evaluate and rank potential parses and sub-parses, and how does the parser cope with the extreme levels of ambiguity introduced by its flexibility parameters? In this chapter we describe our investigations on how to balance flexibility and efficiency in the context of two different robust parsers — a GLR parser and a left corner Chart parser — both based on a unification-augmented context-free grammar formalism. We demonstrate how the combination of a beam search together with ambiguity packing and statistical disambiguation provide a flexible framework for achieving a good balance between robustness and efficiency in such parsers. Our investigations are based on experimental results and comparative performance evaluations of both parsers using a grammar for the spoken language ESST (English Spontaneous Scheduling Task) domain.

J.-C. Junqua and G. van Noord (eds.), Robustness in Languages and Speech Technology, 239–269.

10.1 Introduction

Large practical NLP applications require robust analysis components that can effectively handle input that is disfluent or extra-grammatical. Regardless of the specific formalism and grammar they employ, NLP systems that process free unrestricted input must be prepared to frequently encounter input that deviates from the coverage of their pre-developed grammar (and other knowledge sources). Traditional parsing algorithms that are designed to parse only completely grammatical input sentences (and fail on even the slightest deviation from the grammar) are therefore unsuitable for such applications. This problem is even more evident in systems that process spoken language. Spontaneous spoken language, whether transcribed by hand or by a speech recognizer, is often highly disfluent, with the meaningful portions of the utterance surrounded by a variety of phenomena that disrupt the grammaticality of the overall input stream. Common types of extra-grammaticality include repetitions, false starts, filled pauses, and idiosyncratic constructions found within a speaker's performance grammar but not covered by a particular system's parsing grammar. Processing such language requires an analysis component that is robust to the various types of disfluencies and can extract the most complete interpretation possible from a given input.

Researchers have approached the robust parsing problem from a variety of different directions, including symbolic (Rosé 1997, Lavie 1995, Aït-Mokhtar & Chanod 1997, Neumann, Backofen, Baur, Becker & Braun 1997, McDonald 1993b, McDonald 1990, Lehman 1989, Hipp 1992), statistical (Bod 1998, Miller, Stallard, Bobrow & Schwartz 1996, Pietra, Epstein, Roukos & Ward 1997, Rosé & Waibel 1997, Goodman 1996, Magerman & Marcus 1990, Sanker & Gorin 1993), and connectionist (Henderson & Lane 1998, Buø 1996, Jain 1991, Jain & Waibel 1990). While statistical and connectionist approaches are inherently robust, and can often be trained automatically from labeled corpora, symbolic parsers are most capable of performing deep and detailed analysis based on linguistic principles. Thus, symbolic parsers are still highly attractive for tasks that require these detailed linguistic analyses, such as Machine Translation (MT) or Intelligent Tutoring.

To avoid failure when faced with extra-grammatical input, many robust approaches to symbolic parsing relax syntax and grammar constraints. In some cases, semantic information is used to compensate for the lack of grammaticality in order to partially or completely drive the analysis process. The earliest approaches to robust parsing involved hand-coded grammar specific heuristics for selecting a subset of analyses to extend when each input word was processed (Hobbs, Appelt, Bear & Tyson 1991, McDonald 1993b, McDonald 1993a, McDonald 1992). With these approaches, the work of making the parser robust and efficient must be redone by hand for every new grammar developed. The use of hand-coded heuristics also constrains these approaches to handle only those cases that are consistent with the simplifying assumptions represented by those heuristics.

Some recent approaches to robust parsing focus instead on shallow or partial parsing techniques (Van Noord 1997, Abney 1995, Aït-Mokhtar & Chanod 1997, Worm 1998). Rather than attempting to construct a parse covering an entire ungrammatical sentence, these approaches attempt to construct analyses for maximal contiguous portions of the input. Thus, they address the computational expense problem by limiting the flexibility of their parsing algorithms. The result is a practical form of robust parsing not involving any grammar specific heuristics, but more shallow in its resulting analysis than could possibly be achieved using other types of flexibility. For certain applications, this level of analysis may certainly be sufficient. However, if a more complete analysis is required, additional post-parsing processing is necessary, where robustness must also be applied.

Perhaps the most complete approach to the problem of robust symbolic parsing is *Minimum Distance Parsing* (MDP) (Lehman 1989, Hipp 1992). When faced with input that is extra-grammatical, the goal of an MDP parser is to find the analysis of a corresponding grammatical input that is closest in meaning to the given input. However, for practical reasons, the parser instead searches for an analysis of an input that deviates as little as possible from the original input. This is often accomplished using a "distance" metric that is based on a variety of possible parser flexibilities, such as deletions, insertions or substitutions of words in the input, with the "closest" distance being determined by using adjustable penalties. MDP was originally developed in the context of parsing programming languages, where the types of ungrammaticality were quite different than those often found in natural language. While full MDP parsers (Hipp 1992, Lehman 1989) have been shown to be effective in small natural language applications with small grammars, the approach does not scale up well to large applications. As demonstrated in our previous work (Rosé 1997), with large coverage grammars, the search space that a full MDP parser must explore expands to magnitudes that render the approach computationally infeasible. Attention has thus shifted to more limited approaches, that can consider a more restricted set of parser flexibilities, but can maintain parsing efficiency. Our research work, described in this paper, is an investigation into the tradeoffs involved in balancing robustness and efficiency in such robust symbolic parsers.

The effectiveness and efficiency of robust symbolic parsers are a direct function of three main factors: (1) Flexibility: what types of disfluencies and deviations from the grammar can the parser handle?; (2) Search: How does the parser search the space of possible interpretations, and what techniques are applied to prune the search space?; and (3) Parse Selection and Disambiguation: What methods and resources are used to evaluate and rank potential parses and sub-parses, and how does the parser cope with the extreme levels of ambiguity introduced by its flexibility parameters? In this paper we describe our investigations on how to balance flexibility and efficiency in the context of two different robust parsers: GLR* — a robust GLR parser, and LCFLEX[1] — a left corner Chart parser. Both parsers are based on a unification-augmented context-free grammar formalism. We demonstrate how the combination of a beam search together with ambiguity packing and statistical disambiguation provide a flexible framework for achieving a good balance between robustness and efficiency in such parsers. Our investigations are based on experimental results and comparative performance evaluations of both parsers using a grammar for the spoken language ESST (English Spontaneous Scheduling Task) domain.

We begin with a brief description of the underlying grammar formalism used by both parsers, and the fundamental parsing principles of each. We then describe the flexibility features of our robust parsers. These are accompanied by detailed investigations into the performance effects of each flexibility feature. In our comparisons between our GLR parser and our LC parser we demonstrate that even with the addition of the types of flexibility we explore, our LC parser maintains its order of magnitude computational advantage over GLR in terms of speed. We also demonstrate that a great deal of robustness can be achieved while keeping the parser's efficiency at a practical level. While ambiguity is an orthogonal issue to robustness, it is an essential consideration for maintaining a practical run-time performance in robust parsers because adding flexibility necessarily increases the amount of ambiguity generated. Thus, section 10.9 discusses our approach to managing ambiguity and disambiguation in our robust parsers. Finally, we present our conclusions and planned future research.

All of the evaluations reported in this paper were conducted on the English Spontaneous Scheduling Domain (ESST) Grammar developed for the JANUS speech translation project (Woszcyna, Coccaro, Eisele, Lavie, McNair, Polzin, Rogina, Rosé, Sloboda, Tomita, Tsutsumi, Waibel, Waibel & Ward 1993, Woszcyna, Aoki-Waibel, Buo, Coccaro, Horiguchi, Kemp, Lavie, McNair, Polzin, Rogina, Rosé, Schultz, Suhm, Tomita & Waibel 1994). The grammar is primarily semantic and consists of approximately 1000 rules with an accompanying lexicon with approximately 3000 lexical entries.

10.2 The Grammar Formalism

Much attention has been given in the last decade to unification based formalisms such as LFG (Kaplan & Bresnan 1982), GPSG (Gazdar, E., Pullum & Sag 1985) and HPSG (Pollard & Sag 1987), as declarative theories for describing the structure of natural language. A fundamental primitive of these formalisms is the unification of feature based representations. Several recent practical systems for structural analysis of natural language have been designed to

1. LCFLEX is freely available for research purposes. Interested persons should contact the first author at rosecp@pitt.edu.

```
(< DECL > < -- > (< NP > < VP >)
        (((x2 agr) = (x1 agr))
         ((x0 subject) = x1)
         ((x2 form) = *finite)
         (x0 = x2)))
```

FIGURE 10.1: An English Grammar Rule of the GLR Parser/Compiler

support unification based grammar specifications within a general CFG framework, that allows the system to incorporate known CFG parsing algorithms as the core of the system's implementation. Two examples of such systems are the Core Language Engine (CLE) (Alshawi (ed.) 1992) and the Alvey Natural Language Tools (ANLT) (Carroll 1993).

The underlying grammar formalism used by both GLR* and LCFLEX is a unification-based grammar formalism that was originally developed for the Generalized LR Parser/Compiler (Tomita 1990, Tomita 1987) at the Center for Machine Translation at Carnegie Mellon University. The formalism supports grammatical specification in an LFG framework, that consists of context-free grammar rules augmented with feature bundles that are associated with the non-terminals of the rules. For each context-free rule, the grammar formalism allows the specification of the feature structure associated with a left-hand side non-terminal of the rule as a function of the feature structures that are associated with the non-terminals on the right-hand side of the rule. At run-time, the parser computes the actual feature values of a particular left-hand side non-terminal as a by-product of the operation that creates the left-hand side non-terminal from the identified constituents on the right-hand side of the rule. Feature structure computation is, for the most part, specified and implemented via unification operations. This allows the grammar to constrain the applicability of context-free rules. The LHS constituent of a context-free rule is created only if the associated feature structure unification associated with the rule constituents is successful as well.

An example of an augmented grammar rule is shown in Figure 10.1. The variable $x0$ in the feature specification part of the rule is bound to the left-hand side non-terminal DECL of the context-free part of the rule. The variables $x1$, $x2$, etc. are respectively bound to the right-hand side constituents of the rule. The feature structures that are computed at parse-time are attached to their corresponding parse node structures. Parsing with such an augmented grammar produces both c-structure and f-structure — a set of one or more parse trees, along with the feature structure associated with each of these trees.

10.3 The GLR* Parser

The GLR* parser is a robust extension of the Generalized LR (GLR) parser developed by Tomita (1986). Tomita's GLR parsing algorithm evolved out of the

LR parsing techniques that were originally developed for parsing programming languages in the late 1960s and early 1970s (Aho & Johnson 1974), and also follows ideas developed by Lang (1974). LR parsers parse the input bottom-up, scanning it from left to right, and producing a rightmost derivation. LR parsers are driven by a table of parsing actions that is pre-compiled from the grammar. Tomita's (1986) GLR parsing algorithm extended the original LR parsing algorithm to the case of non-LR languages, where the parsing tables contain entries with multiple parsing actions. The algorithm deterministically simulates the non-determinism introduced by the conflicting actions in the parsing table by efficiently pursuing in a pseudo-parallel fashion all possible actions. The primary tool for performing this simulation efficiently is the *Graph Structured Stack* (GSS). The GSS is a data-structure that efficiently represents the multiple parsing stacks that correspond to different sequences of parsing actions. Two additional techniques, *local ambiguity packing* and *shared parse forests,* are used in order to efficiently represent the various parse trees of ambiguous sentences when they are parsed. Local ambiguity packing collects multiple sub-analyses of the same category type, all of which derive the same substring, into a single structure. Further parsing actions may then refer to the single structure that represents the entire collection of sub-parses, rather than to each of the packed sub-parses separately. Shared packed forests allow common sub-parses of different analyses to be shared via pointers.

The primary type of flexibility used in GLR* is skipping over words and segments of the input string. Two types of skipping are handled by the parser. The parser can skip over an arbitrary prefix or suffix of the input by allowing complete analyses that do not span the entire input string. We refer to this feature of the parser as *inter-analysis word skipping*. Furthermore, the parser can also skip over sequences of words internal to an analysis. We refer to this feature as *intra-analysis word skipping*. While words skipped in both cases can be treated similarly, this distinction allows testing the parser in a mode that permits one type of skipping yet not the other. Our previous investigations demonstrate that inter-analysis skipping is much more critical than intra-analysis skipping (Rosé & Lavie 1997), and some well known parsers have relied entirely upon this form of skipping (Mayfield, Gavaldà, Ward & Waibel 1995, Mayfield, Gavaldà, Seo, Suhm, Ward & Waibel 1995, Ward 1989). Experiments comparing these two forms of skipping are described later in the paper.

GLR* utilizes a free skipping approach constrained by a beam search technique free of any grammar specific heuristics. The beam search keeps the parser's performance within reasonable bounds. Thus, the work for making the parser robust and efficient once completed never needs to be done again. Instead of embodying any simplifying assumptions about which analyses to extend upon processing each input word, the size of the beam bounds the growth of the number of analyses pursued in parallel. Thus, any analysis licensed by the grammar by skipping any portion of the input is, in principle, obtainable via this approach, since the beam size can always be adjusted to a sufficiently large setting. Furthermore, the approach requires no backtracking. However, the GLR parsing architecture itself has been demonstrated to be an order of magnitude less efficient than chart-based parsing architectures (Van Noord 1997). Thus, GLR*'s run-time performance suffers because of inefficiencies in the framework in which it was developed.

10.4 The LCFlex Parser

LCFLEX is a flexible left-corner parser designed for efficient, robust interpretation. Its approach to robustness was originally inspired by our earlier GLR* parser. LCFLEX extended the robustness capabilities of GLR* and supports various types of flexibility including word skipping, non-terminal inserting, and flexible unification. While the additional "flexibility features" offered by LCFLEX could have just as well been implemented within GLR*, our previous experience with GLR* demonstrates that these additional degrees of flexibility render GLR* intractable (Rosé & Lavie 1997). One attractive feature of LCFLEX is that its "flexibility features" can easily be controlled by setting global parameters. Thus, it is possible to tailor the parser's behavior to the specific needs of individual applications where it is used. The more flexibility allowed within the parser, the greater the robustness, but also the more ambiguity and computational expense, both in terms of time and space. Ideally, LCFLEX should be set up in order to achieve the most desirable robustness versus efficiency trade-off. Because this trade-off is application/grammar/genre dependent, determining the most appropriate setting requires some experimentation in the current version of LCFLEX. One of our current research objectives is to semi-automate this process. LCFLEX's basic architecture is based on the left-corner parsing algorithm described in Rosenkrantz & Lewis (1970) and Carroll (1993). It uses the reflexive and transitive closure of the left-corner relation (Van Noord 1997) both for top-down filtering and bottom-up prediction in order to limit ambiguity and thus enhance efficiency.

LCFLEX utilizes a beam skipping technique similar to the one previously developed within the GLR* parser. Like GLR*, LCFLEX searches for the maximal parsable subsequence of a sentence. Thus, it is a grammar-independent approach to robust parsing of full sentences rather than a partial or shallow parsing technique (Abney 1995, Aït-Mokhtar & Chanod 1997, Worm 1998). LCFLEX is implemented within the left-corner parsing architecture, which is an order of magnitude faster than the GLR architecture. Even with its added flexibility, it maintains the computational advantage of chart-based parsers over LR based parsers. Additionally, it incorporates the same efficiency measures that were a part of the GLR* parser. It uses a standard ambiguity packing algorithm augmented with a pruning step similar to that introduced in Lavie (1995). This pruning step eliminates packed analyses that cover a strictly subsumed subset of the portion of the sentence covered by another packed analysis. For ambiguity resolution, LCFLEX uses an adapted version of the statistical disambiguation approach discussed in Lavie (1995) and Carroll & Briscoe (1993).

In addition to the flexibility that was an integral part of the GLR* parser, LCFLEX embodies several other types of flexibility. In addition to skipping, it also has the ability to consider insertion of missing words and categories, which can also be thought of as partial rule matching. Whereas general insertion has been demonstrated to be impractical in realistic sized applications (Rosé 1997, Rosé & Lavie 1997), used sparingly it can be useful in applications such as language tutors and grammar checkers that must not only arrive at an analysis for an extra-grammatical input sentence, but also determine the cause of the extra-grammatical (Schneider & McCoy 1998). LCFLEX can be set to allow specific combinations of insertions. It also has the ability to perform flexible unification. Where complete flexible unification is too inefficient, setting aside a small set of features as "soft-features" can be useful particularly in language learning environments where certain syntactic features tend to be acquired later than others. LCFLEX's final flexibility feature is its two-level parsing mode that allows it to construct a partial analysis at a constrained level of flexibility and then combine the fragments of the partial parse by running the parser again at a far less constrained level of flexibility.

10.5 Word Skipping

Because spoken language in particular is littered with repetitions, false starts, filled pauses, and other noise surrounding islands of parsable language, we have found word skipping to be the most essential form of robustness. Thus, word skipping is featured prominently both in GLR* and in our more recent LCFLEX parser. In this section we describe how our word skipping approach constrained by a beam search technique is implemented in these two alternative parsing architectures, thus demonstrating the architecture independence of the approach. Our end-to-end translation evaluation demonstrates that our beam skipping approach effectively maximizes the parser's robustness while keeping the parser's run time performance practical. Furthermore, we demonstrate that LCFLEX's implementation of beam skipping maintains the computational advantage of LC parsers over GLR parsers.

10.5.1 Word Skipping in GLR*

GLR* accommodates intra-analysis skipping by manipulating the underlying stack structure of the parser. Whereas the standard GLR parser attaches (shifts) each processed input word to the state at the top of the stack structure, GLR* allows processing the new word also from states that are internal to the stack structure. Shifting an input symbol from an internal state is equivalent to skipping the words of the input that were encountered after the parser reached that state and prior to the current word that is being shifted. For a detailed description of how this process is performed, the reader is referred to Lavie (1995).

Allowing the parser to consider all possible internal skipping combinations is computationally infeasible due to the explosion in ambiguity. Thus, GLR* uses a beam search technique to limit the set of words that can be skipped throughout the parsing process. Since the purpose of GLR* is to find maximal (or close to maximal) parsable subsets of the input, applying an adequate beam should allow the flexibility necessary to recover the desired parse. The beam search restricts the parser to pursuing a "beam" of a fixed size k of parsing actions, which are selected according to a criterion that locally minimizes the number of words skipped. There exists a direct tradeoff between the amount of search (and word skipping) that the parser is allowed to pursue and the time and space performance of the parser itself. For a given grammar, one must determine the smallest possible setting of the beam parameter that allows the parser to find the desired parses in an overwhelming majority of cases.

10.5.2 Word Skipping in LCFlex

Like GLR*, LCFLEX is able to skip portions of its input sentences in order to search for the maximal portion for which it can construct an analysis. It allows the user to control how freely it is permitted to skip in searching for the maximal parsable subset of an input utterance. Like GLR*, LCFLEX incorporates both inter-analysis skipping and intra-analysis skipping, and these two forms of skipping may be used either separately or together.

Inter-analysis skipping in LCFLEX allows an analysis to begin or end at *any* vertex in the chart. In this way, the parser constructs analyses for every contiguous portion of the input where it is possible to do so. Inter-analysis skipping is accomplished in LCFLEX by modifying its prediction mechanisms. Normally active edges are created wherever their corresponding left hand side category is predicted by LCFLEX's top-down and bottom-up prediction mechanisms. When inter-analysis skipping is enabled, the goal category (or categories) are predicted at each vertex in the chart in addition to the categories predicted based on the active edges ending at those vertices. This allows full analyses to begin at any vertex in the chart. When the whole utterance has been processed, analyses can be extracted from the chart by searching for inactive edges with one of the goal categories as their category. When inter-analysis skipping is turned on, analyses that begin and end at any vertex are returned, rather than only the analyses that begin with the first word and end with the last word. From the extracted list of analyses it is possible to return the subset of largest analyses that span the maximum subset of the utterance. In this way, LCFLEX can perform like other well-known partial parsers.

Often inter-analysis skipping is inadequate because it fails to handle the disfluencies that appear *within* an analysis. To handle these cases, the second type of skipping, *intra-analysis skipping*, allows the parser to skip words *within* an analysis. Intra-analysis skipping in LCFLEX is accomplished by copying references to active edges forward in the chart, in effect skipping over all of the words in between the current vertex and the vertices where the edges were copied from. Before each word is parsed, active edges from some number of previous vertices are copied forward to the current vertex. The exact number of previous vertices from which active edges are copied is determined by the parser's two different beams, described below, that bound the number of analyses pursued by the parser in parallel. An alternative to copying active edges forward would have been to search backwards in the chart for active edges to extend each time an inactive edge is inserted. This would have appeared superficially more similar to the approach taken in GLR*. However, this makes it necessary to do the work of skipping over the same portion of the sentence each time an inactive edge of a particular category is inserted into the chart beginning at the same vertex. Approaching the problem of intra-analysis skipping by copying active edges forward allows the parser to do the work of skipping each subset of the sentence only once rather than multiple times. Note also that intra-analysis skipping could have also been accomplished by copying active edges forward in the chart, allowing us to handle both types of skipping with a single mechanism. However, our intra-analysis skipping approach introduces far less ambiguity than copying active edges forward in the chart. Additionally, our experience has revealed that even the noisiest input sentences tend to be composed of mostly grammatical islands surrounded by noise. If we treat these two types of skipping separately, we can achieve the same robustness with a lower threshold on active edge copying and thus less computational cost.

We have experimented with two types of beams in LCFLEX to control the amount of skipping. The first type of beam, *maximum of n beam*, allows the user to specify how many words can be skipped within an single analysis. The second type of beam, *n-best beam* allows the user to specify a number that constrains the number of analyses it pursues in parallel, as in GLR*. It is not a true n-best beam in the sense that it limits the absolute number of analyses pursued in parallel. Instead, it approximates this by bounding the number of active edges that are allowed to end at any one vertex in the chart. These two different beams can be used separately or together. Thus, LCFLEX allows a great deal of control over its skipping behavior.

10.5.3 Evaluating Word Skipping

We conducted a series of empirical tests to evaluate the effect of the word skipping flexibility allowed by the parser on its runtime performance. Our evaluation here demonstrates that LCFLEX maintains the order of magnitude computational advantage of left-corner parsers over GLR parsers. For both GLR* and LCFLEX, the amount of word skipping flexibility is controlled via the parser's search beam. We compared the alternative settings of the two parsers based both on parse times and on translation quality of text produced automatically by our generation component from the parser output. In all cases, LCFLEX

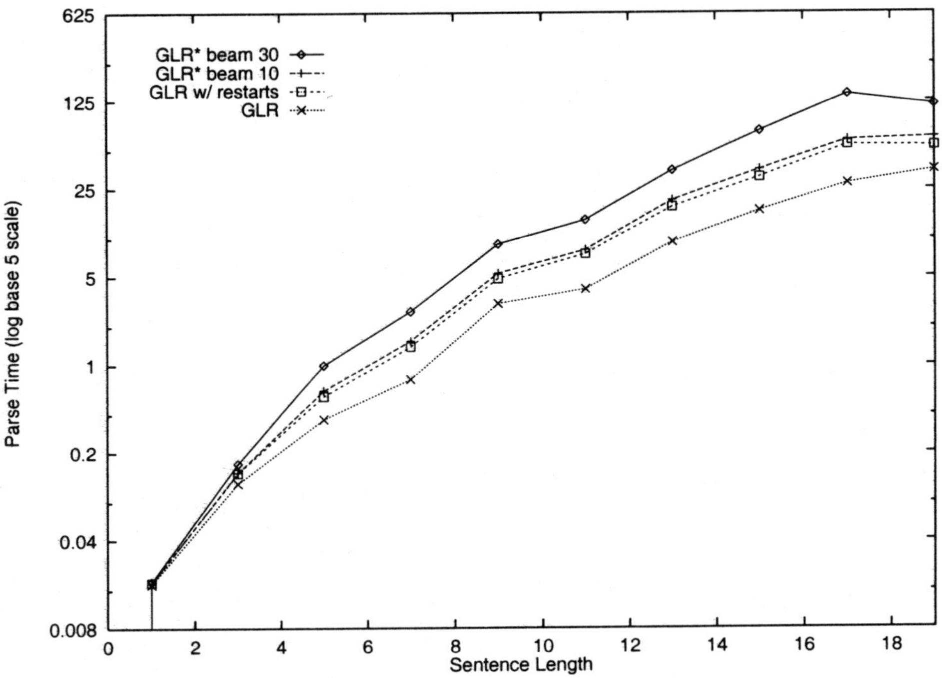

FIGURE 10.2: Avg Parse Time vs. Sentence Length for GLR*, Various Skipping Settings

TABLE 10.1: Translation Quality with Alternative Parsers

	Bad	OK	Perfect	Total Acceptable
No Robustness	47.16	11.67	41.17	52.84 (60.39)
Restarts	32.83	21.00	46.17	67.17 (76.77)
GLR* beam 10	31.50	21.00	47.50	68.50 (78.29)
GLR* beam 30	30.33	19.50	50.17	69.67 (79.62)
LCFLEX 2 Level	29.00	20.50	50.50	71.00 (81.14)
LCFLEX Skip 3	26.67	20.00	53.33	73.33 (83.80)

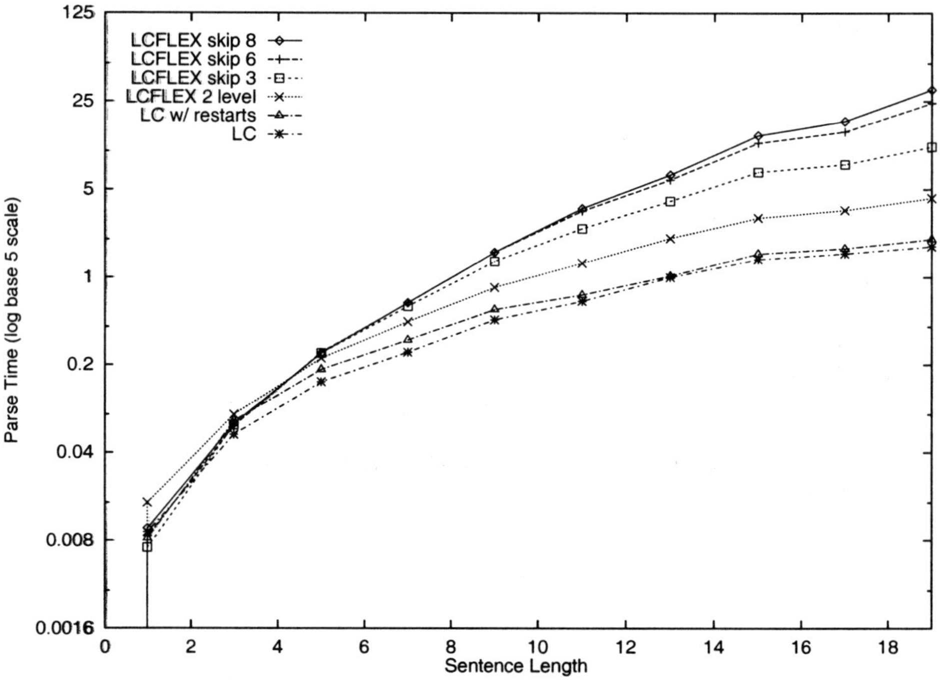

FIGURE 10.3: Avg Parse Time vs. Sentence Length for LCFLEX, Various Skipping Settings

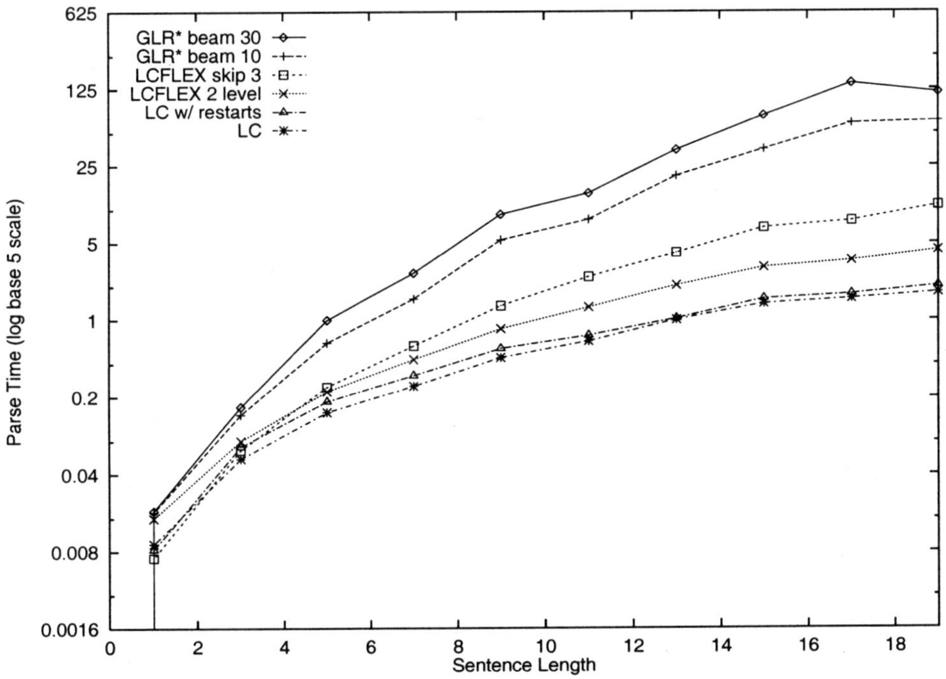

FIGURE 10.4: Avg Parse Time vs. Sentence Length for LCFLEX compared with GLR*

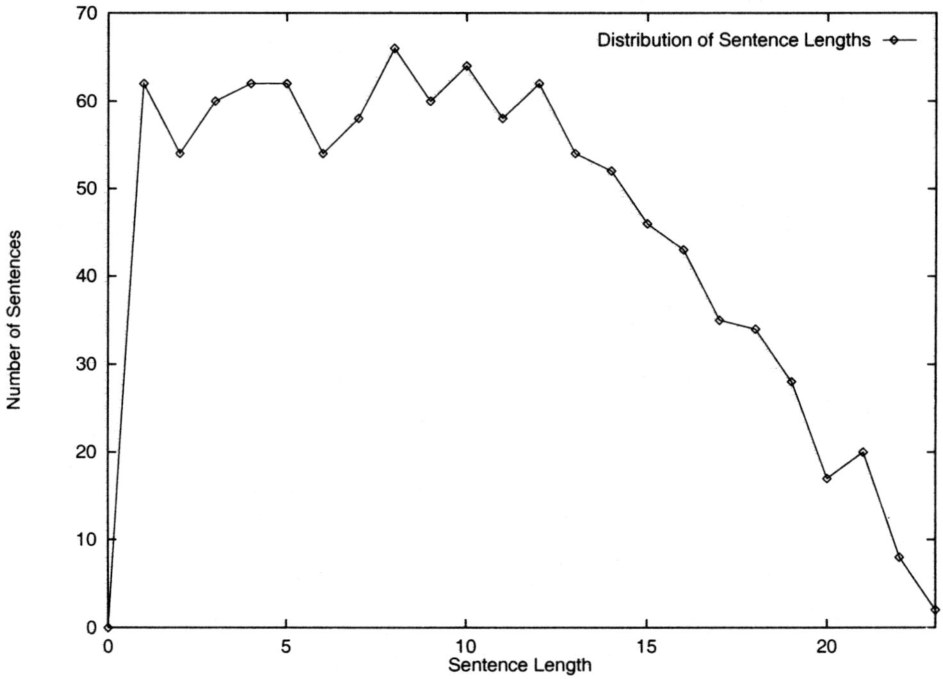

FIGURE 10.5: Distribution of Sentence Lengths

performs at least an order of magnitude more efficiently than GLR* set to a roughly equivalent flexibility setting. The complete test set contains 607 sentences of transcribed spontaneous speech from the ESST task. The distribution of sentence lengths is found in Figure 10.5. Of these 607 sentences, a subset of 200 sentences was selected for grading. These 200 sentences were contiguous sentences from five dialogues in our ESST corpus. We selected this set of contiguous sentences in order to ensure a realistic distribution of sentence lengths and in-domain vs. out-of-domain sentences.

Each translation was graded by three independent judges, receiving a grade of Perfect, OK, or Bad. For each strategy, the number of sentences assigned each grade was totaled for each grader. These totals were then averaged to arrive at the percentages presented in Table 10.1. Of the 200 sentences in the test set, 25 were determined to be out of domain. Thus, the percentages in parentheses in the final column are percentages of acceptable translations out of the in-domain sentences.

In the case of GLR*, we tested the parser in four different settings: (1) "GLR": with no skipping allowed — this is equivalent to the original (non-robust) GLR parser; (2) "GLR with restarts": allowing only inter-analysis skipping (of prefixes and suffixes of the input string); (3) "GLR* beam 10": allowing inter-analysis skipping with the parser using a search beam of size 10; and (4) "GLR* beam 30": as in (3), but with a search beam of 30. Average parse times as a function of sentence length for all four tests appear in Figure 10.2. As expected, greater levels of skipping flexibility result in increasing parse times. It is interesting to note that GLR* with a beam of 10 is only slightly more time consuming than just allowing restarts, while already allowing a significant amount of intra-analysis skipping. In fact, our experiments indicated that GLR* with a beam of 10 was usually flexible enough to find a correct analysis when such an analysis can be recovered with word skipping (see also the translation quality evaluation results in Figure 10.1). A beam of 30 already results in significantly longer parse times, approaching infeasible levels for online parsing with sentences of length greater than 15 words.

For LCFLEX, we conducted a similar set of experiments, with varying amounts of skipping flexibility. In LCFLEX, the amount of skipping flexibility is controlled by a global skip parameter, which determines the total number of words allowed to be skipped in the process of obtaining an analysis. We tested the parser with five different settings: (1) "LC": no skipping allowed — equivalent to a non-robust left-corner parser; (2) "LC with restarts": allowing only inter-analysis skipping (of prefixes and suffixes of the input string); (3) "LCFLEX skip 3": a maximum of 3 words can be skipped within an analysis; (4) "LCFLEX skip 6": similar, with a maximum of 6 skipped words; and (5) "LCFLEX skip 8": similar, with a maximum of 8 skipped words. In addition to our beam comparison tests, we also tested LCFLEX in its two level parsing mode discussed below in Section 10.8 in order to demonstrate the increased efficiency possible with that mode.

Our purpose in testing different beam settings for both parsers was to demonstrate that LCFLEX maintains the computational advantage of left-corner chart parsers over GLR parsers even as the flexibility in both parsers is increased. Because GLR*'s beam is based on the nature of its Graph Structured Stack, which is not a part of LCFLEX, it is impossible to constrain the two parsers to skip exactly the same amount or in exactly the same places. However, in practice, setting LCFLEX's maximum skip beam to 3 causes it to closely mimic GLR*'s beam of 30. Setting its maximum skip beam to 6 or 8 causes it to allow more skipping than GLR* with a beam of 30. While it would have been possible to re-implement GLR*'s beam to control skipping based on number of words skipped, as in LCFLEX, our informal investigations indicate that controlling GLR*'s skipping in this way causes it to parse far less efficiently than with its current beam. Thus, our current results are sufficient to demonstrate that LCFLEX does maintain the computational advantage of left-corner parsers over GLR parsers.

Average parse times as a function of sentence length for all tests appear in Figure 10.3. As expected, greater levels of skipping flexibility result in increasing parse times. However, for LCFLEX, inter-analysis skipping does not result in a very significant increase in parse time, while already providing a good deal of robustness. Two level parsing performs slightly slower while achieving another sizable increase in robustness. While a skip parameter value of 3 does incur a significant time cost, parse times remain feasible for even relatively long sentences. Increasing the skip parameter above 3 did not improve the overall translation quality produced.

Note that with all of the above settings, LCFLEX is at least an order of magnitude faster than GLR* in comparable settings. (See Figure 10.4.) LCFLEX with a skip value of 8 is about similar in speed to GLR with no skipping at all.

10.6 Inserting in LCFlex

While in our experience, skipping is the most important flexibility feature for efficient robust parsing, under certain circumstances it is still not enough. Such is the case when a speaker uses an abbreviated form that is not grammatical, for example when a determiner is dropped in informal language. (i.e., "cat at nine o'clock" to indicate the location of an approaching cat). While such constructions can be modeled within the grammar, doing so leads to an unnecessary explosion in ambiguity. In the case of determiner dropping, two analyses would always be constructed for every noun phrase normally requiring a determiner, one including the determiner and one omitting it. Even in cases where a correct analysis is constructed, the incorrect one (without the determiner) would be constructed as well, doubling the number of analyses that contain a representation for the associated noun phrase. If these types of ungrammaticalities are handled within the parser instead, for instance by allowing the parser

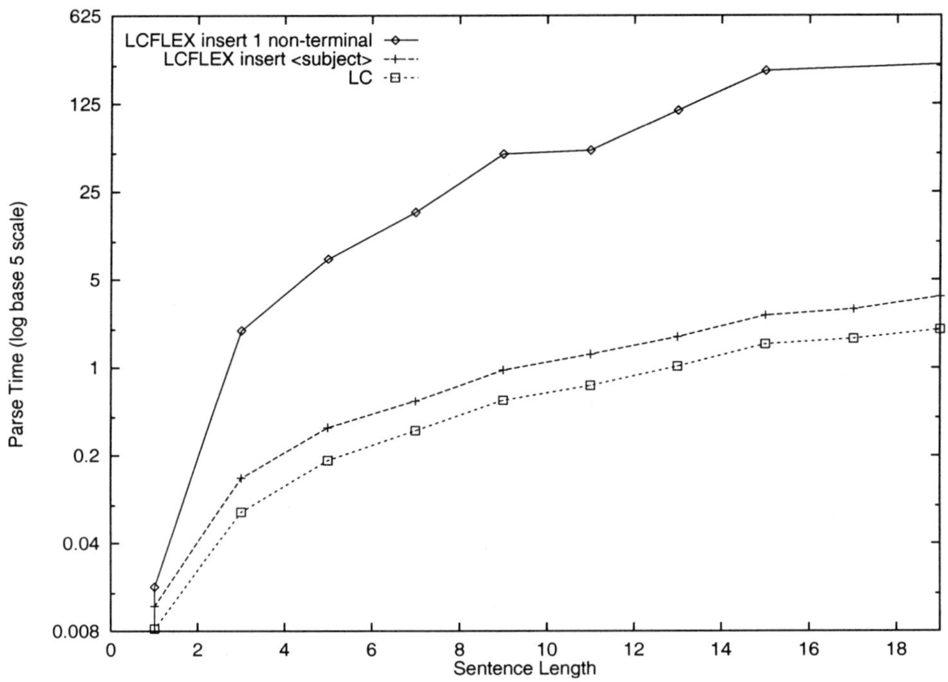

FIGURE 10.6: Avg Parse Time vs. Sentence Length for LCFLEX, Various Insertion Settings

to insert words or non-terminals in order to complete a partial rule match, then two analyses will still be constructed for the noun phrase in question, but the one with the greater error can eventually be pruned. Because the parser is aware of the fact that the analysis without a determiner is less correct than the one with a determiner, it can prevent the explosion of ambiguity that arises from modeling this type of ungrammaticality within the grammar.

10.6.1 LCFlex's Limited Insertions

While insertions have been constrained in previous approaches by limiting the number of inserted non-terminals, as with Lehman's prioritized agenda mechanism (Lehman 1989), LCFLEX allows the user to control the number of occurrences of each specific non-terminal that may be inserted. Thus, the user may limit insertions to just a small number of a handful of categories. We conducted a set of experiments to evaluate the effect of word and category insertions on the runtime performance of the LCFLEX parser. As mentioned earlier, word and category insertion is a very costly form of flexibility, as it dramatically increases the search space of analyses that must be considered by the parser. While it has been demonstrated in Rosé (1997) that even limited numbers of general insertions are infeasible with a practical sized grammar, we demonstrate here that LCFLEX's style of limited insertions does not demand too high of a price in speed.

Insertions in LCFLEX are primarily controlled by setting a global parameter and then recompiling the grammar. When insertions are turned on, the global variable that controls this behavior is an association list of non-terminals paired with integers or the symbol `*any*`. This indicates that any combination of the specified maximum number of the specified non-terminals may be inserted into an analysis. For example, `((< np > 1) (< det > *any*))` specifies that at most one noun phrase and any number of determiners may be inserted. `*or*` may be used to specify a disjunction of constraints. Thus, `(*or* ((< np > 1) (< det > *any*)) ((< copula > 1)))` indicates that either at most one copula or any determiner and at most one noun phrase may be inserted.

One example where inserting is necessary in our ESST task is where speakers drop the subject as in, "Don't have any two hour slots." Because our ESST grammar does not allow the speaker to drop the subject in constructions such as this one, this sentence would ordinarily not be able to parse. Figure 10.7 displays the correct c-structure for this construction according to our ESST grammar. This sentence can only parse by inserting a $< subject >$ non-terminal before the $< aux >$ non-terminal under the $< *subject - aux >$ non-terminal. When a non-terminal is inserted, a "dummy" feature structure is inserted into the analysis in place of the one that would have normally been constructed by the parser for the corresponding input words. The dummy feature structure has the feature `dummy` with a value of +. This dummy feature ensures that inserting a dummy feature structure will not cause unification

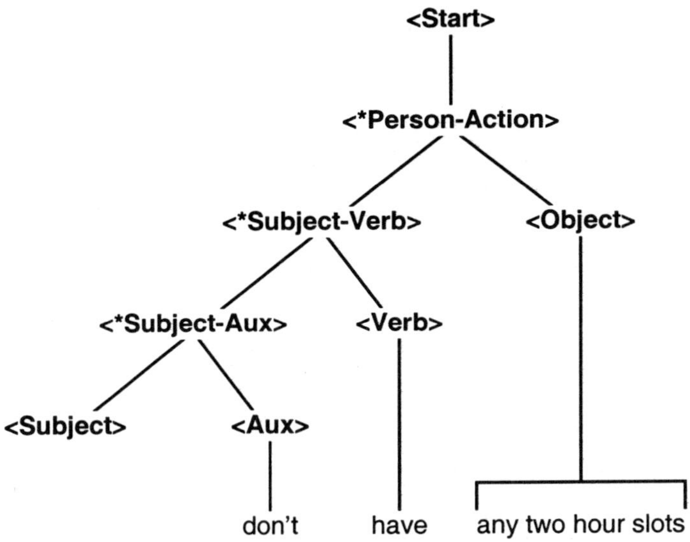

RULE: $(< *SUBJECT - AUX > (< SUBJECT > < AUX >))$

CHILDREN:
$< SUBJECT >$ ((DUMMY +))
$< AUX >$ ((ROOT DO) (NEGATIVE +))

RESULT:
$< *SUBJECT - AUX >$ ((SENTENCE-TYPE *STATE) (VERB-FORM2 BASE)
(AUX DO) (DUMMY +))

RULE: $(< *SUBJECT - VERB > (< *SUBJECT - AUX > < VERB >))$

CHILDREN:
$< *SUBJECT - AUX >$ ((VERB-FORM2 BASE) (SENTENCE-TYPE *STATE)
(AUX DO) (DUMMY +))
$< VERB >$ ((VERB-FORM BASE) (ROOT HAVE))

RESULT:
$< *SUBJECT - VERB >$ ((VERB ((ROOT HAVE))) (ADVERB *DUMMY*)
(NEGATIVE +) (SENTENCE-TYPE *STATE)
(AUX DO) (DUMMY +))

FIGURE 10.7: Insertion Example

constraint equations to fail. Whenever a unification equation refers to an empty slot within a dummy feature structure, the value that is returned is `*dummy*`. Constraint equations involving a `*dummy*` value never fail. Figure 10.7 shows the result of two unifications involving dummy feature structures. In the first rule application, The dummy feature structure is unified with the result so that the resulting feature structure is also a dummy feature structure since it contains the `dummy` feature with value +. Additional constraint equations in the unification augmentation part of the rule, not shown, cause other features of the dummy feature structure to be assigned values. In the second rule application, the value of the `adverb` feature of the dummy feature structure is passed up to the result. Since the `adverb` slot of the dummy feature structure is empty, the value `*dummy*` is assigned. Thus, subsequent constraint equations testing the value of the `adverb` feature of the resulting feature structure will not fail.

Allowing the parser to insert a single non-terminal symbol only slightly decreases the parser's efficiency. To demonstrate this, we evaluated the effect of inserting one particular grammar category, $< subject >$, and compared it with allowing the insertion of any one category into the analysis. We picked the $< subject >$ category because we have evidence of a few sentences, such as the example above, in which the insertion of this particular category is in fact necessary for recovering the correct analysis. Average parse times as a function of sentence length for these tests appear in Figure 10.6. As can be seen, the insertion of the $< subject >$ category does result in a noticeable, but tolerable, effect on runtime performance. However, the insertion of even one arbitrary category results in infeasible parse times.

10.6.2 Inserting Non-Terminals in an LC Framework

Insertions affect every aspect of how the parsing algorithm proceeds, including the conditions under which active edges are created, extended, and considered complete. Because its impact on the operation of the parser is so pervasive, inserting comes at a greater computational expense than skipping. Normally active edges corresponding to rules are created when an inactive edge of the category of the associated rule's first right hand side daughter are inserted into the chart. With unconstrained insertions, active edges would be created whenever an inactive edge of the category of any of the associated rule's right hand side daughters is inserted into the chart. If insertions are constrained, either in number or in terms of which non-terminals are allowed to be inserted, then rather than doing this for every right hand side category, the parser would only do this for the category of every right hand side daughter such that the constraints on insertions allow the preceding right hand side daughters to be inserted. Thus, when insertions are enabled, top-down predictions are computed with a modified version of the left-corner relation that is computed not only between the left hand side category of a rule and the category of the first right hand side daughter, but instead between the left hand side category and the category of any right hand side daughter that can be the "trigger" for creating an active edge.

Just as one or more non-terminal symbols may be inserted at the beginning of a rule match, one or more non-terminals may be inserted at the end. Thus, insertions also affect the conditions under which an edge may be considered complete, and thus become inactive. As long as inserting the remaining "needed" categories does not cause the constraints on insertions (if any) to be broken, an edge may become inactive. However, if the "needed" list is non-empty at this point, the corresponding active edge remains in the chart in case any of the remaining "needed" categories may be constructed and thus make it possible to create a similar inactive edge with a smaller insertion error.

Non-terminals may also be inserted in the middle of rule matches. Thus, inserting also affects the way in which active edges are extended. Normally, an active edge ending at vertex i with x as the next needed category will only be extended when an inactive edge of category x is inserted in the chart beginning at vertex i. With insertions enabled, an active edge may be extended when an inactive edge of category x is inserted beginning at vertex i as long as x is any one of its "needed" categories and the constraints on insertion allow the insertion of the categories preceding x on the "needed" list.

10.7 Selective Flexible Feature Unification in LCFlex

Skipping and inserting lend flexibility to the manner in which LCFLEX constructs c-structures for sentences from the context-free backbone portion of its LFG-like grammar formalism. Similarly, its selective flexible unification adds flexibility to the manner in which the grammar formalism's unification augmentations are evaluated. LCFLEX's unification augmentations serve the dual purpose of enforcing grammatical constraints such as agreement and building a feature structure representation for the sentence. To the extent that feature unification is used in any grammar to eliminate undesired analyses allowed by the context-free portion of the grammar, flexible unification can be used to allow these less desirable analyses in the case where it is impossible to construct a parse otherwise.

10.7.1 Why Selective Flexible Feature Unification

Because the unification augmentations of our LFG-like grammar formalism are used to eliminate a large number of incorrect analyses, unconstrained flexible unification would cause a massive explosion of ambiguity. We address this problem primarily by allowing users to specify which features LCFLEX should treat as "soft features" and in which order they should be evaluated. Our evaluation demonstrates that setting aside a small number of features as "soft features" does not dramatically decrease the parser's overall efficiency.

LCFLEX allows its users to specify which features to set aside as "soft features" by setting a global variable. When flexible unification is turned on, the

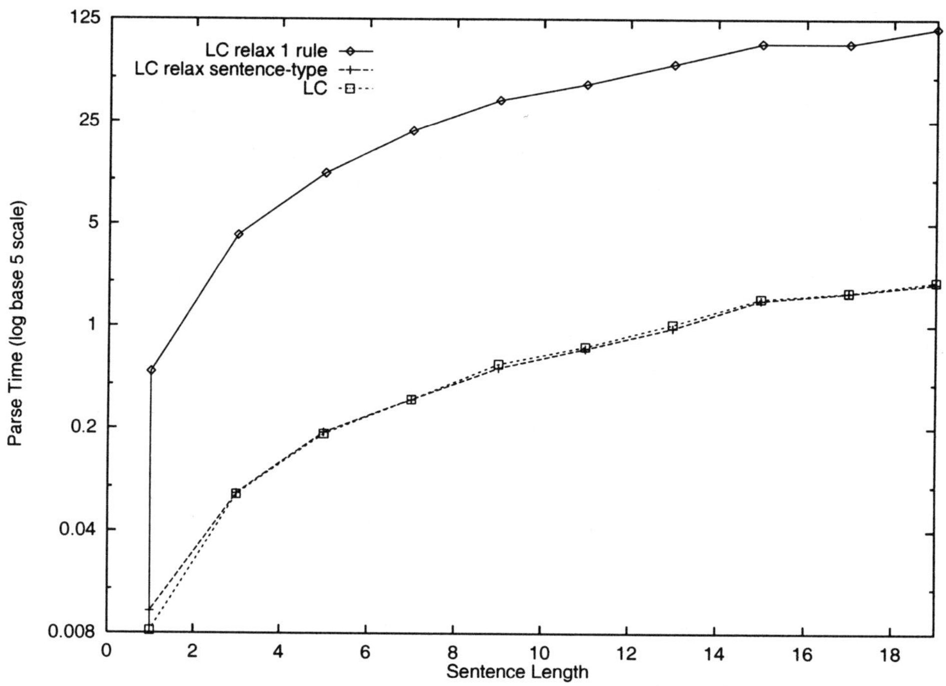

FIGURE 10.8: Avg Parse Time vs. Sentence Length for LCFLEX, Various Flexible Unification Settings

```
RULE: (< SENTENCE > (< WH - WORD > < *SUBJ - BE - LOOK >))

CHILDREN:
< WH - WORD > ((SENTENCE-TYPE2 *QUERY-IF)
               (SENTENCE-TYPE *QUERY-REF)
               (WHAT ((FRAME *HOW))))
< *SUBJ - BE - LOOK > ((VERB ((ROOT LOOK)))
                       (SENTENCE-TYPE *STATE)
                       (TYPE TEMPORAL)
                       (WHEN ((SPECIFIER (*MULTIPLE* DEFINITE
                                                      FOLLOWING))
                              (NAME WEEK)
                              (FRAME *SPECIAL-TIME))))
RESULT:
< SENTENCE > ((FRAME *HOW) (NO-WHEN-0 +)
              (A-SPEECH-ACT *SUGGEST)
              (SENTENCE-TYPE *QUERY-REF)
              (WHEN ((SPECIFIER (*MULTIPLE* DEFINITE
                                             FOLLOWING))
                     (NAME WEEK)
                     (FRAME *SPECIAL-TIME))))
```

FIGURE 10.9: Flexible Unification Example

global variable that controls this behavior stores a list of lists of soft features. Each list of features represents a separate unification stage. At parse time, when an edge becomes complete and the unification part of the rule is fired, it fires in stages according to the division of features in this list of lists. With each successive stage, the resulting feature structure becomes further specified with the result of unifications involving the associated features. The parser fires successive stages until it either finishes the complete set of stages or one stage fails. At that point, it returns the result from the last successful unification stage. If it fails in one of its unification stages, it also returns an indication of which stage failed.

An example where flexible feature unification is useful is the sentence "How the week after look?", as displayed in Figure 10.9. The problem with this sentence is that the auxiliary "does" has been dropped. Our ESST grammar allows "How" to be attached to a yes/no question in order to form a wh-question. Without the "does", the text "the week after look" still parses as a sentence, but it is analyzed as a statement rather than a question. Thus, the rule that would normally attach the "How" in order to form a wh-question fails when it tests the `sentence-type` feature. This sentence can be handled by relaxing the `sentence-type` feature. Relaxing a single feature such as `sentence-type` only slightly decreases the parser's efficiency, as demonstrated in Figure 10.8. In this evaluation, relaxing a single specific feature was compared with allowing a single rule application to fail during unification for any analysis. In the case where a rule application failed, a dummy feature structure was returned as the result from the unification. Notice how much more efficient relaxing one specific feature is than allowing a single rule application to fail. This demonstrates the effectiveness of selective flexible feature unification in increasing the parser's flexibility without compromising its efficiency.

One problem with allowing features to "fail soft" is that it may cause an infinite recursion in the grammar. Thus, in addition to allowing the user to specify which features should be treated as "soft features", LCFLEX allows the user to specify how many rules may "fail soft" within a single parse. This is accomplished by setting a separate global variable. By limiting the set of "soft features" and possibly the number of rules allowed to "fail soft" within a single analysis, flexible unification can be handled in an efficient manner.

10.7.2 Implementing Selective Flexible Feature Unification

Selective flexible unification is accomplished in LCFLEX by modifying the way the grammar's unification augmentations are compiled into lisp code. When the grammar is compiled, and each grammar rule's unification augmentation is compiled into a lisp function, each constraint equation is placed into one of several stages. Those constraint equations not involving any of the soft features are placed in the first stage. Those involving features from the first list but not involving any features from the subsequent lists are placed in the second stage, and so on. Each list of constraint equations is compiled separately into lisp code. This lisp code is then placed in an overarching `case` statement such that which set of constraint equations is applied is dependent upon the stage. This `case` statement is then placed in a lisp function associated with the grammar rule from which the original list of constraint equations was taken. At parse time, the stage is first set to 0 and the function is evaluated. Thus, all of the constraint equations not involving any soft features are evaluated. If that unification fails, no result is returned. The unification is said to have failed. If it does not fail, however, a result will be returned regardless of what happens with unification at subsequent stages. The stage is then set to 1 and the previous result is passed back into the lisp function. As the constraint equations associated with stage 2 are applied, they have the affect of elaborating the result from the previous stage that was passed in. If this unification fails, the previous result is returned with an associated flexible unification error code of 1 that gets inserted into the resulting edge's `unif` slot. If it does not fail, the stage is set to 2 and process cycles again. This continues until either unification fails at some stage or the function has been evaluated for every stage. The most complete result is returned.

10.8 Two Level Parsing in LCFlex

Two level parsing further enhances LCFLEX's efficiency by allowing flexibility to be applied to a sentence on an "as needed" basis. The simple idea behind it is to parse a sentence first at a constrained level of flexibility, and then to combine the best partial analyses constructed in the first pass at a higher level of flexibility. The largest partial analyses are extracted from the chart using a greedy "shortest path" algorithm. In the second stage, inactive edges corresponding to the best partial analyses constructed in the first pass are inserted in the second pass the same way that edges corresponding to lexical entries were inserted in the first pass. Thus, the effective "sentence length" in the second stage is shorter than that in the first stage. Because the additional flexibility is only applied to the places in the analysis where it is really needed (i.e., that would not parse at the lower level of flexibility), the overall parse time and number of edges created is smaller with two level parsing than with one level parsing.

We conducted a preliminary test of LCFLEX's two level parsing mode. In the first stage, we enabled inter-analysis skipping and intra-analysis skipping with a maximum skip limit of 1. In the second stage we increased the maximum skip limit to 6. The run time performance and translation quality for this two level parsing experiment were already reported above in Section 10.5.3.

10.9 Managing Ambiguity

Increasing the parser's flexibility also increases the amount of ambiguity produced by it. Consequently, an increase in flexibility in general slows down the parser and makes selecting the best analysis a more challenging task. Thus, while managing ambiguity is an orthogonal issue to robustness, it is intimately related. Any practical robust parser must manage ambiguity effectively. Both in GLR* and LCFLEX we approach the efficiency issue with local ambiguity packing and pruning, and the selection issue with statistical disambiguation. Specifics of GLR*'s ambiguity packing, pruning, and statistical disambiguation are described in depth elsewhere (Lavie 1995). Thus, we focus here on how these techniques are incorporated into LCFLEX.

10.9.1 Ambiguity Packing and Pruning

Because LCFLEX is a chart parser, it keeps track of the parse trees constructed in the course of the parsing process via inactive edges stored in its chart. Whenever a grammar rule is successfully applied, a corresponding inactive edge is created. That inactive edge has the rule's left hand side category as its category, its first child's starting position as its starting position, and its last child's ending position as its ending position. In order to make parsing as efficient as possible, LCFLEX attempts to pack together all edges of the same category that have the same starting and ending positions. Thus, it attempts to pack together all analyses of the same span of text that result in the same category.

Packing edges together reduces the number of rule applications the parser must attempt, thus reducing both the parser's time and space requirements. However, the resulting packed edges have much more complex feature structures than those of non-packed edges. And rule applications involve evaluating unification equations over the feature structures of the child edges. Thus, while the parser will have fewer rule applications with ambiguity packing turned on, each rule application involving packed edges will take longer to complete than a corresponding rule application on non-packed child edges. Increasing the parser's flexibility along any dimension exacerbates this problem since it increases the number of analyses of the same category that span the same segment of text. Thus, while ambiguity packing alone results in a significant savings in time in traditional non-robust parsers, it is not sufficient in robust parsers even with limited flexibility.

This problem is addressed in LCFLEX, as it was in GLR*, by pruning local analyses that will almost definitely result in global analyses of lesser quality. While pruning is attractive because it directly addresses the efficiency issue, it is risky in unification augmented grammar formalism's such as LCFLEX's because it is not clear on a local level which analyses will be eliminated through unification in subsequent rule applications. Thus, it is dangerous to base decisions about which analyses to prune only on the statistical score of the context free portion of the analysis. At the same time, it is not reliable to prune based on the magnitude of the error level (i.e., how many words skipped, how many non-terminals inserted, how many flexible unifications) of local analyses since an analysis with a lower error locally may only be able to be included in analyses that have a higher global error level than other local analyses with a locally higher error level.

LCFLEX prunes based on a heuristic adapted from GLR*. Analyses are pruned that have errors that strictly subsume errors in other analyses of the same category over the same span of text. In GLR*, analyses were pruned that covered a strict subset of words that were covered by another analysis of the same category spanning the same text. Since LCFLEX allows flexibility along more dimensions than GLR*, it needs a correspondingly more complex pruning heuristic that takes into account not only skipping but also insertions and unification relaxations. However, the same principle of eliminating analyses with errors that strictly subsume errors of otherwise similar analyses can be extended to this more complex case.

Pruning in LCFLEX takes place in three stages. In the first stage, analyses are divided into sets of analyses that have the same insertion and flexible unification errors. Within these sets, pruning takes place based on skipping as in GLR*. Then the remaining analyses are combined into a single set which is then divided into subsets again, this time grouping those analyses that have the same skipping and inserting errors. This time, analyses within each subset are

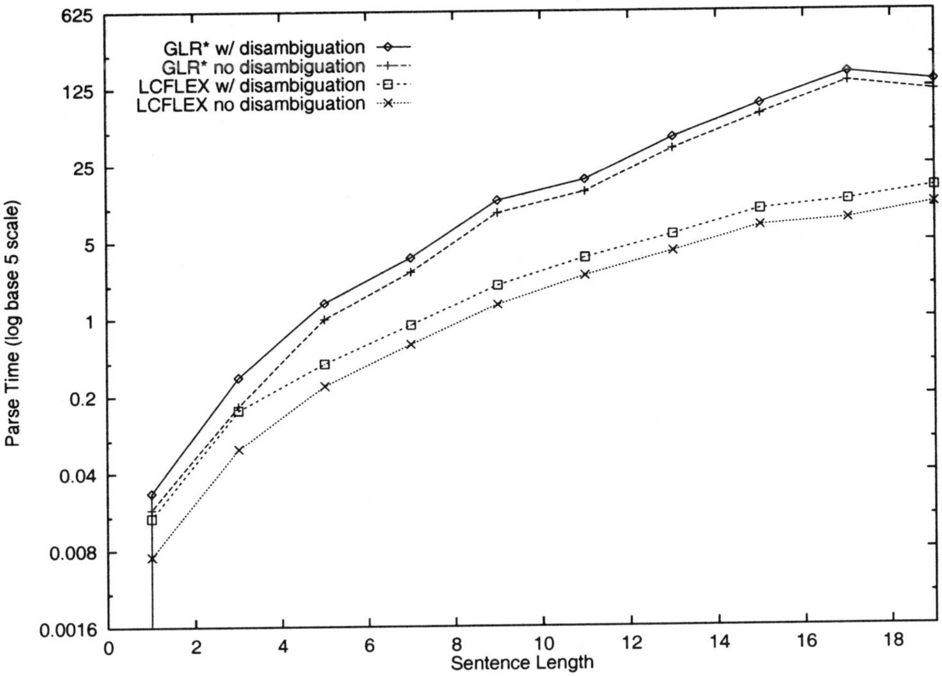

FIGURE 10.10: Parse Times with and without Statistical Disambiguation

pruned if they contain flexible unification errors that subsume those of other analyses within the same subset. Finally, the remaining analyses are divided into subsets once again, this time according to skipping and flexible unification errors. Analyses with insertion errors that subsume those of other analyses within the same subset are then pruned.

10.9.2 Statistical Disambiguation

As in GLR*, statistical disambiguation is used in LCFLEX to select a single analysis from a packed list of analyses. Traditionally, probabilistic context free grammars assign probabilities to grammar rules. Recently, researchers have argued (Carroll & Briscoe 1993, Lavie 1995) that it is advantageous to assign probabilities to transitions between parser actions instead because it lends context-sensitivity to the score assigned to each rule application. So far that principle has only been implemented in the LR framework where probabilities can be associated with parser actions within individual states. Thus, the probability assigned to each parser action depends upon the state the parser was in when the action was performed. A similar context sensitivity is achieved in LCFLEX by assigning probabilities to bigrams of rule applications. Thus, the probability of a rule application is dependent upon the previous rule application.

Sequences of rule applications are extracted from analyses similarly to how action sequences are extracted from analyses in GLR* (Lavie 1995). A postfix procedure traverses the parse tree such that the extracted sequence of rule applications corresponds to the order in which the rules were applied to build the parse tree. Rule transition bigrams are trained in LCFLEX the same way action probabilities are trained in GLR*. A training corpus is used that contains sentences paired with correct feature structures. Each sentence is parsed. Then the list of produced analyses are searched for the one that produces the correct feature-structure. The sequence of rule applications is then extracted from this analysis. The counts for each of the rule application bigrams in the sequence are then incremented. After the entire corpus has been processed, the counts are smoothed using the Good Turing method. Finally, these counts are converted into probabilities.

While statistical disambiguation has been demonstrated to be an effective approach to selecting from multiple alternative analyses, it comes at a significant cost in terms of run time performance. Figure 10.10 displays parse times with and without statistical disambiguation for GLR* with a beam of 30 and LCFLEX with a maximum skip limit of 3.

10.10 Conclusions and Future Directions

This paper described our investigations into how to effectively balance robustness and computational efficiency in a general unified parsing framework for unification-based large-scale grammars. Our investigations were based on empirical evaluations conducted with two general robust parsers: GLR* — a robust GLR parser, and LCFLEX — a left-corner chart parser. We focused on a spontaneous spoken language analysis task, where disfluent and extra-grammatical input is extremely common. Our investigation shows that flexible skipping, controlled via a beam search, is the most effective mechanism for achieving robustness at an acceptable level of computational cost. LCFLEX can tolerate relatively high levels of skipping (and its resulting high levels of ambiguity), while still maintaining online parsing times. While the other types of flexibility investigated — category insertion and flexible unification — can help recover a correct parse in certain cases, their computation cost is very high. Our experiments show that these types of flexibility should be applied sparingly, with only a selective set of insertion categories and features. The specific appropriate sets of insertion categories and features for relaxation are grammar and domain dependent. Our flexible robust parsing framework, however, is general enough to work with any specified sets. One area of future research is the development of automatic methods for identifying appropriate insertion categories and features for relaxation, using machine learning techniques.

Another interesting direction for future research is multi-level parsing. The current LCFLEX parser supports two-level parsing that allows it to construct a partial analysis at a more constrained level of flexibility and then combine the fragments of the partial parse by running the parser again at a far less constrained level of flexibility. The experiments we reported here demonstrated that the two-level mode can achieve an almost comparable level of robustness at a significantly lower computational cost. However, we suspect that the specific types and levels of flexibility that should be applied at each parsing stage are domain and grammar dependent. While our framework is general enough to support any division of labor between the parsing levels, we would like to explore methods for finding the optimal settings for a given grammar and domain automatically. We also would like to investigate whether adding additional parsing levels can result in further gains in efficiency.

As emphasized in this paper, effective parse scoring and disambiguation is essential in robust parsers such as GLR* and LCFLEX. The framework for parse scoring and disambiguation in our current parsers does not model the various types of flexibility in a general uniform way. We would thus like to develop a parse scoring and disambiguation model that explicitly accounts for the various flexibility operations as well as for preferences for particular structures on both the constituent and feature levels. The relative fitness scores can then be applied to constrain and prune the parser search, as well as for parse selection and disambiguation. The relative preference (or cost) for applying certain types of flexibility should also be accounted for via the parse scoring model. A more unified parse evaluation framework would also allow us to better apply machine learning techniques, in order to automatically derive the set of optimal parameters for any specific domain or application.

Acknowledgments

Special thanks are due to all of those patient individuals who used LCFLEX during its beta testing phase and sent us many helpful bug reports. David Schneider was the first to license a copy of LCFLEX for a Written English as a Second Language tutoring system. His research on detecting syntactic errors provided the inspiration for LCFLEX's limited insertions and selective flexible unification. In particular we'd like to thank the students in our Robust Parsing seminar class for providing us with stimulating discussions and getting involved in extending LCFLEX's capabilities. Finally, we'd like to thank the many other people who have encouraged us to continue to develop this work, especially the members of CIRCLE[2] who's vision for dialogue-based tutoring systems inspires us to continue to work towards the goal of robust and efficient natural language understanding. Our deepest gratitude belongs to our families, in particular Eric, Rachel, and Bob, without who's loving support this work would not be possible.

This research was supported in part by the Office of Naval Research, Cognitive and Neural Sciences Division (Grants N00014-91-J-1694 and N00014-93-I-0812) and the National Science Foundation (Grant IRI-94-57637).

---

2. CIRCLE is the NSF funded center for intelligent tutoring housed at Carnegie Mellon University and the University of Pittsburgh.

References

Abney, S. (1995). Partial parsing via finite-state cascades, *in* J. Carroll (ed.), *Workshop on Robust Parsing; Eight European Summer School in Logic, Language and Information*, pp. 8–15.

Aho, A. V. & Johnson, S. C. (1974). LR parsing, *Computing Surveys* **6**(2): 99–124.

Aït-Mokhtar, S. & Chanod, J. (1997). Incremental finite-state parsing, *Proceedings of the Fifth Conference on Applied Natural Language Processing*, pp. 72–79.

Alshawi (ed.), H. (1992). *The Core Language Engine*, MIT Press, Cambridge, MA.

Bod, R. (1998). Spoken dialogue interpretation with the DOP model, *Proceedings of COLING/ACL-98*, pp. 138–144.

Buø, F. D. (1996). FEasPar — A feature structure parser learning to parse spoken language, *Proceedings of COLING-96*, pp. 188–193.

Carroll, J. A. (1993). *Practical Unification-Based Parsing of Natural Language*, PhD thesis, University of Cambridge, Computer Laboratory.

Carroll, J. & Briscoe, T. (1993). Generalized probabilistic LR parsing of natural language (corpora) with unification-based grammars, *Computational Linguistics* **19**(1): 25–59.

Gazdar, G., E., K., Pullum, G. & Sag, I. (1985). *Generalized Phrase Structured Grammar*, Blackwell, Oxford, UK.

Goodman, J. (1996). Parsing algorithms and metrics, *Proceedings of the 34th Annual Meeting of the Association for Computational Linguistics*, pp. 177–183.

Henderson, J. & Lane, P. (1998). A connectionist architecture for learning to parse, *Proceedings of COLING/ACL-98*, pp. 531–537.

Hipp, D. R. (1992). *Design and Development of Spoken Natural-Language Dialog Parsing Systems*, PhD thesis, Dept. of Computer Science, Duke University.

Hobbs, J. R., Appelt, D. E., Bear, J. & Tyson, M. (1991). Robust processing of real-world natural-language texts, *Technical report*, SRI International.

Jain, A. N. (1991). *PARSEC: A Connectionist Learning Architecture for Parsing Speech*, PhD thesis, School of Computer Science, Carnegie Mellon University.

Jain, A. N. & Waibel, A. H. (1990). Incremental parsing by modular recurrent connectionist networks, *in* D. S. Tourertzky (ed.), *Advances in Neural Information Processing 2*, Morgan Kaufman Publishers, pp. 346–371.

Kaplan, R. & Bresnan, J. (1982). Lexical-functional grammar: A formal system for grammatical representation, *The Mental Representation of Grammatical Relations*, MIT Press, pp. 173–281.

Lang, B. (1974). Deterministic techniques for efficient non-deterministic parsers, *Proceedings of 2nd Colloquium on Automata, Languages and Programming*, Lecture Notes in Computer Science, Springer Verlag, Saarbruken, Germany, pp. 255–269.

Lavie, A. (1995). *A Grammar Based Robust Parser For Spontaneous Speech*, PhD thesis, School of Computer Science, Carnegie Mellon University.

Lehman, J. F. (1989). *Adaptive Parsing: Self-Extending Natural Language Interfaces*, PhD thesis, School of Computer Science, Carnegie Mellon University.

Magerman, D. M. & Marcus, M. P. (1990). Parsing a natural language using mutual information statistics, *Proceedings of AAAI*, pp. 984–989.

Mayfield, L., Gavaldà, M., Seo, Y.-H., Suhm, B., Ward, W. & Waibel, A. (1995). Parsing real input in JANUS: A concept-based approach to spoken language translation, *Proceedings of the Theoretical and Methodological Issues in Machine Translation*, pp. 196–205.

Mayfield, L., Gavaldà, M., Ward, W. & Waibel, A. (1995). Concept-based speech translation, *Proceedings of IEEE International Conference on Acoustics, Speech and Signal Processing (ICASSP'95)*.

McDonald, D. (1990). Robust partial-parsing through incremental, multi-level processing: Rationales and biases, *in* P. S. Jacobs (ed.), *Proceedings of the AAAI Spring Symposium on Text-Based Intelligent Systems: Current Research in Text Analysis, Information Extraction, and Retrieval.* A technical report from the GE Research and Development Center, Schnectedy NY, no 90CRD198.

McDonald, D. (1992). An efficient chart-based algorithm for partial-parsing of unrestricted texts, *Proceedings of the 3rd Conference on Applied Natural Language Processing*, pp. 193–200.

McDonald, D. (1993a). Efficiently parsing large corpora. Submitted to the *ACL Workshop on Very Large Corpora: Academic and Industrial Perspectives*.

McDonald, D. (1993b). The interplay of syntactic and semantic node labels in partial parsing, *Proceedings of the Third International Workshop on Parsing Technologies*, pp. 171–185.

Miller, S., Stallard, D., Bobrow, R. & Schwartz, R. (1996). A fully statistical approach to natural language interfaces, *Proceedings of the 34th Annual Meeting of the Association for Computational Linguistics*, pp. 55–61.

Neumann, G., Backofen, R., Baur, J., Becker, M. & Braun, C. (1997). An information extraction core system for real world German text processing, *Proceedings of the Fifth Conference on Applied Natural Language Processing*.

Pietra, S., Epstein, M., Roukos, S. & Ward, T. (1997). Fertility models for statistical natural language understanding, *Proceedings of the 35th Annual Meeting of the Association for Computational Linguistics*, pp. 168–173.

Pollard, C. & Sag, I. (1987). *Information Based Syntax and Semantics: Vol. 1 — Fundamentals*, University of Chicago Press, Chicago, IL.

Rosé, C. P. (1997). *Robust Interactive Dialogue Interpretation*, PhD thesis, School of Computer Science, Carnegie Mellon University.

Rosé, C. P. & Lavie, A. (1997). An efficient distribution of labor in a two stage robust interpretation process, *Proceedings of the Second Conference on Empirical Methods in Natural Language Processing*, pp. 26–34.

Rosé, C. P. & Waibel, A. (1997). Recovering from parser failures: A hybrid statistical/symbolic approach, *in* J. Klavans & P. Resnik (eds), *The Balancing Act: Combining Symbolic and Statistical Approaches to Language Processing*, The MIT Press, pp. 157–179.

Rosenkrantz, D. J. & Lewis, P. M. (1970). Deterministic left corner parsing, *Procedings of the IEEE Conferenc eof the 11th Annual Symposium on Switching and Automata Theory*, pp. 139–152.

Sanker, A. & Gorin, A. (1993). Adaptive language acquisition in a multi-sensory device, *IEEE Transactions on Systems, Man, and Cybernetics* .

Schneider, D. & McCoy, K. F. (1998). Recognizing syntactic errors in the writing of second language learners, *Proceedings of COLING/ACL 98*, pp. 1198–1204.

Tomita, M. (1986). *Efficient Parsing for Natural Language: A Fast Algorithm for Practical Systems*, Kluwer Academic Publishers.

Tomita, M. (1987). An efficient augmented context-free parsing algorithm, *Computational Linguistics* **13**(1–2): 31–46.

Tomita, M. (1990). The generalized LR parser/compiler — version 8.4, *Proceedings of International Conference on Computational Linguistics (COLING'90)*, Helsinki, Finland, pp. 59–63.

Van Noord, G. (1997). An efficient implementation of the head-corner parser, *Computational Linguistics* **23**(3): 425–456.

Ward, W. (1989). Understanding spontaneous speech, *Proceedings of the DARPA Speech and Natural Language Workshop*, pp. 137–141.

Worm, K. (1998). A model of robust processing of spontaneous speech by integrating viable fragments, *Proceedings of COLING-ACL 98*, pp. 1403–1407.

Woszcyna, M., Aoki-Waibel, N., Buo, F. D., Coccaro, N., Horiguchi, K., Kemp, T., Lavie, A., McNair, A., Polzin, T., Rogina, I., Rosé, C. P., Schultz, T., Suhm, B., Tomita, M. & Waibel, A. (1994). JANUS 93: Towards spontaneous speech translation, *Proceedings of the International Conference on Acoustics, Speech, and Signal Processing*, pp. 345–349.

Woszcyna, M., Coccaro, N., Eisele, A., Lavie, A., McNair, A., Polzin, T., Rogina, I., Rosé, C. P., Sloboda, T., Tomita, M., Tsutsumi, J., Waibel, N., Waibel, A. & Ward, W. (1993). Recent advances in JANUS: A speech translation system, *Proceedings of the ARPA Human Languages Technology Workshop*, p. 1295.

Text, Speech and Language Technology

1. H. Bunt and M. Tomita (eds.): *Recent Advances in Parsing Technology.* 1996 ISBN 0-7923-4152-X
2. S. Young and G. Bloothooft (eds.): *Corpus-Based Methods in Language and Speech Processing.* 1997 ISBN 0-7923-4463-4
3. T. Dutoit: *An Introduction to Text-to-Speech Synthesis.* 1997 ISBN 0-7923-4498-7
4. L. Lebart, A. Salem and L. Berry: *Exploring Textual Data.* 1998 ISBN 0-7923-4840-0
5. J. Carson-Berndsen, *Time Map Phonology.* 1998 ISBN 0-7923-4883-4
6. P. Saint-Dizier (ed.): *Predicative Forms in Natural Language and in Lexical Knowledge Bases.* 1999 ISBN 0-7923-5499-0
7. T. Strzalkowski (ed.): *Natural Language Information Retrieval.* 1999 ISBN 0-7923-5685-3
8. J. Harrington and S. Cassiday: *Techniques in Speech Acoustics.* 1999 ISBN 0-7923-5731-0
9. H. van Halteren (ed.): *Syntactic Wordclass Tagging.* 1999 ISBN 0-7923-5896-1
10. E. Viegas (ed.): *Breadth and Depth of Semantic Lexicons.* 1999 ISBN 0-7923-6039-7
11. S. Armstrong, K. Church, P. Isabelle, S. Nanzi, E. Tzoukermann and D. Yarowsky (eds.): *Natural Language Processing Using Very Large Corpora.* 1999 ISBN 0-7923-6055-9
12. F. Van Eynde and D. Gibbon (eds.): *Lexicon Development for Speech and Language Processing.* 2000 ISBN 0-7923-6368-X; Pb: 07923-6369-8
13. J. Véronis (ed.): *Parallel Text Processing.* Alignment and Use of Translation Corpora. 2000 ISBN 0-7923-6546-1
14. M. Horne (ed.): *Prosody: Theory and Experiment.* Studies Presented to Gösta Bruce. 2000 ISBN 0-7923-6579-8
15. A. Botinis (ed.): *Intonation.* Analysis, Modelling and Technology. 2000 ISBN 0-7923-6605-0
16. H. Bunt and A. Nijholt (eds.): *Advances in Probabilistic and Other Parsing Technologies.* 2000 ISBN 0-7923-6616-6
17. J.-C. Junqua and G. van Noord (eds.): *Robustness in Languages and Speech Technology.* 2001 ISBN 0-7923-6790-1

KLUWER ACADEMIC PUBLISHERS – DORDRECHT / BOSTON / LONDON